ELECTRONIC CIRCUIT DESIGN

From Concept to Implementation

ELECTRONIC CIRCUIT DESIGN

From Concept to Implementation

NIHAL KULARATNA

CRC Press
Taylor & Francis Group
Boca Raton London New York

CRC Press is an imprint of the
Taylor & Francis Group, an **informa** business

CRC Press
Taylor & Francis Group
6000 Broken Sound Parkway NW, Suite 300
Boca Raton, FL 33487-2742

First issued in paperback 2019

© 2008 by Taylor & Francis Group, LLC
CRC Press is an imprint of Taylor & Francis Group, an Informa business

No claim to original U.S. Government works

ISBN-13: 978-0-8493-7617-7 (hbk)
ISBN-13: 978-0-367-38726-6 (pbk)

Library of Congress Cataloging-in-Publication Data

Kularatna, Nihal.
    Electronic circuit design : from concept to implementation / Nihal Kularatna.
        p. cm.
    Includes bibliographical references and index.
    ISBN 978-0-8493-7617-7 (alk. paper)
    1. Electronic circuit design. I. Title.

TK7867.K78 2008
621.3815--dc22
                                               2007048859

Visit the Taylor & Francis Web site at
http://www.taylorandfrancis.com

and the CRC Press Web site at
http://www.crcpress.com

This book is dedicated to Sir Arthur C. Clarke, who departed us recently.

With loving thanks to my wife Priyani, and the two daughters, Dulsha and Malsha, who always tolerate my addiction to tech-writing and electronics.

Contents

Preface ix
About the Author xi
Acknowledgments xiii
Contributors xvii

1   **Review of Fundamentals** 1

2   **Design Process** 57
Shantha Fernando

3   **Design of DC Power Supply and Power Management** 77

4   **Preprocessing of Signals** 205

5   **Data Converters** 277

6   **Configurable Logic Blocks for Digital Systems Design** 343
Morteza Biglari-Abhari

7   **Digital Signal Processors** 363

8   **An Introduction to Oscillators, Phase Lock Loops, and Direct Digital Synthesis** 413
Sujeewa Hettiwatte, Coauthor

9   **System-on-a-Chip Design and Verification** 427
Chong-Min Kyung

Index 469

Preface

Electronics is probably one of the few subjects where the "knowledge half-life" of a professional is very short. Today it is probably only 3 to 4 years. Designing electronic systems today requires a unique combination of (1) fundamentals; (2) research and development directions in the latest semiconductors and passive components; (3) nitty-gritty aspects within the "mixed signal world;" (4) access to component manufacturers' data sheets, design guidelines, and development environments; and, most importantly, (5) a timely and practical approach to overall aspects of a design project.

Classen's rule—"the usefulness of a product is proportional to log (technology)"—reminds us that engineers have to add a lot of technology and engineering teamwork to get more user-friendliness into electronics. With portability and miniaturization becoming buzzwords in electronics, design engineers have to concentrate on many additional aspects in the core of a design. Some of these are power supply design, packaging, thermal design, and reliability. In the new millennium, with very large-scale integrated (VLSI) circuits and system-on-a-chip (SOC) technologies maturing, designers have many options when designing electronic systems.

My own career of more than 32 years in different environments such as aviation, telecommunications, and power electronics, with a long midcareer in research and development, has given me the opportunity to look at the world of electronics in a broader perspective. A budding electronics engineer with a degree-level qualification takes a few years to appreciate the breadth of the subject while learning the depth in limited specialized areas. The breadth and depth of the subject together are necessary to produce a commercially viable product or system, giving consideration to time to market (TTM).

This work attempts to address several areas of analog and mixed signal design, including power supply design, signal conditioning, essentials of data conversion, and signal processing, while summarizing a large amount of information from theory texts, application notes, design bulletins, research papers, and technology magazine articles. In a few chapters I had the assistance of experts in different subject areas, as chapter authors.

Because of page limitations, I had to summarize a large amount of useful subject matter extracted from more than 500 technical publications. I suggest that readers refer to the cited references for more details. I would also appreciate your assistance in notifying me of any errors found in the book.

Nihal Kularatna
Auckland
New Zealand

About the Author

Former CEO of the Arthur C. Clarke Institute for Modern Technologies (ACCIMT) in Sri Lanka, **Nihal Kularatna** is an electronics engineer with more than 30 years of experience in professional and research environments. He is the author of two Electrical Measurement Series books for the IEE (London), titled *Modern Electronic Test and Measuring Instruments* (1996) and *Digital and Analogue Instrumentation: Testing and Measurement* (2003), and two Butterworth (USA) titles, *Power Electronics Design Handbook: Low Power Components and Applications* (1998) and *Modern Component Families and Circuit Block Design* (2000). He coauthored *Essentials of Modern Telecommunications Systems* for Artech House Publishers (April 2004).

From 1976 to 1985 he worked as an electronics engineer responsible for navigational aids and communications projects in civil aviation and digital telephone exchange systems. In 1985 he joined the ACCIMT as a research and development engineer and earned a principal research engineer status in 1990; he was appointed CEO/Director in 2000. From 2002 to 2005 he was a senior lecturer at the Department of Electrical and Electronic Engineering, University of Auckland, New Zealand. He has participated in many specialized training programs with equipment manufacturers, universities, and other organizations in the United States, United Kingdom, France, and Italy. He was an active consultant for two U.S. companies, including the Gartner Group, and many Sri Lankan organizations. He is currently a member of the expert reviewer panel of the Foundation for Research, Science and Technology, New Zealand.

He is currently active in research in transient propagation and power conditioning in power electronics, embedded processing applications for power electronics, and smart sensor systems. He has contributed more than 60 papers to academic and industry journals and international conference proceedings. He was the principal author of the McGraw-Hill (Datapro) report "Sri Lanka Telecoms—An Industry and Market Analysis" (1997).

A Fellow of the IEE (London), a Senior Member of IEEE (USA), and an honors graduate from the University of Peradeniya, Sri Lanka, during his research career in Sri Lanka, he was the winner of a Presidential Award for Inventions (1995), Most Outstanding Citizens Award (1999, Lions Club), and a TOYP Award for academic accomplishment (Jaycees) in 1993. He is a fellow of the Institution of Engineers, Sri Lanka.

He is currently employed as a senior lecturer in the Department of Engineering, University of Waikato, New Zealand. His hobby is gardening cacti and succulents.

Acknowledgments

Since my graduation in 1976 from the University of Peradeniya, Sri Lanka, I have spent more than 30 years in electronics and associated fields. Since 2002 I've had a full-time university teaching career in New Zealand, which has given me the time to think of the differences between industry and academia. My 16-year career (1985–2002) at the Arthur C. Clarke Institute for Modern Technologies (ACCIMT) in Sri Lanka gave me the opportunity to work closely with engineers from other countries where electronics technology has had rapid progress. Colleagues, project partners, students, friends and family always encouraged me to be closely involved with the world of electronics and enjoy the opportunities. There are too many people to mention by name, but I thankfully acknowledge them all with a very grateful heart.

This book attempts to provide a reasonable link between the theoretical knowledge domain and the valuable practical information domain from the technology developers. The broad approach in this work is to understand the complete systems and appreciate their interfacing aspects, embracing many digital circuit blocks coupled with the mixed signal circuitry in complete systems.

A large amount of published material from industry and academia has been used in this book. The work and organisations that deserve strong acknowledgments are:

1. Many published text books for the material in Chapter 1.
2. Many published articles in the *Power Electronics Technology* magazine (Penton Media) for Chapter 3.
3. Analog Devices, Inc., for a large amount of material in Chapters 4 and 5.
4. Industry magazines such as *EDN, Electronic Design, Test & Measurement World* and many IEEE/ IEE publications.

I am very thankful to the tireless attempts by the following for creating figures, word processing, etc.:

1. Chandrika Weerasekera, and Jayathu Fernando, ACCIMT staff members from Sri Lanka, who spent a lot of their private time coordinating with me across different time zones.
2. Pawan Shestra and David Nicholls, from our own technical staff at the University of Waikato, and Heidi Eschmann from the department administration.

3. Postgraduate student Zhou Weiqian and undergraduate student Ben Haughey.

4. Thiranjith Weerasinghe, Kasun Talwatte, and Dilini Attanayake, my friends and neighbours.

5. Dulsha, Malsha, and Rajith, from the family.

Without the above persons it would have been impossible for me to deal with writing the 5½ chapters of the book and the overall manuscript preparation.

My special gratitude is extended to my chapter authors Shantha Fernando (Shantha, after many years since our ACCMT days you made me feel part of a good team work again across the Tasman sea!), Morteza Biglari-Abhari, Sujeewa Hettiwatte and Prof. Chong-Min Kyung of KAIST, Korea, for their great team work with regards to the contents of the book.

I am also very indebted to our postgraduate student Chandani Jinadasa for her assistance in checking my proof reading comments.

For the copyright permission for certain contents in the book I am very thankful to the following:

1. David Morrison, Editor-in-Chief of the *Power Electronics Technology Magazine*, Penton Media, USA

2. Jon Titus (formerly with *EDN Magazine*), Kasey Clark and Maury Wright of the Editorial Group of the *EDN Magazine*

3. Mark David of *Electronic Design* magazine, USA

4. Rich Fassler of Power Integrations Inc., USA

5. Steve West of On Semiconductor, USA

6. John Hamburger of Linear Technology Inc., USA

7. Bill Hohl and Katherine Souter of ARM Limited, UK

8. Mike Phipps and Hamish Rawnsley of Altera

9. Bill Hutchins, Jaqueline Eischmann and Carol Popvich of Microchip Technologies, USA

10. Yvette Huygen of Synopsys Inc., USA

11. James Bryant from Analog Devices Inc, European Headquarters.

12. Deborah Sargent and Rick Nelson of *Test and Measurement World Magazine*, USA

13. Joseph P. Hayton and colleagues of Elsevier Science & Technology Books

It was an extremely pleasant experience to work with the staff of the CRC Press on this book, which occurred during a very busy time for my family due to my daughter's wedding. I am particularly thankful to publisher Nora Konopka for her understanding and the support to get the project moving smoothly. Jessica Vakili, Katherine Colman, and the other members of Editorial are gratefully acknowledged for their support in collecting my manuscript in several stages, allowing me to balance my time between work, family and book writing, particularly during our stay in Sri Lanka for the wedding. I am very grateful to Robert Sims and the Production staff for their assistance in solving difficult problems during the production of the work. I am very glad to mention that the Editorial and Production group of the CRC Press is a very understanding team, and it was a great pleasure to work with them.

Since moving to Hamilton, New Zealand, to work at the University of Waikato, I am privileged to work with a very friendly team of colleagues who give me lots of encouragement for my work. I particularly appreciate the spectacularly beautiful area in New Zealand and the friendly kiwi atmosphere in which I have been able to complete the final stages of this book.

To Sir Arthur Clarke, Patron of the Clarke Institute, I was inspired by you during the 16 years of my work at the ACCIMT. Thank you so much for your encouragement in my work.

Last, but not least, my special thanks go to Priyani, who made this work possible by taking over my family commitments —looking after the family needs and taking the full responsibility of planning a wedding in Sri Lanka.

Nihal Kularatna
November 2007

Contributors

Dr. Morteza Biglari-Abhari Department of Electrical and Computer Engineering, University of Auckland, Auckland, New Zealand

Shantha Fernando Advanced Technology Centre NSW Police Force, Australia

Dr. Sujeewa Hettiwatte Department of Electrical and Electronic Engineering, School of Engineering, Auckland University of Technology, Auckland, New Zealand

Prof. Chong-Min Kyung Department of Electrical Engineering and Computer Science, Korea Advanced Institute of Science and Technology, Daejeon, Republic of Korea

1

Review of Fundamentals

CONTENTS

1.1 Introduction 2
1.2 Ohm's Law, Kirchoff's Laws, and Equivalent Circuits 3
1.3 Time and Frequency Domains 3
   1.3.1 The Fourier Transform 5
1.4 Discrete and Digital Signals 6
   1.4.1 Discrete Time Fourier Transform 7
   1.4.2 Discrete Fourier Transform 7
   1.4.3 Fast Fourier Transform 8
1.5 Feedback and Frequency Response 9
   1.5.1 Gain Desensitization 10
   1.5.2 Noise Reduction 10
   1.5.3 Reduction in Nonlinear Distortion, Bandwidth Extension,
        and Input/Output Impedance Modification by Feedback 11
1.6 Loop Gain and the Stability Problem 13
   1.6.1 The Nyquist Plot 17
   1.6.2 Poles and Zeros, S-Domain, and Bode Plots 17
   1.6.3 Bode Plots and Gain and Phase Margins 20
1.7 Amplifier Frequency Response 22
   1.7.1 BJT Equivalent Circuits 22
   1.7.2 BJT Small Signal Operation and Models 26
   1.7.3 High-Frequency Models of the Transistors and the
        Frequency Response of Amplifiers 26
       1.7.3.1 Low-Frequency Response 28
       1.7.3.2 High-Frequency Response 30
       1.7.3.3 Use of Short-Circuit and Open-Circuit Time
            Constants for the Approximate Calculations of ω_L
            and ω_H 30
1.8 Transistor Equivalent Circuits, Models, and Frequency Response
    of Common Emitter/Common Source Amplifiers 32
   1.8.1 Calculation of the Low-Frequency 3-dB Corner Frequency,
        ω_L 32
   1.8.2 Calculation of the High-Frequency 3-dB Corner
        Frequency, ω_H 34

1.9   Noise in Circuits 39
       1.9.1   Noise in Passive Components 39
       1.9.2   Effect of Circuit Capacitance 41
       1.9.3   Noise in Semiconductors and Amplifiers 41
       1.9.4   Circuit Noise Calculations and Noise Bandwidth 44
       1.9.5   Noise Figure and Noise Temperature 47
1.10   Passive Components in Circuits 49
       1.10.1 Resistors 50
       1.10.2 Capacitors 51
       1.10.3 Inductors 53
       1.10.4 Passive Component Tolerances and Worst-Case Design 53
References 55

1.1   Introduction

Circuit design can be considered an art based on the fundamental concepts we learn in electrical and electronic engineering. With the unprecedented advancement of semiconductors, today a designer has many choices of components. Although keeping track of all the new integrated circuits (ICs) appearing on the market is a difficult task, it may be particularly useful if design challenges include miniaturization of the overall product. Passive components such as resistors, capacitors, inductors, and transformers need to be mixed effectively and optimally with semiconductor components in building a particular circuit. In this exercise of "design and development," the designer needs to work with the delicate balance between the real or the analog world, where signals can take any value within a given range, and the digital world, where we make use of processors, memories, and other peripheral devices to accurately process information. To summarize the need for the delicate balance required, it may be appropriate to cite Jim Williams, a well-known linear circuit designer: "Wonderful things are going on in the forgotten land between ONE and ZERO. This is real electronics." Within the past quarter century, a new domain of semiconductors has appeared that links the analog and digital worlds. That is the world of mixed signal electronics, where analog-to-digital conversion, and vice versa, occurs. In this book, emphasis is placed on the analog world of electronics together with the mixed signal domain of design.

In dealing with the challenges we face in the process of design and development, a few essential fundamentals need to be reviewed. This chapter reviews the essentials so that designers can comfortably link theory and practice. The reader can find details related to theory and analysis in standard textbooks used for undergraduate and postgraduate courses.

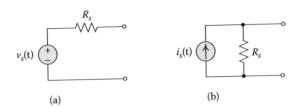

FIGURE 1.1
Useful forms of equivalent circuits: (a) Thevenin form; (b) Norton form.

1.2 Ohm's Law, Kirchoff's Laws, and Equivalent Circuits

Ohm's law relates the voltage and current in a circuit element with the familiar relationship $V = IR$. From this relationship, and then applying Kirchoff's laws to more complex circuits, a design engineer has a precise tool set to analyze a complete circuit with both active and passive components. Deriving from these basic laws the most commonly used equivalent circuits, such as the Thevenin and Norton forms, allows us to simplify many complex circuit blocks, provided that we use reasonable assumptions to simplify each case. For example, a transducer such as a microphone may be simplified as a voltage source in series with a resistor, giving us the familiar Thevenin or Norton form as in Figure 1.1. Similarly, a regulated direct current (DC) power supply may be represented by its Thevenin equivalent to explain the essential characteristics defined in its specifications (discussed later in Chapter 2). One essential reminder regarding the use of Thevenin's equivalent is that it does not allow us to calculate the dissipation within a circuit. Another is that if we replace the voltage or current sources with their internal resistances to calculate the equivalent resistance, R_s, these sources need be independent ones. Another reminder is to take care when calculating R_s in circuits where dependent voltage or current circuits exist.

1.3 Time and Frequency Domains

All electrical signals can be described as a function of either time or frequency. When we observe signals as a function of time, they are called time domain measurements. Sometimes we observe the frequencies present in signals, in which case they are called frequency domain measurements. The word spectrum refers to the frequency content of any signal. All practical components we use as building blocks perform according to design specifications only within a limited frequency range. Therefore, we can define a bandwidth for each circuit block and an overall bandwidth of operation for the entire product.

When signals are periodic, time and frequency are simply related; namely, one is the inverse of the other. Then we can use the Fourier series to find the spectrum of the signal. For nonperiodic signals, a Fourier transform is used to obtain the spectrum. However, performing a Fourier transformation involves integration over all time, that is, from $-\infty$ to $+\infty$. Because this is not practicable, we approximate the Fourier transform by a discrete Fourier transform (DFT), which is performed on a sampled version of the signal. The computational load for direct DFT, which is usually computed in a processor subsystem, increases rapidly with the number of samples and sampling rate. As a way around this problem, Cooley and Tukey invented the fast Fourier transform (FFT) algorithm in 1954. With the FFT, computational loads are significantly reduced.

Let $f(t)$ be an arbitrary function of time. If $f(t)$ is also periodic, with a period T, then $f(t)$ can be expanded into an infinite sum of sine and cosine terms. This expansion, which is called the Fourier series of $f(t)$, may be expressed in the form

$$f(t) = \frac{a_0}{2} + \sum_{n=1}^{\infty}\left[a_0 \cos\left(\frac{2\pi nt}{T}\right) + b_n \sin\left(\frac{2\pi nt}{T}\right) \right],\qquad (1.1)$$

where Fourier coefficients a_n and b_n are real numbers independent of t and which may be obtained from the following expressions:

$$a_n = \frac{2}{T}\int_0^T f(t)\cos\left(\frac{2\pi nt}{T}\right)dt, \text{ where } n = 0, 1, 2, \ldots,\qquad (1.2a)$$

$$b_n = \frac{2}{T}\int_0^T f(t)\sin\left(\frac{2\pi nt}{T}\right)dt, \text{ where } n = 0, 1, 2, \ldots.\qquad (1.2b)$$

Further, $n = 0$ gives us the DC component, $n = 1$ gives us the fundamental, $n = 2$ gives us the second harmonic, and so on.

Another way of representing the same time function is

$$f(t) = \frac{a_0}{2} + \sum_{n=1}^{\infty} d_n \cos\left(\frac{2\pi nt}{T} + \phi_n\right),\qquad (1.3)$$

where $d_n = \sqrt{a_n^2 + b_n^2}$ and $\tan\phi_n = -b_n/a_n$. The parameter d_n is the magnitude and ϕ_n is the phase angle.

Equation (1.3) can be rewritten as

$$f(t) = \sum_{n=-\infty}^{+\infty} c_n e^{j2\pi nt/T},$$

(1.4)

where

$$c_n = \frac{1}{T} \int_0^T f(t) e^{-j2\pi nt/T} dt \text{ and } n = 0, \pm 1, \pm 2.$$

The series expansion of Equation 1.4 is referred to as the complex exponential Fourier series. The c_n are called the complex Fourier coefficients.

According to this representation, a periodic signal contains all frequencies (both positive and negative) that are harmonically related to the fundamental. The presence of negative frequencies is simply a result of the fact that the mathematical model of the signal given by Equation 1.4 requires the use of negative frequencies. Indeed, this representation also requires the use of complex exponential functions, namely $e^{j2\pi nt/T}$, which have no physical meaning either. The reason for using complex exponential functions and negative frequency components is merely to provide a complete mathematical description of a periodic signal, which is well suited for both theoretical and practical work.

1.3.1 The Fourier Transform

The transformation from the time domain to the frequency domain and back again is based on the Fourier transform and its inverse. When the arbitrary function $f(t)$ is not necessarily periodic, we can define the Fourier transform of $f(t)$ as

$$F(f) = \int_{-\infty}^{\infty} f(t) e^{-j2\pi ft} dt.$$

(1.5)

The time function $f(t)$ is obtained from $F(f)$ by performing the inverse Fourier transform:

$$f(t) = \int_{-\infty}^{\infty} F(f) e^{j2\pi ft} df.$$

(1.6)

Thus, $f(t)$ and $F(f)$ form a Fourier transform pair. The Fourier transform is valid for both period and nonperiod functions that satisfy certain minimum conditions. All signals encountered in the real world easily satisfy these

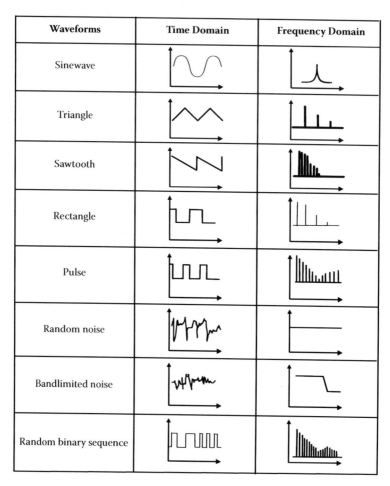

FIGURE 1.2
Signal observation with oscilloscope and spectrum analyzer.

conditions. These conditions, known as the Dirichlet conditions, determine whether a function is Fourier expandable [1].

An oscilloscope displays the amplitude of a signal as a function of time, whereas a spectrum analyzer represents the same signal in the frequency domain. The two types of representations (which form Fourier transform pairs) are shown in Figure 1.2 for some common signals encountered in practice.

1.4   Discrete and Digital Signals

A discrete signal is either discrete in time and continuous in amplitude, or discrete in amplitude and continuous in time. Discrete signals can occur in charge-coupled device (CCD) arrays or switched capacitor filters. A digital signal, however, is discrete in both time and amplitude, such as those signals encountered in digital signal processing (DSP) applications.

1.4.1 Discrete Time Fourier Transform

The discrete time Fourier transform (DTFT) maps a discrete time function $h[k]$ into a complex function $H(e^{j\omega})$, where inverse transforms are

$$H\left(e^{j\omega}\right) = \sum_{k=-\infty}^{\infty} h[k] e^{-jk\omega} \tag{1.7}$$

$$h[k] = \frac{1}{2\pi} \int_{-\pi}^{\pi} H\left(e^{j\omega}\right) d\omega. \tag{1.8}$$

Note that the inverse transform, as in the Fourier case, is an integral over the real frequency variable. A complex inversion integral is not required. The DTFT is useful for manual calculations but not for computer calculations because of continuous variable ω [2]. The role of the DTFT in discrete time system analysis is very much the same as the role the Fourier transform plays for continuous time systems.

1.4.2 Discrete Fourier Transform

An approximation to the DTFT is the discrete Fourier transform (DFT). The DFT maps a discrete time sequence of N-point duration into an N-point sequence of frequency domain values. Formally, the defining equations of the DFT are [2]

$$H\left(e^{j\frac{2\pi n}{N}}\right) = H[n] = \sum_{k=0}^{N-1} h[k] e^{j\frac{2\pi nk}{N}} \quad \text{for } k, n = 0, 1, \dots, N-1 \tag{1.9}$$

$$h[k] = \frac{1}{N} \sum_{n=0}^{N-1} H[n] e^{j\frac{2\pi nk}{N}} \quad \text{for } k, n = 0, 1, \dots, N-1. \tag{1.10}$$

These equations are the DTFT relationships with the frequency discretized to N points spaced $2\pi/N$ radians apart around the unit circle, as in Figure 1.3. In addition, the time duration of the sequence is limited to N points.

The DFT uses two finite series of N points in each of the definitions. Together they form the basis for the N points DFT. The restriction to N points

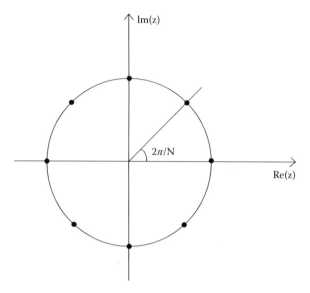

FIGURE 1.3
Unit circle with $N = 8$.

and reduction from integral to summation occur because we are only evaluating the DTFT at N points in the Z domain. The value of N is usually determined by constraints in the problems or the resources for analysis. The most popular values of N are in powers of 2. However, many algorithms do not require this. N can be almost any integer value [2].

1.4.3 Fast Fourier Transform

The FFT is not a transform but an efficient algorithm for calculating the DFT. The FFT algorithms remove the redundant computations involved in computing the DFT of a sequence by exploiting the symmetry and periodic nature of the summation kernel, the exponential term, and the signal. For large values of N, FFTs are much faster than performing direct computation DFTs. Table 1.1 shows the approximate computation loads for various sizes of directly calculated DFTs and the DFT computed using the FFT decimation-in-time algorithm. Note that as N grows beyond 16, the savings are quite dramatic [2].

For details on practical FFT algorithms, in particular, the decimation-in-time algorithm, see Lynn and Fuerst [3]. In modern digital storage scopes, the FFT function is a common feature, and most of these techniques and algorithms are used within the processor subsystem of the scope (see Kularatna [4], Chapters 6 and 9).

TABLE 1.1

Comparison of Approximate Computational Loads
for Direct DFT and Decimations–in-Time FFT

Number of Points	Multiplications		Additions	
	DFT	FFT	DFT	FFT
4	16	4	12	8
8	64	12	56	24
16	256	32	240	64
32	1024	80	992	160
64	4096	192	4032	384
128	16,384	448	16,256	896
256	65,536	1024	65,280	2048
512	262,144	2304	261,632	4608
1024	1,048,576	5120	1,047,552	10,240
2048	4,194,304	11,264	4,192,256	22,528
4096	16,777,216	24,576	16,773,120	49,152

1.5 Feedback and Frequency Response

Harold Black, an electronics engineer with Western Electric, invented the feedback amplifier in 1928. Since then, the technique has been so widely used that it is almost impossible to think of electronic circuits without some form of feedback.

Feedback can be either negative or positive. In amplifier design, negative feedback is commonly used to achieve one or more of the following:

- Desensitize the gain—make the gain less sensitive to variations in the value of circuit components.
- Reduce nonlinear distortion—make the output gain constant and independent of the signal level.
- Reduce the effect of noise—make improvements to the signal-to-noise ratio (SNR).
- Control the input/output impedances.
- Extend the bandwidth.

Figure 1.4, which is a signal flow diagram where x indicates either voltage or current signals, indicates the concept of applying feedback to an amplifier. With the addition of the feedback block, the basic amplifier with an open-loop gain of A now receives a modified input signal $x_i = x_s - x_f$. The feedback

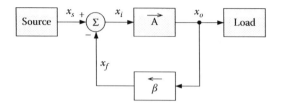

FIGURE 1.4
General structure of the feedback amplifier.

block with a feedback factor of β makes $x_f = \beta x_o$, where x_i is the input into the open-loop amplifier, x_o is the output of the open-loop amplifier, and x_f is the feedback signal. With the assumptions that the transmission of the signals through the basic amplifier is only in the forward direction and the transmission through the feedback block is only in the backward direction, it can be shown that the overall gain of the feedback system A_f is given by the relationship

$$A_f = \frac{x_o}{x_s} = \frac{A}{1+A\beta}. \tag{1.11}$$

Quantity $A\beta$ is called the loop gain and quantity $(1 + A\beta)$ is called the amount of feedback. In a practical design, open-loop gain $A\beta$ is large, $A\beta \gg 1$, then from Equation 1.11, it follows that $A_f \approx 1/\beta$, which is a very practical and useful property in design, because the close loop gain is now entirely dependent on the feedback network.

1.5.1 Gain Desensitization

From the above discussion, we can further derive many useful attributes of the feedback system in Figure 1.4. For example, we can show that the percentage change in A_f (due to variations in some circuit parameters) is smaller than the percentage change in A by the amount of feedback $(1 + A\beta)$. This is given by the relationship

$$\frac{dA_f}{A_f} = \frac{1}{(1+A\beta)} \frac{dA}{A}. \tag{1.12}$$

1.5.2 Noise Reduction

Negative feedback can be employed to reduce the noise or interference in an amplifier and hence improve the SNR. This situation is summarized in Figure 1.5. If we refer the total noise in the amplifier to its input by a noise source of value V_n, we can show that the SNR is given by V_s/V_n for the case in Figure 1.5a. Now, if the system is modified by adding a theoretically

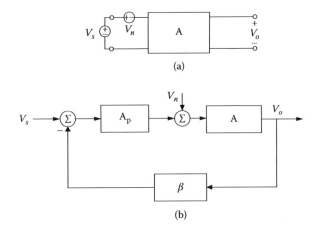

FIGURE 1.5
Concept of increasing the SNR of a circuit: (a) basic amplifier indicating the case of input referred noise; (b) with an additional noise-free amplifier at the front end for SNR improvement.

noise-free amplifier with gain A_p and a feedback block with a feedback factor of β, the total output V_o is given by Figure 1.5b:

$$V_o = V_s \frac{AA_p}{1 + AA_p\beta} + V_n \frac{A}{1 + AA_p\beta}. \tag{1.13}$$

Thus, the modified SNR becomes

$$SNR_{new} = \frac{V_S}{V_n} A_p. \tag{1.14}$$

This is an improvement by a factor of A_p. For details, see Sedra and Smith [5].

1.5.3 Reduction in Nonlinear Distortion, Bandwidth Extension, and Input/Output Impedance Modification by Feedback

Another useful application of feedback is the linearization of transfer characteristics of an amplifier. For a discussion on this, see Chapter 8 in Sedra and Smith [5]. Similarly, for a low-pass filter, by applying feedback, the bandwidth of the circuit can be increased at the expense of a lower midband gain. To illustrate the case, consider a single-pole amplifier with a midband gain of A_M and a 3-dB upper cutoff frequency of ω_H, where the transfer function is given by

$$A(s) = \frac{A_M}{1 + s/\omega_H}. \tag{1.15}$$

With a feedback circuit with a feedback factor of β added to the circuit, commencing from the relationships in Equation 1.11 and Equation 1.15, we can show that the new transfer function is given by

$$A_f(s) = \frac{A_M/(1+A_M\beta)}{1+s/\omega_H(1+A_M\beta)}.$$ (1.16)

This simply illustrates the basis for increasing or decreasing the bandwidth at the expense of decreasing or increasing the midband gain. Similar discussions are possible for other types of amplifiers [5]. Also, it reminds us that the gain bandwidth product for a single-pole low-pass circuit is constant.

Feedback can be utilized to modify the input and output impedances of amplifier stages. At this point it may be appropriate to discuss the four common types of amplifiers: voltage, current, transconductance, and transresistance. To ease the design process, for each case we can insert a feedback stage and suitable practical simplifications such as (1) an open-loop amplifier, which acts as a unilateral circuit (where no reverse feedback occurs through the amplifier) and (2) a β circuit (no forward transfer of signals occurs from the input to output side of the amp through the β circuit). Figure 1.6 shows the four topologies with the feedback networks. We give different names in each case, depending on the way we mix and sample the signals at the input and output sides, respectively. For example, in the case of a voltage amplifier, we

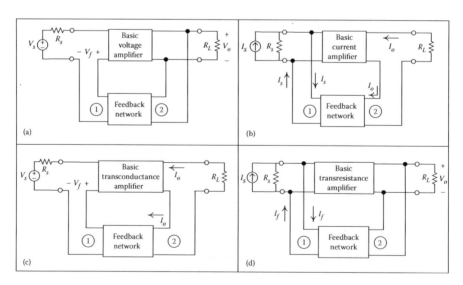

FIGURE 1.6
The four basic feedback topologies: (a) voltage-sampling series mixing (series-shunt) topology; (b) current-sampling shunt mixing (shunt-series) topology; (c) current-sampling series-mixing (series-series) topology; (d) voltage-sampling shunt-mixing (shunt-shunt) topology.

mix a voltage sample with the input (hence a series configuration at input) by sampling the output voltage (hence a shunt sampling case). Thus, this case is called "series-shunt" feedback. It may be worth noting that the cases come from the basic requirement for mixing identical types of signals, voltages, or currents. (Two ideal voltage sources can be mixed in series only, whereas two ideal current sources can be mixed in parallel only.)

Although a complete treatment of the four different cases is beyond the scope of this chapter, Table 1.2 summarizes the four cases. For details, see Sedra and Smith [5].

Figure 1.7 indicates the use of a two-port h parameter feedback circuit with a voltage amplifier and derivation of the A circuit (a modified version of the basic amplifier) and the β circuit for the series-shunt feedback. In a practical design, one needs to simplify the case to have a unilateral β circuit, as shown in Figure 1.7b, after neglecting the parameter h_{21}. Similarly, for the transconductance amplifier, Figure 1.8 indicates the derivation of the A circuit and β circuit where series-series feedback is used.

The ideal structure of a series-shunt feedback amplifier is shown in Figure 1.9, consisting of a unilateral open-loop amplifier (the A circuit) and an ideal voltage sampling series mixing feedback network (the β circuit). The A circuit has an input resistance R_i, a voltage gain A, and an output resistance R_o. It is assumed that the source and load resistances have been included in the A circuit. Furthermore, note that the β circuit does not load the A circuit; that is, connecting the β circuit does not change the value of A, which is defined as V_o/V_i. Similarly, for the series-series feedback case, the ideal structure and the equivalent circuit are shown in Figure 1.10.

1.6 Loop Gain and the Stability Problem

In Figure 1.4, the open-loop gain is a function of frequency, and the feedback factor β is dependent on frequency in a general case. This situation can be indicated by the generalized case for the feedback with a closed-loop transfer function $A_f(s)$ as in Equation 1.17:

$$A_f(s) = \frac{A(s)}{1 + A(s)\beta(s)}. \qquad (1.17)$$

If the amplifier is assumed to be a direct-coupled case with a constant DC gain of A_0 with the poles and zeros occurring in the high-frequency band, the loop gain $A(s)\beta(s)$ becomes a constant $(A_0\beta_0)$ at low frequencies when the feedback factor $\beta(s)$ at low frequencies reduces to a constant value (β_0). This is a common situation we assume in design work for simplicity and convenience

TABLE 1.2

Different Amplifier Types, Feedback, and the Modification of Input and Output Impedances

Basic Amplifier Type	Gain	Type of Feedback	Parameters Used for Two-Port Network-Based Analysis	Assumptions for Simplifications	Modified Values	
					Input Impedance	Output Impedance
Voltage amplifier	$A_v = v_o/v_s$	Series-shunt	h parameters	$\left|h_{21}\right|_{\substack{feedback\\network}} \ll \left|h_{21}\right|_{\substack{basic\\amplifier}}$ $\left|h_{12}\right|_{\substack{basic\\amplifier}} \ll \left|h_{12}\right|_{\substack{feedback\\network}}$ (unilateral condition)	$R_i^*(1 + A\beta)$	$R_O/(1 + A\beta)$
Current amplifier	$A_i = i_o/i_s$	Shunt-series	g parameters	$\left|g_{21}\right|_{\substack{feedback\\network}} \ll \left|g_{21}\right|_{\substack{basic\\amplifier}}$ $\left|g_{12}\right|_{\substack{basic\\amplifier}} \ll \left|g_{12}\right|_{\substack{feedback\\network}}$	$R_i/(1 + A\beta)$	$R_O(1 + A\beta)$
Transconductance amplifier	$G_m = i_o/v_s$	Series-series	z parameters	$\left|z_{21}\right|_{\substack{feedback\\network}} \ll \left|z_{21}\right|_{\substack{basic\\amplifier}}$ $\left|z_{12}\right|_{\substack{basic\\amplifier}} \ll \left|z_{12}\right|_{\substack{feedback\\network}}$ (unilateral condition)	$R_i^*(1 + A\beta)$	$R_O(1 + A\beta)$
Transresistance amplifier	$R_m = v_o/i_s$	Shunt-shunt	y parameters	$\left|y_{12}\right|_{\substack{basic\\amplifier}} \ll \left|y_{12}\right|_{\substack{feedback\\network}}$ $\left|y_{21}\right|_{\substack{feedback\\network}} \ll \left|y_{21}\right|_{\substack{basic\\amplifier}}$	$R_i/(1 + A\beta)$	$R_O/(1 + A\beta)$

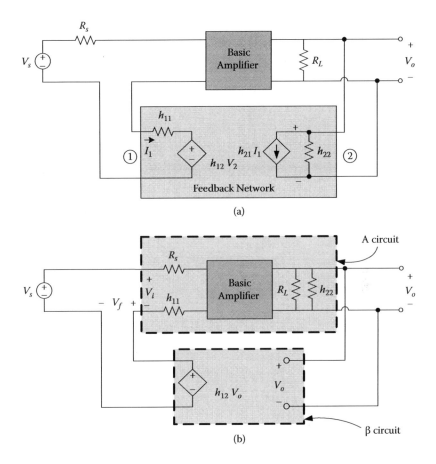

FIGURE 1.7
Derivation of the A circuit and β circuit for the series-shunt feedback amplifier: (a) the circuit in Figure 1.6a with the feedback represented by h parameters; (b) the circuit in Figure 1.7a, neglecting h_{21} and modifying the basic amplifier to form the A circuit.

and to achieve a negative feedback condition. However, at higher frequencies this need not be the case, and by substituting $s = j\omega$,

$$A_f(j\omega) = \frac{A(j\omega)}{1 + A(j\omega)\beta(j\omega)}.$$ (1.18)

This clearly indicates that $A(j\omega)\beta(j\omega)$ is a complex number $L(j\omega)$ that can be represented by

$$L(j\omega) \equiv |A(j\omega)\beta(j\omega)| e^{j\phi(\omega)}.$$ (1.19)

This indicates that at higher frequencies, loop gain $L(j\omega)$ can be positive or negative, depending on the phase angle ϕ. To appreciate this fact, consider

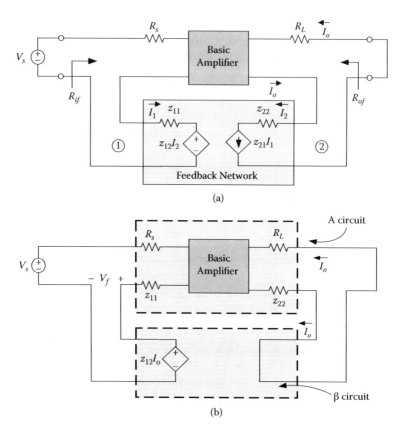

FIGURE 1.8
Derivation of the A circuit and the β circuit for a series-series feedback amplifier: (a) the circuit in Figure 1.6c with the feedback represented by z parameters; (b) the same circuit neglecting z_{21} and modifying the basic amplifier to form the A circuit.

the frequency at which $\phi(\omega)$ becomes 180°. At this frequency, ω_{180}, $A(j\omega)\beta(j\omega)$, or the loop gain will be a negative real number. As per Equation 1.18, at this frequency the feedback will be positive. If at $\omega = \omega_{180}$ the loop gain is less than unity, the closed-loop gain will be greater than the open-loop gain $A(j\omega)$. Nevertheless, the feedback amplifier will be stable.

In the special case where the loop gain becomes −1 at $\omega = \omega_{180}$, the amplifier closed-loop gain will be infinite, with the practical situation of output rising to a very high value for no appreciable input signal, which represents the case of an oscillator with a frequency of oscillation equal to ω_{180}. This is discussed further in Chapter 8.

When the magnitude of the loop gain is greater than unity at ω_{180} it is not very obvious from Equation 1.18, but the circuit will oscillate with the amplitude growing gradually until the circuit nonlinearities limit the amplitude and the loop gain becomes unity.

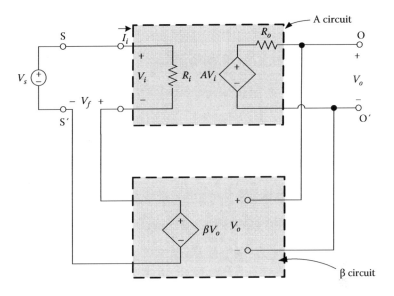

FIGURE 1.9
The series-shunt feedback amplifier.

1.6.1 The Nyquist Plot

Based on the preceding discussion, a formalized approach for testing the stability is the Nyquist plot, which is simply a polar plot of loop gain with frequency as a parameter. Figure 1.11 depicts a Nyquist plot where the radial distance is $|A\beta|$ and the angle is the phase angle φ. The solid line indicates positive frequencies and the dotted line indicates negative frequencies, which forms a mirror image of the plot for the positive and negative ω. The Nyquist plot intersects the negative real axis at the frequency ω_{180}, indicating that if the intersection occurs to the left of the point $(-1,0)$, loop gain $|A\beta| > 1$, making the amplifier unstable. However, if the plot intersects the negative real axis to the right of $(-1,0)$ the amplifier will be stable. In summary, if the Nyquist plot encircles the point $(-1,0)$, the amplifier will be unstable, which is a simplified version of the Nyquist criterion. For the full theory, see Haykin [6].

1.6.2 Poles and Zeros, S-Domain, and Bode Plots

The frequency response of an amplifier can be analyzed by representing the gain as a function of the complex frequency s. In the s domain analysis, a capacitor of value C is represented by an equivalent impedance of $1/sC$, and an inductance of value L is represented by Ls. Using common circuit analysis techniques, a voltage transfer function $T(s)$ is derived as $T(s) \equiv V_o(s)/V_i(s)$.

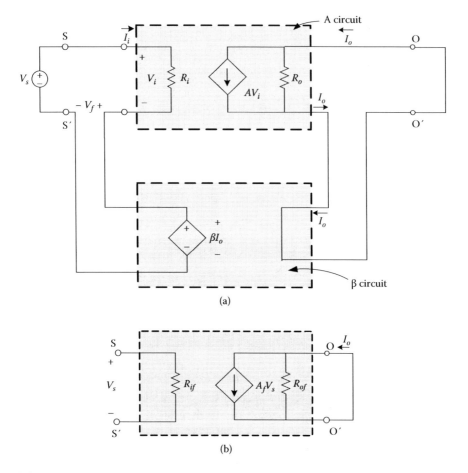

FIGURE 1.10
The series-series feedback amplifier: (a) ideal structure; (b) equivalent circuit with modified input and output impedances.

Once this function is derived, by replacing s with $j\omega$ we can evaluate its frequency behavior.

In general, the transfer function $T(s)$ in its own form (without substituting $s = j\omega$) can reveal many useful details about the stability of the circuit. Function $T(s)$ can be expressed in many different forms, including

$$T(s) = a_m \frac{(s-z_1)(s-z_2)(s-z_3)\ldots(s-z_m)}{(s-p_1)(s-p_2)(s-p_3)\ldots(s-p_n)} \tag{1.20}$$

and

$$T(s) = \frac{a_m s^m + a_{m-1} s^{m-1} + a_{m-2} s^{m-2} + \ldots + a_0}{s^n + b_{n-1} s^{n-1} + b_{n-2} s^{n-2} \ldots + b_0}. \tag{1.21}$$

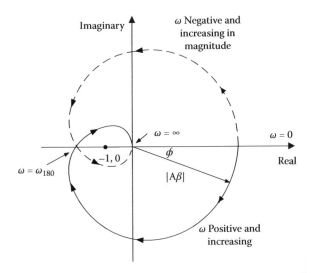

FIGURE 1.11
The Nyquist plot of an unstable amplifier.

In Equation 1.20, $z_1, z_2, \ldots, z_m$ are called transfer function zeros or transmission zeros, and $p_1, p_2, \ldots, p_n$ are called transfer function poles or the natural modes of the network. The n of the transfer function is called the order of the network. The equivalent form of the transfer function in Equation 1.21, where coefficients a and b are real numbers, gives us the condition that the poles or zeros must occur as conjugate pairs; that is, if, for example, a pole or a zero occurs at $3 + 2j$, there should be another pole or zero at $3 - 2j$. A zero that is pure imaginary ($\pm j\omega_z$) causes the transfer function to be exactly zero at $\omega = \omega_z$.

For a system to be stable, its poles should lie in the left half of the s plane. A pair of complex-conjugate poles on the $j\omega$ axis gives rise to sustained oscillations. Poles on the right-hand side of the s plane give rise to growing oscillations, which will be limited by nonlinearities in a practical system. Figure 1.12 indicates the three possibilities.

From the closed-loop transfer function of Equation 1.17, we see that the poles of the feedback amplifier can be obtained by solving the equation

$$1 + A(s)\beta(s) = 0, \tag{1.22}$$

which is called the characteristic equation. This indicates that feedback to a system changes its poles. In an amplifier with an open-loop transfer function characterized by a single pole, the open-loop transfer function is

$$A(s) = \frac{A_o}{1 + \dfrac{s}{\omega_p}}. \tag{1.23}$$

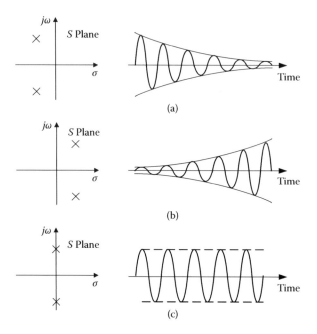

FIGURE 1.12
Relationship between pole location and transient response: (a) poles on the left-hand side of the imaginary axis; (b) poles on the right-hand side of the imaginary axis; (c) poles on the imaginary axis.

The closed-loop transfer function is given by

$$A_f(s) = \frac{A_o / (1 + A_o\beta)}{1 + s / \omega_p(1 + A_o\beta)}. \tag{1.24}$$

Feedback moves the pole along the negative real axis to a frequency ω_{Pf}:

$$\omega_{Pf} = \omega_P(1 + A_o\beta) \tag{1.25}$$

This process is illustrated in Figure 1.13. With feedback, the new pole shifts to the left-hand side of the original pole. Figure 1.13b indicates its effect on the Bode plot, where at low frequencies the gain drops by an amount equal to $20\log(1 + A_o\beta)$ and the two curves coincide at high frequencies. Figure 1.13b indicates that applying feedback extends the bandwidth of the amplifier at the expense of loop gain. Because the pole of the closed-loop amplifier never enters the right-hand side of the s plane, the single-pole amplifier is unconditionally stable.

1.6.3   Bode Plots and Gain and Phase Margins

From Section 1.6 we know that we can determine whether a feedback amplifier is or is not stable by examining its loop gain $A\beta$ as a function of frequency.

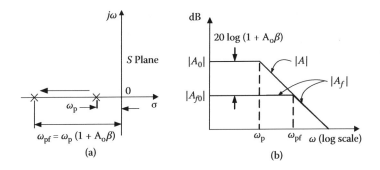

FIGURE 1.13
Effect of feedback on a single-pole amplifier: (a) pole location; (b) the frequency response (Bode magnitude plot).

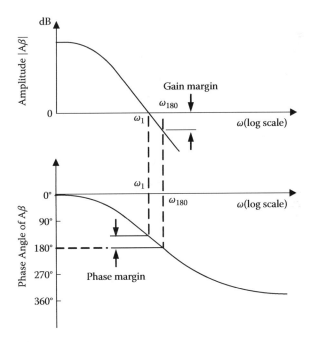

FIGURE 1.14
Bode plots for the loop gain illustrating the gain and phase margin.

The most effective and easiest way to do this is through the use of a Bode plot of $A\beta$, such as the one in Figure 1.14. This feedback amplifier will be stable, because at the frequency of the 180° phase shift, ω_{180}, the magnitude of the loop gain is less than unity or negative in decibel (dB) terms. The difference between the value of $|A\beta|$ at ω_{180} and unity is called the gain margin (usually expressed in decibels). This important parameter represents the amount by which the loop gain can be increased while stability is maintained. Feedback amplifiers are usually designed to have sufficient gain margins to allow for the inevitable changes in loop gain with physical parameters such as temperature, relative humidity, and the age of components.

Another way to investigate stability and its degree is to examine the Bode plot at the frequency for which $|A\beta| = 1$, which is the location at which the magnitude plot crosses the 0-db line. If at this point the phase angle is less (in magnitude) than 180°, then the amplifier is stable. This situation is illustrated in Figure 1.14. The difference between the phase angle at this frequency and 180° is called the phase margin. In other words, if at the frequency of unity loop gain magnitude the phase lag is in excess of 180°, the amplifier will be unstable. An alternative and much simpler approach for investigating stability is to construct the Bode plots for the open-loop gain and, assuming that the β is independent of frequency, plot the horizontal line corresponding to 20log(1/β) as a horizontal straight line on the same amplitude plot, looking at the difference between the two curves. For details, see Chapter 8 and Appendix E of Sedra and Smith [5].

1.7   Amplifier Frequency Response

As a review, let's summarize a few important equivalent circuit models for bipolar junction transistors (BJTs) and metal oxide semiconductor field effect transistors (MOSFETs) and then briefly discuss the frequency response considerations of simple transistor amplifier circuits. Comprehensive discussions of this subject can be found in many textbooks, including Sedra and Smith [5]. It is important to note that for transistors we use different circuit models for large-signal and small-signal operations. Data sheet parameters clearly distinguish these two subsets.

1.7.1   BJT Equivalent Circuits

The following sections review transistor models and equivalent circuits in low-frequency and high-frequency regions. Because there are many books and other publications that provide details on BJTs as well as FETs, the purpose of the section is to remind the reader of how and which models are useful for different applications. Sedra and Smith [5] is very comprehensive in the treatment of these devices within the discrete as well as the IC domains. Ayers [7] is a good source of information related to digital circuit implementations.

An NPN BJT's transfer characteristics can be simplified as shown in Figure 1.15. This i_C-v_{BE} characteristic can be summarized by the exponential relationship

$$i_C = I_s e^{v_{BE}/V_T} , \qquad (1.26)$$

where v_{BE} is the instantaneous base-emitter voltage; V_T is the thermal voltage, which is approximately 25 mV at 25°C; and I_S is the scale current, which

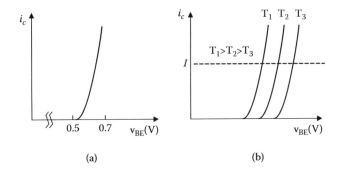

FIGURE 1.15
Characteristics of the transistor: (a) i_C-v_{BE} plot; (b) temperature behavior indicating -2 mV/°C at a constant emitter current.

is dependent on the process parameters and the dimensions of the transistor. Similar exponential plots exist for i_E-v_{BE} and i_B-v_{BE} but with different scale currents, I_S/α and I_S/β, respectively. Because the constant of exponential characteristic, $1/V_T$, is high (≈ 40), the curve rises very sharply. For v_{BE} smaller than 0.5 V, the current is negligibly small. Also, over most of the normal current range, v_{BE} lies in the range of 0.6 to 0.8 V. For a PNP transistor circuit, the i_C-v_{EB} characteristic will look identical to that of Figure 1.15, except that v_{BE} is replaced with v_{EB}.

In one of the most commonly used configurations, the common emitter configuration, i_C is dependent on the value of the collector voltage; this is shown in Figure 1.16. In this common emitter mode, at low values of v_{CE}, as the collector voltage goes below that of the base by more than 0.4 V, the collector-base junction becomes forward biased and the transistor leaves the active mode and becomes saturated. In the active region, i_C-v_{CE} curves are

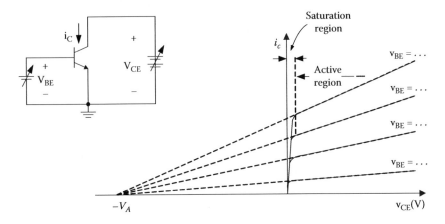

FIGURE 1.16
Common emitter characteristics indicating the practical behavior of transistors at different V_{BE} values.

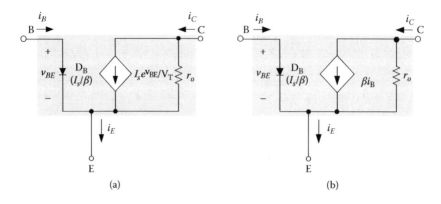

FIGURE 1.17
Large signal model of the transistor (a) based on v_{BE} as the control input; (b) based on i_B as the control input.

reasonably straight and have nonzero slope values. In fact, when these are extrapolated, they meet at a point on the negative axis, at $v_{CE} = -V_A$. The voltage V_A (a positive number) is a parameter of the particular BJT and is called the Early voltage. Typical values for V_A are in the range of 50 to 100 V. The linear dependence of i_C on v_{CE} can be accounted for by assuming that I_S remains constant and including a factor of $1 + (v_{CE}/V_A)$ in Equation 1.26. The nonzero slope of the i_C-v_{CE} straight lines indicates that the output resistance looking into the collector is not infinite. This gives us the relationship

$$r_o = (V_A + V_{CE})/I_C. \tag{1.27}$$

It is rarely necessary to include the dependence of i_C on v_{CE} in DC bias calculations or analysis. However, the parameter r_o can have a significant effect on the small signal gain of the transistor amplifiers.

Based on the above characteristics, a large signal equivalent circuit model for the BJT is depicted in Figure 1.17. In this figure we have assumed that the transistor gain β is constant for all operational conditions of the transistor. However, this assumption is not accurate, and in practical design we usually define two different β values, namely the large signal β or DC β and the small signal β or AC β.

Common emitter characteristics with base current as the control input are shown in Figure 1.18; compare this to the case of v_{BE} used as the input control parameter in Figure 1.16. The operating point Q in the graph allows us to define these two parameters, β_{DC} and β_{AC}, respectively, as

$$\beta_{DC} \equiv I_{CQ}/I_{BQ} \tag{1.28a}$$

$$\beta_{ac} \equiv \frac{\Delta i_C}{\Delta i_B}\bigg|_{v_{CE}=\text{constant}}. \tag{1.28b}$$

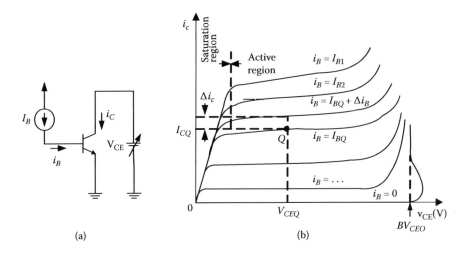

FIGURE 1.18
Common emitter characteristics showing the definition of h_{FE} and h_{fe} and also the primary breakdown limits at the v_{CE} value of BV_{CEO}. (a) Simplified circuit; (b) i_c versus V_{CE} characteristics with breakdown phenomenon at high V_{CE} values.

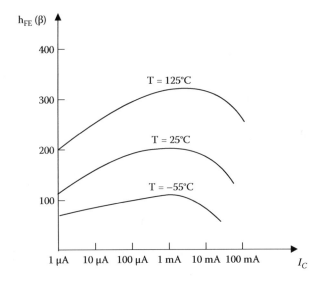

FIGURE 1.19
Dependence of h_{FE} for the DC collector current.

In data sheets, β_{DC} is referred to as h_{FE} and β_{AC} is referred to as h_{fe}, representing the two port h parameter equivalent circuits used. Figure 1.19 indicates the typical dependence of h_{FE} on the quiescent DC over a scale of several orders. In Figure 1.18 another important limiting parameter, collector-emitter breakdown voltage BV_{CEO}, is also indicated (on data sheets this parameter is sometimes referred to as sustaining voltage [LV_{CEO}]). This is the limiting value at $i_B = 0$ where the transistor collector current suddenly rises; under general operating conditions, reaching this collector-emitter voltage value should be avoided.

1.7.2   BJT Small Signal Operation and Models

In most amplifier configurations, a transistor circuit is biased around a quiescent point, and we try to operate the amplifier around that point for better linearity. In such situations, instead of the large signal models shown in Figure 1.17, small signal models are used. The parameters based on the hybrid-π model are shown in Figure 1.20d and Figure 1.20e. Note that these parameters are based on a circuit with DC sources as bias voltage supplies, later reduced to the two cases separately for DC bias analysis and AC signal components only, as in Figure 1.20a to Figure 1.20c. Figure 1.20f shows linear operation of the transistor under small signal conditions.

For the small signal models of the transistor, it can be shown that

$$g_m = I_c/V_T \tag{1.29a}$$

$$r_\pi = \beta/g_m = V_T/I_B. \tag{1.29b}$$

A more in-depth analysis of these topics with other models can be found in Sedra and Smith [5].

1.7.3   High-Frequency Models of the Transistors and the Frequency Response of Amplifiers

Let us begin this review with the transfer function of a capacitively coupled amplifier. As indicated in Figure 1.21, the amplifier gain is almost constant over a wide frequency range called the midband. In this frequency range, all capacitances such as coupling, bypass, and device internal values are considered to have negligible effects. At the high-frequency end of the spectrum, the gain drops because of the effect of device internal capacitances, whereas the low-frequency end is mostly governed by the bypass and the coupling capacitances. The extent of the midband is defined by the two frequencies ω_L and ω_H, where the gain drops by 3 dB below the midband. The amplifier bandwidth (BW) is usually defined as

$$BW = \omega_H - \omega_L. \tag{1.30}$$

In many cases where $\omega_L \ll \omega_H$ and $BW \approx \omega_H$ and a figure of merit for an amplifier, the gain-bandwidth product is defined as

$$GB \equiv A_M\omega_H, \tag{1.31}$$

where A_M is the magnitude of the midband gain.

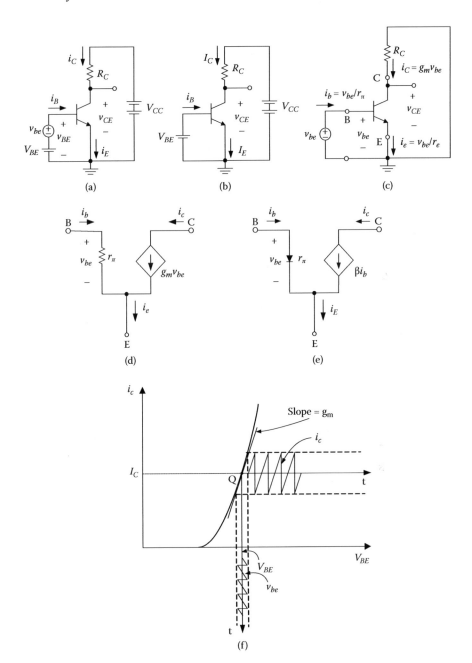

FIGURE 1.20

Common emitter transistor circuit for small signal analysis and bias calculations with the small signal hybrid-π equivalent circuits: (a) conceptual circuit to illustrate the amplifier with DC bias; (b) signal source v_{BE} eliminated for DC (bias) analysis; (c) small signal circuit without DC sources; (d) small signal model as a voltage-controlled current source (transconductance amplifier); (e) small signal model as a current-controlled current source; (f) relationship showing the small signal operation about a quiescent point Q defining the parameter g_m for the transistor.

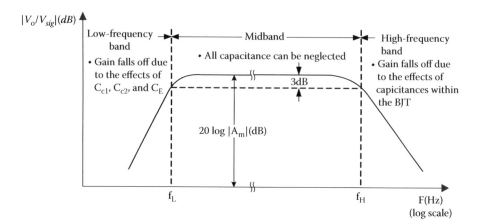

FIGURE 1.21
Frequency response of capacitively coupled amplifiers.

The amplifier gain as a function of the complex frequency s can be written in the form

$$A(s) = A_M F_L(s) F_H(s), \tag{1.32}$$

where $F_L(s)$ and $F_H(s)$ are functions that account for the dependence of the frequency in the low-frequency and high-frequency bands, respectively. For frequencies much above ω_L, the value of the function $F_L(s)$ approaches unity; similarly for the case of frequencies much lower than ω_H, the function $F_H(s)$ approaches unity. Thus, we can write the following approximate relationships for three different frequency bands:

$$A(s) \approx A_M, \text{ for the frequency range } (\omega_L \ll \omega \ll \omega_H) \tag{1.33a}$$

$$A_L(s) \approx A_M F_L(s), \text{ for the frequency range } \omega < \omega_L \tag{1.33b}$$

$$A_H(s) \approx A_M F_H(s), \text{ for the frequency range } \omega > \omega_H. \tag{1.33c}$$

Table 1.3 shows the characteristics and assumptions used in design and analysis for first approximations in circuits.[*]

1.7.3.1 Low-Frequency Response

In the low-frequency band, the function $F_L(s)$ takes the general form

$$F_L(s) = \frac{(s+\omega_{Z1})(s+\omega_{Z2})\dots(s+\omega_{ZnL})}{(s+\omega_{P1})(s+\omega_{P2})\dots(s+\omega_{PnL})}, \tag{1.34}$$

---

[*] In this discussion we disregard the frequency variations of transistor barameters and treat them as frequency independent variables.

TABLE 1.3

Characteristics and Assumptions for Three Frequency Bands of a Capacitively Coupled Amplifier

Frequency Range	Effect of Capacitors	Transfer Functions
0 to f_L	Coupling and bypass capacitors are in effect	$F_H(s) \approx 1$ $A(s) \approx A_M F_L(s)$
f_L to f_H	No capacitors are effective	$F_L(s) \approx 1; F_H(s) \approx 1$ $A(s) \approx AM$
Over f_H	Device internal capacitors are in effect	$F_L(s) \approx 1$ $A(s) \approx A_M F_H(s)$

where $\omega_{P1}, \ldots, \omega_{PnL}$ are positive numbers representing the frequencies of the n_L low-frequency poles and $\omega_{Z1}, \ldots, \omega_{ZnL}$ are positive, negative, or zero numbers representing the n_L zeros. It should be noted that as $s = j\omega$ approaches midband frequencies, $F_L(s)$ approaches unity. The designer's interest is usually in the low frequencies close to the midband, because it may be necessary to estimate and sometimes modify the value of the 3-dB frequency ω_L. In practical situations, the zeros may be at much lower frequencies than ω_L as to be of little importance in determining ω_L. Also, usually one of the poles (for example, ω_{P1}) has a much higher frequency than all other poles. If that is the case, for low values of ω close to the midband, $F_L(s)$ can be approximated by the transfer function of a first-order high-pass network:

$$F_L(s) = s/(s + \omega_{P1}). \tag{1.35}$$

In this case, the low-frequency response of the amplifier is dominated by the pole at $s = -\omega_{P1}$ and the lower 3-dB frequency is

$$\omega_L \approx \omega_{P1}. \tag{1.36}$$

In general, if this dominant pole approximation holds, it becomes a simple matter to determine the value of ω_L. Otherwise, one has to develop the complete Bode plot for $F_L(s)$ and thus determine the value of ω_L. As a rule of thumb, the dominant pole approximation can be made if the highest frequency pole is separated from the nearest pole or zero by at least a factor of four (that is, at least two octaves). If a dominant pole does not exist, an approximate formula can be derived for ω_L in terms of the poles and zeros. For the case of two poles and two zeros in the low-frequency band where

$$F_L(s) = \frac{(s+\omega_{Z1})(s+\omega_{Z2})}{(s+\omega_{P1})(s+\omega_{P2})}$$

by simple calculation of $|F_L(j\omega)^2|$ and equating it to half at $\omega = \omega_L$, it can be shown that

$$\omega_L \approx \sqrt{\omega_{P1}^2 + \omega_{P2}^2 - 2\omega_{Z1}^2 - 2\omega_{Z1}^2} \ . \tag{1.37}$$

This relationship can be extended to any number of poles and zeros, and if one of the poles, P1, is dominant, then Equation 1.37 reduces to Equation 1.36.

1.7.3.2   High-Frequency Response

In the low-frequency band, the function $F_H(s)$ can be expressed in the general form

$$F_H(s) = \frac{(1+s/\omega_{Z1})(1+s/\omega_{Z2})...(1+s/\omega_{ZnH})}{(1+s/\omega_{P1})(1+s/\omega_{P2})...(1+s/\omega_{PnH})}, \tag{1.38}$$

where $\omega_{P1}, \ldots, \omega_{PnL}$ are positive numbers representing the frequencies of the n_L high-frequency real poles and $\omega_{Z1}, \ldots, \omega_{ZnL}$ are positive, negative, or infinite numbers representing the n_H high-frequency zeros. Similar to the case of the low-frequency transfer function, as $s = j\omega$ approaches midband frequencies, $F_H(s)$ approaches unity. In this case also, the designer is interested only in the area that is close to the midband and many times will attempt to modify the value of the upper 3-dB frequency ω_H. In many cases, the zeros are either at infinity or at very high values, and the function can be simplified to the case of a first-order low-pass network, where

$$F_H(s) = \frac{1}{(1+s/\omega_{P1})}. \tag{1.39}$$

Under this simplification,

$$\omega_H \approx \omega_{P1} \tag{1.40}$$

and for the case of two poles and two zeros,

$$\omega_H = \frac{1}{\sqrt{\dfrac{1}{\omega_{P1}^2} + \dfrac{1}{\omega_{P2}^2} + \ldots - \dfrac{2}{\omega_{Z1}^2} - \dfrac{2}{\omega_{Z2}^2}}}. \tag{1.41}$$

Similar to the case of the low-frequency transfer function, if one of the poles is dominant, Equation 1.40 is valid as a first approximation.

1.7.3.3   Use of Short-Circuit and Open-Circuit Time Constants for the Approximate Calculations of ω_l and ω_H

If the poles and zeros of the transfer function can be determined easily, then the values of ω_L and ω_H can be determined easily using the methods described above. However, in many practical circuits it is not easy to determine poles and zeros. In such cases, the following common technique is helpful.

Let's consider the high-frequency response first. The factors in $F_H(s)$ of Equation 1.38 can be rewritten in the following form:

$$F_H(s) = \frac{1 + a_1 s + a_2 s^2 + \ldots + a_{nH} s^{nH}}{1 + b_1 s + b_2 s^2 + \ldots + b_{nH} s^{nH}}, \qquad (1.42)$$

where coefficients a and b are related to zero and pole frequencies, respectively. Specifically, the coefficient b_1 can be written as

$$b_1 = \frac{1}{\omega_{P1}} + \frac{1}{\omega_{P2}} . + \ldots + \frac{1}{\omega_{PnH}} . \qquad (1.43)$$

Assuming that there are n_H number of capacitors in the high-frequency equivalent circuit, the value of b_1 can be computed by summing a set of individual time constants called the open-circuit time constants. Open-circuit time constants are calculated by open-circuiting all other capacitors except a single capacitor, C_i, and then calculating the resistance seen by this capacitor as R_{io}. Once each time constant is calculated as $R_{io}C_i$, b_1 is exactly given by

$$b_1 = \sum_{i=1}^{n_H} R_{io}C_i . \qquad (1.44)$$

Specifically, if zeros are not dominant and if one of the poles is dominant, then Equation 1.42 reduces to

$$b_1 \approx 1/\omega_{P1}, \qquad (1.45)$$

and the approximate 3-dB frequency ω_H is given by

$$\omega_H \approx \frac{1}{\left[\displaystyle\sum_{i=1}^{n_H} R_{io}C_i \right]}. \qquad (1.46)$$

Even though it is sometimes hard to identify if there is a dominant pole or not, the relationship in Equation 1.46 provides good results [5].

Similarly, we can use the short-circuit time constants to determine the lower 3 dB frequency, ω_L, by rewriting Equation 1.34 as

$$F_L(s) = \frac{s^{nL} + d_1 s^{nL-1} + \ldots}{s^{nL} + e_1 s^{nL-1} + \ldots}, \qquad (1.47)$$

where coefficients d and e are related to the zero and pole frequencies, respectively, and the coefficient e_1 is given by

$$e_1 = \omega_{P1} + \omega_{P2} + \ldots + \omega_{PnL} = \sum_{i=1}^{n_L} \frac{1}{C_i R_{is}}, \tag{1.48}$$

where all other capacitors except C_i are set to short-circuit conditions, and then calculating the resistance R_{is} as seen by C_i.

Based on similar approximations, such as in the case of a high-frequency 3-dB value, this gives the approximate value for ω_L for the case where a dominant pole exists as

$$\omega_L \approx \sum_{i=1}^{n_L} \frac{1}{C_i R_{is}}. \tag{1.49}$$

In the design of practical amplifiers, these approaches to calculating ω_H and ω_L are very useful, particularly when there is dominant pole present. For more details, see Gray and Searle [8].

1.8 Transistor Equivalent Circuits, Models, and Frequency Response of Common Emitter/Common Source Amplifiers

This section discusses the applications of models and equivalent circuits to practically realize the parameters of a BJT- or FET-based amplifier. Figure 1.22a depicts the common emitter configuration with three capacitors: C_{C1}, C_{C2} (the coupling elements), and the bypass capacitor C_E. Operation is based on a single DC power rail. In this analysis we assume that the transistor β is finite and constant.

1.8.1 Calculation of the Low-Frequency 3-dB Corner Frequency, ω_L

The circuit in Figure 1.22b uses an improved version of small-signal equivalent circuit in Figure 1.20d while short-circuiting the V_{CC} rail to the ground. By using general circuit analysis techniques, we can determine the transfer function of the circuit and hence derive the poles and zeros. However, this may be a little too complicated for the busy circuit designer to get a rough estimate of ω_L, and, instead, we can use the method of short-circuit time constants, described earlier. The determination of ω_L proceeds as follows. First, we set V_{Sig} to zero and then set C_E and C_{C2} to infinity and find the resistance R_{C1} seen by C_{C1}. From Figure 1.22b, with C_E set to infinity, we find that

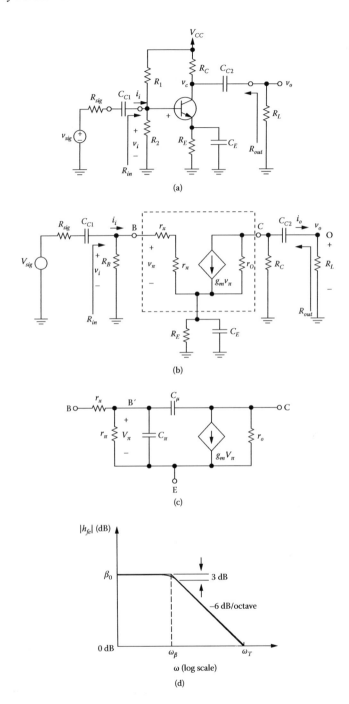

FIGURE 1.22
Common emitter transistor amplifier with three capacitors: (a) circuit; (b) equivalent circuit for small signal analysis; (c) transistor equivalent circuit at high frequencies; (d) transistor DC gain at different frequencies.

$$R_{C1} = [R_B//(r_x + r_\pi)] + R_{sig}. \tag{1.50a}$$

where, $R_B = R_1//R_2$.

Next, we set C_{C1} and C_{C2} to ∞ and determine the resistance R_{E1} seen by C_E. Assuming that the value of r_O is large, we can show that

$$R_{E1} = R_E // \frac{r_\pi + r_x + (R_B//R_S)}{\beta + 1}. \tag{1.50b}$$

Finally, we set both C_{C1} and C_E to ∞ and obtain the resistance seen by C_{C2} as

$$R_{C2} = R_L + (R_C//r_O). \tag{1.50c}$$

By combining the three short-circuit time constants for the three capacitors,

$$\omega_L \approx \frac{1}{C_{C1}R_{C1}} + \frac{1}{C_E R_{E1}} + \frac{1}{C_{C2}R_{C2}}. \tag{1.50d}$$

At this point, we should note that the zero introduced by C_E is at a value of s that makes $Z_E = 1/(1/R_E + sC_E)$ infinite:

$$s_Z = -1/(C_E R_E), \tag{1.50e}$$

and this zero is usually much lower than ω_L, justifying the approximations involved in the method used.

Given a desired value for ω_L, Equation 1.50d can be used for design as follows:

1. Because R_{E1} is usually the smallest of the three values R_{C1}, R_{E1}, and R_{C2}, we select a value for C_E so that $1/C_E R_{E1}$ is the dominant term on the right-hand side of Equation 1.50d. (This makes C_E the dominant low-frequency pole.)
2. We can make it so that $1/C_E R_{E1} = 0.8\omega_L$.
3. The remaining 20% of ω_L is then split equally between the other two terms.
4. Finally, the practical values for capacitors are determined so that the actual ω_L value is equal to or smaller than the specified low-frequency corner frequency.

1.8.2   Calculation of the High-Frequency 3-dB Corner Frequency, ω_H

The following section discusses the approach to obtain the high-frequency equivalent circuit, with some simplifications using Miller's theorem. Compared to Figure 1.20d, the equivalent circuit in Figure 1.22c has two internal capacitances of significance, C_μ and C_π. Also note that the resistor r_x is added to show the resistance between the physical base terminal and the fictitious internal terminal. Another useful property of this model is to explain the frequency-dependent behavior of the $|h_{FE}|$ as indicated in data sheet values. Figure 1.22d shows this, and details can be found in Sedra and Smith [5].

Looking at Figure 1.23a, it is easy to see that the analysis of a particular circuit could be tedious and time consuming. A useful technique based on Miller's theorem is to simplify the circuit to the case of Figure 1.23c. Figure 1.24 shows the application of Miller's theorem to arrive at the Miller's equivalent circuit for an impedance connecting the input and output side of a circuit. Referring to Figure 1.24a, the impedance Z connecting the two nodes labeled 1 and 2 can be replaced by two independent resistances, Z_1 and Z_2, where

$$Z_1 = Z/(1 - K) \tag{1.51a}$$

$$Z_2 = Z/(1 - 1/K). \tag{1.51b}$$

In this situation it is assumed that the voltage at node 2 is related to the voltage at node 1 by

$$V_2 = KV_1. \tag{1.51c}$$

One important assumption in this simplification is that while using this technique, the rest of the circuit remains unchanged. For details, see Sedra and Smith [5].

The circuit in Figure 1.23a can be simplified in two stages indicated in Figure 1.23b and Figure 1.23c and then analyzed using the techniques discussed in Section 1.7 to derive the two corner frequencies. Details of this analysis are beyond the scope of this chapter; see Sedra and Smith [5] for details.

We can extend the analysis to a common source MOSFET amplifier, as shown in Figure 1.25a. The high-frequency equivalent circuit model for the MOSFET is shown in Figure 1.25b. In deriving the high-frequency performance, we can assume that all capacitors such as C_{C1}, C_{C2}, and C_S act as perfect short circuits, and the circuit becomes equivalent to Figure 1.25c. Using Thevenin's equivalent at the input side and replacing the parallel combination of the three resistances at the output side, we can further simplify the circuit to the case in Figure 1.25d. Miller's theorem allows us to simplify the case in Figure 1.25c to the case in Figure 1.25d. In Figure 1.25c, the capacitance between the drain and gate forms an impedance similar to Z in Figure 1.24a, which will be equal to $1/sC_{gd}$. If we neglect the current flowing in the capacitor, the output voltage, v_o, will be approximately equal to $-g_m R'_L v_{gs}$. This gives us the amplification factor, $K = -g_m R'_L$, as in Figure 1.24b, and this allows us to represent the effect of Z using two elements given by

$$Z_1 = \frac{1}{(1 + g_m R'_L)sC_{gd}} \tag{1.51d}$$

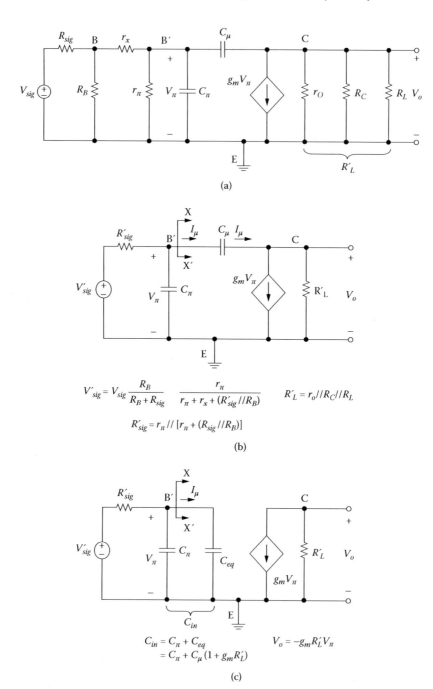

FIGURE 1.23
Calculation of high-frequency response of the common emitter amplifier: (a) equivalent circuit for small signal operation; (b) simplified at both the input and output side; (c) further simplification based on Miller's theorem.

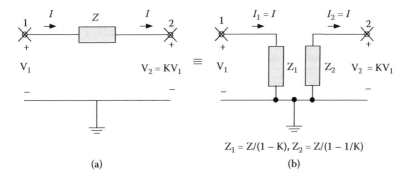

$$Z_1 = Z/(1 - K), Z_2 = Z/(1 - 1/K)$$

(a)   (b)

FIGURE 1.24
Miller's theorem for simplifications of circuits such as the high-frequency model of the BJT or FET:
(a) circuit to be simplified; (b) simplified case.

and

$$Z_2 = \frac{1}{1 - \left(\dfrac{1}{-g_m R_L}\right)} \times \frac{1}{sC_{gd}}.$$ (1.51e)

However, if the gain is relatively large compared to one, the effect of Z_1 will be a capacitance of value $g_m R'_L C_{gd}$ parallel with C_{gs} in Figure 1.25d. The effect of Z_2, appearing on the output side of the circuit (which is nearly equal to C_{gd}) will be negligible for first-order approximations. (The same concepts apply to the BJT example in Figure 1.23.) This indicates that there is a multiplication effect on C_{gd} that ultimately appears at the input side as $(1 + g_m R'_L)C_{gd}$. This is known as the Miller effect, and the multiplication factor is known as the Miller multiplier. Based on the discussions in previous sections, we can show that the approximate upper corner frequency of the circuit is given by

$$f_H = \frac{1}{2\pi \left(C_{gs} + \left(1 + g_m R'_L\right)C_{gd}\right)R'_{sig}}.$$ (1.51f)

In practical circumstances, as R_G is fairly large compared to the signal source resistance, we can see that the upper corner frequency is governed by the signal source resistance and the Miller effect. The analysis done here is a very approximate one based on a single time constant circuit at the input side. For a more rigorous analysis, see Chapter 6 of Sedra and Smith [5]. The lower corner frequency of the circuit f_L is based on three poles because of capacitors C_{C1}, C_s, and C_{C2}, and this can be calculated by the methods discussed in Section 1.8.1. A detailed analysis can be found in Chapter 6 of Sedra and Smith [5].

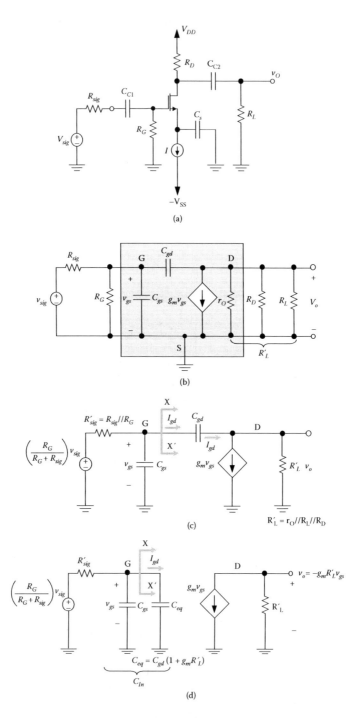

FIGURE 1.25
Common source amplifier: (a) circuit; (b) high-frequency equivalent circuit; (c) simplification
using Thevenin's equivalent circuit; (d) applying Miller's theorem to simply the circuit.

1.9 Noise in Circuits

In the broadest sense, noise can be defined as any unwanted disturbance that obscures or interferes with a desired signal. The existence of noise is basically due to the fact that electrical charge is not continuous but, rather, is carried in discrete amounts equal to the electron charge. Noise is a totally random signal and consists of frequency components that are random in both amplitude and phase. Although the long-term root mean square (RMS) value of noise can be measured, the exact amplitude at any instant of time cannot be predicted. In practical design environments, noise is important, as it can limit the resolution of a sensor or the dynamic range of the system. Much noise has a Gaussian or normal distribution of instantaneous amplitude with time. To a good engineering approximation, common electrical noise lies within the plus or minus three times the RMS value of the noise waveform.

There are three fundamental categories of noise in circuits: device noise, emitted or radiated noise, and conducted noise. Device noise can be in three different forms: thermal noise, low-frequency $(1/f)$ noise, and shot noise.

1.9.1 Noise in Passive Components

Thermal noise (or Johnson noise) is caused by the random thermally excited vibration of the charge carriers in a conductor. In every conductor at a temperature above absolute zero, the electrons are in random motion, and this vibration depends on the temperature. The available noise power (P_n) is proportional to the absolute temperature and the bandwidth of the measuring system. P_n is given by

$$P_n = kT\Delta f, \qquad (1.52)$$

where k is the Boltzman constant (1.38×10^{-23}), T is the absolute temperature, and Δf is the noise bandwidth of the measuring system. There is equal noise power in each hertz of the bandwidth, irrespective of the center frequency. For example, power in the bandwidth from 1 to 2 Hz is the same as from 1000 to 1001 Hz. This results in thermal noise being called "white noise." From the relationship in Equation 1.50 for a resistance of R, the equivalent RMS noise voltage, E_n, can be written as

$$E_n = \sqrt{4kTR\Delta f}\ . \qquad (1.53)$$

The above relationship defines a useful parameter, the noise-voltage spectral density, which is the equivalent noise voltage per 1-Hz bandwidth. Some useful facts in designing circuits are that the noise voltage spectral density of a 50-Ω resistor is approximately 0.9 nV/$\sqrt{\text{Hz}}$ and for a 1-kΩ resistor

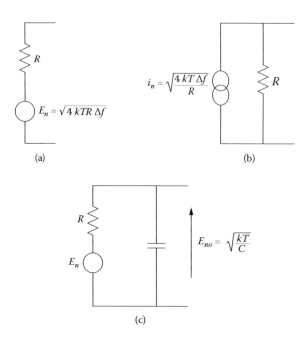

FIGURE 1.26
Thermal noise in a resistor: (a) noise as a voltage source; (b) noise as a current source; (c) noise in a practical resistor shunted by its parasitic capacitance.

is approximately 4 nV/√Hz. Figure 1.26a and Figure 1.26b represent alternatives for circuit equivalents for thermal noise.

If a DC flows through a resistor, in addition to the thermal (or Johnson) noise, an excess noise is generated. This is also referred to as current noise. This is due to the flow of current through a discontinuous medium such as a carbon composition resistor, where carbon granules and a binding medium are combined. This manufacturing process can cause microarcs due to variations in conductivity. Excess noise is usually a $1/f$ power spectrum, and it dominates at lower frequencies; at higher frequencies, thermal noise dominates. For a resistor, an experimentally determined equation for excess noise voltage (E_{ex}) is

$$E_{ex}^2 = \frac{K_{ex} I_{DC}^2 R^2}{f},\tag{1.54}$$

where K_{ex} is a constant dependent on the manufacturing process and I_{DC} is the DC through the resistor. For more details, see Chapter 9 of Motchenbacher and Fitchen [9]. Excess noise in a resistor becomes significant at low frequencies only when there is significant voltage drop across the resistor. Excess noise in a resistor can be measured in terms of a noise index, expressed in decibels. The noise index is the RMS noise in microvolts in the resistor per volt of DC drop across the resistor in every 10 units of frequency. For more details, see Motchenbacher and Connelly [9].

1.9.2 Effect of Circuit Capacitance

Generally, capacitors and inductors do not generate any device noise. Equation 1.53 predicts that an open circuit or an infinite resistance generates an infinite noise voltage. This is not practically observed, because there is always some shunt capacitance that limits the voltage. Figure 1.26 shows the case of a resistor shunted by its capacitance. In this case it can be shown [9] that total RMS noise at the output is given by

$$E_{no} = \sqrt{\int_0^\infty [E_{no}^2(f)]df} = \sqrt{\frac{kT}{C}}. \tag{1.55}$$

This relationship indicates that in practical circumstances, total noise energy of a resistor is limited by temperature and capacitance.

1.9.3 Noise in Semiconductors and Amplifiers

Different sources of noise are present in semiconductor devices. Significant types include shot noise, flicker noise, burst (or popcorn) noise, and avalanche noise. All these are in addition to the thermal noise discussed in the previous section.

Shot noise is always associated with a DC flow and is present in diodes, bipolar transistors, and MOS transistors. As the DC in a semiconductor is always based on electrons and holes, it can be shown that the resulting noise current, i_{ns}, has a mean square value of

$$\overline{i_{ns}^2} = 2qI_D\Delta f, \tag{1.56}$$

where I_D is the DC. Equation 1.56 is valid until the frequency becomes comparable to $1/\tau$, where τ is the transit time through the depletion region. For most devices, τ is extremely small, and Equation 1.54 is accurate well into the gigahertz region [10]. Note that shot noise current is independent of temperature. The junction diode small-signal equivalent circuit with noise is shown in Figure 1.27. The equivalent resistance, r_d, is considered a noiseless element, as it represents the dynamic resistance of the diode. For more details, see Gray et al. [10].

Similar to the case of excess noise in resistors, flicker noise is another noise source in semiconductors. The origin of flicker noise varies, but it is caused mainly by traps associated with contamination and crystal defects. These traps capture and release carriers in a random fashion, and the time

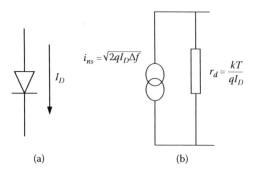

(a)   (b)

FIGURE 1.27
Small-signal equivalent of junction diode with noise.

constants associated with the process give rise to a noise signal with energy concentrated at low frequencies. This is always associated with a flow of DC and displays a spectral density of the form

$$\overline{i_{nf}^2} = K_f \frac{I^a}{f^b} \Delta f , \tag{1.57}$$

where I is the DC, K_f is a constant for a particular device, a is a constant in the range from 0.5 to 2, and b is a constant approximating unity. Because b is close to unity, the noise spectral density has an approximate $1/f$ frequency dependence. For this reason, flicker noise is sometimes called $1/f$ noise. Compared to shot noise and thermal noise, which have well-defined mean square values that can be expressed in terms of current flow, resistance, and well-known physical constants, flicker noise contains an unknown constant K_f. This constant not only varies by orders of magnitude from one device type to the next, but it can also vary widely for different transistors or ICs from the same wafer. Hence, statistically derived values from the process are used to predict the average or typical flicker noise performance for ICs from a given process [10].

Another type of low-frequency noise in ICs is burst noise or popcorn noise, where the noise is superimposed on a number of discrete levels (two or more) and lies in the audio range. This can produce a popping sound if fed to a loudspeaker and carries the wave shape and the spectral density shown in Figure 1.28. The spectral density of the burst noise is given by

$$i_{nb} = K_b \frac{I^c}{1+\left(\dfrac{f}{f_{cb}}\right)^2} \Delta f , \tag{1.58}$$

where K_b is the constant for a particular device carrying current of I, c is a constant in the range of 0.5 to 2, and f_{cb} is a particular frequency for a given

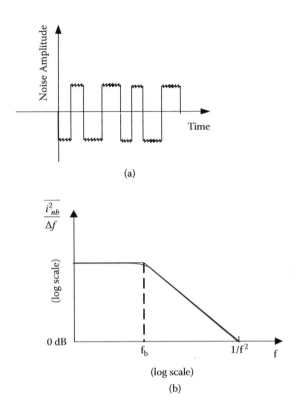

FIGURE 1.28
Burst noise: (a) typical waveform; (b) spectral density.

noise process. Burst noise can occur with different time constants and can produce multiple humps in the spectrum [10]. Figure 1.29 shows a combination of burst noise and flicker noise.

Avalanche noise is another form of a noise produced in zener or avalanche diodes. This is a cumulative effect of a large series of noise spikes and is associated with the DC through the device. The magnitude is difficult to predict, and it depends on the device structure and the uniformity of the silicon crystal [10,11]. A typical value for $\overline{v_n^2}/\Delta f$ could be 10^{-14} V²/Hz, and this could be equivalent to the thermal noise voltage of a 600-kΩ resistor and hence significant in value.

Based on the above discussion on noise sources, for basic semiconductor components one can develop small-signal equivalent circuits for diodes, BJTs, and MOS transistors, as detailed in Table 1.4. In the BJT noise model, r_b and r_c are physical resistors, but the relationships in Table 1.4 have taken only the significant contribution of r_b, as r_c is a relatively small series resistor. Note that r_π and r_o are fictitious resistors, and hence they have no noise contribution in the model. Avalanche noise in BJTs is negligible if V_{CE} is kept at least about 5 V below the breakdown voltage; thus, it is not shown in the model. The base current noise spectrum (without the burst noise component)

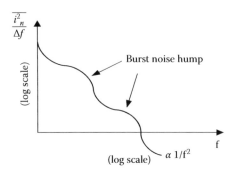

FIGURE 1.29
Combined spectral density due to burst and flicker noise.

is shown in Figure 1.30. The shot noise and flicker noise asymptotes meet at a corner frequency f_a that can be anything from 100 Hz to 10 MHz, depending on the processing. For details, see Gray et al. [10] and Peterson [12].

In MOS devices, because the channel material is resistive, thermal noise is significant, and flicker noise is also contributory. The relationship in Table 1.4 is usually valid for only long-channel devices, and for short-channel devices (less than 1-μm channel length), thermal noise is two to five times larger [13]. The relationships in each case in Table 1.4 are all independent of each other. In high-frequency long-channel MOS devices, another correlated gate noise component is added and the magnitude of correlation is about 0.39 [14]. For short-channel devices, this component could be higher [15].

For noise performance of transistor combinations and operational amplifiers, Gray et al. [10] provide a comprehensive discussion, and Israelsohn [16,17] provides a useful summary.

1.9.4   Circuit Noise Calculations and Noise Bandwidth

The device equivalent circuits discussed in previous sections can be used for the calculation of noise performance. First, the methods for circuit calculations with noise sources must be established. In general, for a noise source as in Figure 1.31a, we establish its input noise spectral density as a graph of power versus frequency, as in Figure 1.31b, and then we derive an equivalent sine wave at the selected frequency with a narrow bandwidth of Δf, as in Figure 1.31b. For the case of a noise current source with a mean square value

$$\overline{v_n^2} = S(f)\Delta f \, , \tag{1.59a}$$

this gives the equivalent RMS amplitude of

$$v_n = \sqrt{S(f)\Delta f} \, . \tag{1.59b}$$

TABLE 1.4

Device Equivalent Models with Noise Sources

Device	Equivalent Circuit	Noise Source Values (Approximate)
Junction diode (forward biased)	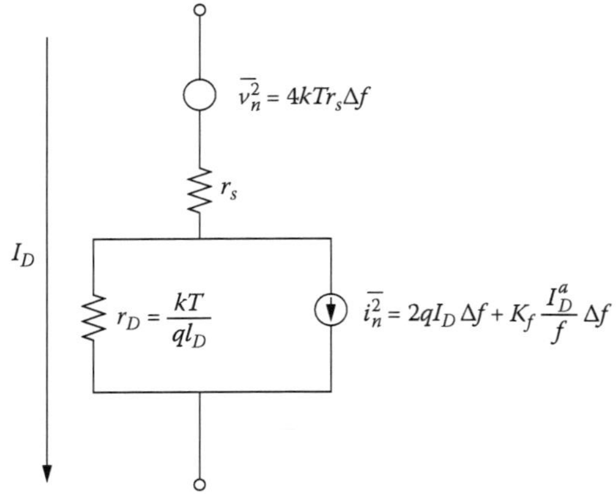 (a)	$\overline{v_n^2} = 4kTr_s\Delta f$ $\overline{i_n^2} = 2qI_D\Delta f + K_f \dfrac{I_D^a}{f}\Delta f$ I_D...Forward diode current
Bipolar transistor	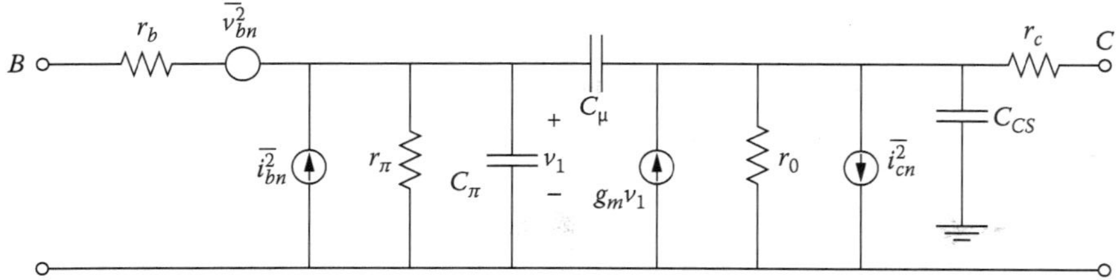 (b)	$\overline{v_{bn}^2} = 4kTr_b\Delta f$ $\overline{i_{cn}^2} = 2qI_C\Delta f$ $\overline{i_{bn}^2} = 2qI_B\Delta f + K_f \dfrac{I_B^a}{f}\Delta f + K_b \dfrac{\dfrac{I_B^c}{f}}{1 + \left(\dfrac{f}{f_{cb}}\right)^2}\Delta f$ I_B...DC current in the base I_C...DC current in the collector

(continued on next page)

In the equivalent circuit (a): $\overline{v_n^2} = 4kTr_s\Delta f$, r_s, $\overline{i_n^2} = 2qI_D\Delta f + K_f \dfrac{I_D^a}{f}\Delta f$, $r_D = \dfrac{kT}{qI_D}$, I_D.

In the equivalent circuit (b): r_b, $\overline{v_{bn}^2}$, $\overline{i_{bn}^2}$, r_π, C_π, C_μ, $+v_1-$, $g_m v_1$, r_0, $\overline{i_{cn}^2}$, r_c, C_{CS}, B, C.

TABLE 1.4 (continued)

Device Equivalent Models with Noise Sources

Device	Equivalent Circuit	Noise Source Values (Approximate)
MOS transistor		$\overline{i_{dn}^2} = 4kT\left(\dfrac{2}{3}g_m\right)\Delta f + K_f \dfrac{I_{DM}^a}{f}\Delta f$ $\overline{i_{gn}^2} = 2qI_{GM}\Delta f$ (for long channel devices) $\overline{i_{gn'}^2} = \dfrac{16}{15}kT\omega^2 C_{gs}^2\Delta f$ (for high-frequency, long channel devices) $C_{gs} = (2/3)C_{ox}WL$ I_{DM}…drain bias current g_m…device transconductance at the operating point C_{gs}…gate–source capacitance C_{ox}…gate capacitance per unit area W…Width of the device channel L……Length of the device channel

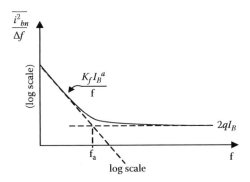

FIGURE 1.30
Base current noise density of a BJT.

Using values similar to the case in Equation 1.59b, we can treat noise sources as familiar sinusoidal waveforms for mathematical treatment. For a given noise source, noise bandwidth is defined as a range of frequency where the noise amplitude is flat (compared to the case of a single-pole amplifier, where frequency rolls off with a 3-dB amplitude at the corner frequency), based on the relationship

$$f_N = \frac{1}{A_{v0}^2} \int_0^\infty |A_v(jf)|^2 df , \qquad (1.59c)$$

where A_{v0} is the low-frequency gain of the circuit transfer function and $A_v(jf)$ is the magnitude of the voltage gain of the circuit at a selected frequency f. Figure 1.31c and Figure 1.31d show the circuit transfer function and the output noise voltage spectral density, respectively. Figure 1.31e shows the concept of noise bandwidth, with a flat noise voltage spectral density over the band-width of f_N.

Based on the relationship in Equation 1.59c, the value of f_N for a low-pass filter with a 3-dB corner frequency of f_1 is $1.57f_1$. For a detailed discussion of this topic, see Motchenbacher and Connelly [9] and Gray et al. [10].

1.9.5 Noise Figure and Noise Temperature

The noise figure is a commonly used method of specifying the noise perfor-mance of a circuit or device. The definition of the noise figure (F) of a circuit is

$$F = [SNR]_{input}/[SNR]_{output}. \qquad (1.60)$$

F is usually expressed in decibels. This parameter approximates the level of degradation of the SNR caused by the circuit. For an ideal noiseless ampli-fier, this figure is unity or 0 dB. A useful alternative representation of Equa-tion 1.58 is

$$F = \text{total output noise/output noise due to source resistance.} \qquad (1.61)$$

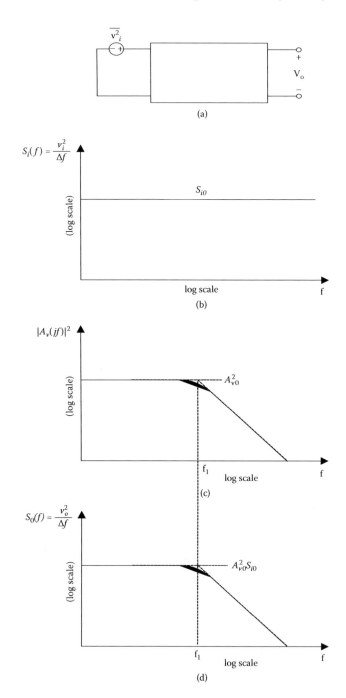

FIGURE 1.31
Circuit noise bandwidth calculation: (a) equivalent input noise voltage generator; (b) equivalent input noise voltage spectral density; (c) circuit transfer function; (d) output noise voltage spectral density; (e) equivalent transfer function indicating the noise bandwidth f_N.

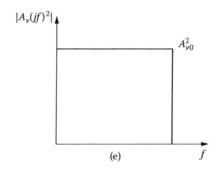

FIGURE 1.31 (continued)

The noise temperature (T_n) of a circuit is defined as the temperature at which the source resistance R_s must be held so that the noise output from the circuit due to R_s equals the noise output due to the circuit itself.

In practical circuits where there are different noise sources, one needs to ascertain the overall effect of all noise sources. If two noise sources of equivalent noise voltages v_{n1} and v_{n2} are connected in series, the resultant noise voltage is given as

$$V_{total} = \sqrt{v_{n1}^2 + v_{n2}^2 + 2C_n v_{n1} v_{n2}} ,$$
(1.62)

where C_n is a correlation coefficient between −1 and 1, including 0. When the noise sources are completely independent (such as in the case of two series resistances), the value of this coefficient is zero. This relationship indicates that for totally independent noise sources connected in series, an RMS value can be estimated [9].

1.10 Passive Components in Circuits

Except in the case of predominantly digital circuit blocks, passive components such as resistors, capacitors, inductors, and transformers are required as basic elements in circuit design. For example, good circuit design practice demands accurate and stable amplifiers, but active devices are by nature unstable, so they need to be tamed with passive components. Feedback is employed in almost all circuit designs to ensure that the circuit performance is a function of the passive rather than the active components. Passive components are neglected in the rush to complete the design of electronic systems. Many designers select passives as an afterthought and choose them from a list of standard components. Although this practice is adequate for some circuits, it does not suffice in the demanding world of high-frequency amplifiers, precision sample-hold circuits, data converters, and other analog and mixed signal circuits. The hardware designer must select adequate passive components to obtain the specified performance in demanding applications.

The first selection criterion for passive components dictates that they should be accurate and stable to ensure proper circuit performance. Second, there are further requirements such as low cost, small size, and surface mounting that must be met to satisfy the broader design specifications. Accuracy normally dictates larger size, so the accuracy requirement and the small-size requirements often conflict. With new families of surface-mount parts entering the commercial domain, the design engineer is constantly searching for accurate and stable passives that meet all design criteria. Responsible and reputable manufacturers will provide designers with application information such as dimensions, electrical specifications, model parameters, performance curves at different frequencies or other operating conditions, tolerances, and reliability data. If a manufacturer is not responsive or is reluctant to release such data, the designer should look for alternative suppliers.

1.10.1 Resistors

There are many different kinds of resistors available for designers, including carbon film, carbon composition, metal film, wire wound, and bulk metal types. For surface-mount designs, surface-mount chip film resistors are available. A resistor is characterized by its resistance value, power dissipation capability, temperature coefficient, and manufacturing tolerances. Tolerances can range from ±10% for carbon composition resistors, to 0.01% to 0.1% for precision metal film resistors, to as low as 0.005% for precision bulk metal types. Temperature coefficients can be as low as ±1 to ±5 ppm for bulk metal types to as high as +400 to −900 ppm for carbon composition resistors. Details are available in Williams [18] and Fowler [19].

A resistor at high frequency is not ideal, and it has an equivalent circuit, as shown in Figure 1.32a. A simplified version of the equivalent circuit is shown in Figure 1.32b, and the overall impedance at different frequencies is depicted in Figure 1.32c. In Figure 1.32a, C_G indicates the stray capacitance from the resistor body to the ground, and C_P indicates the capacitance across the terminals, whereas L_L indicates the inductance of each lead. In summary, these models should remind a designer that a simple resistor at a DC will not be an ideal resistor at higher frequencies. In power circuit designs and pulsed operations, other important parameters come into play [18]. The limiting element voltage (LEV) is such a parameter and specifies the maximum applicable voltage difference across the terminals of a resistor. This will be the case for high-value resistances, because calculations according to the maximum allowed power dissipation will make higher voltages possible across the device [18]. Pulse ratings are applied when the resistor is used in a situation like surge protection in telecommunications circuits [18]. In design calculations, such as derived voltage reference sources using high-precision reference sources and resistor dividers, tolerance values and temperature coefficients can significantly alter the expected output at different temperatures.

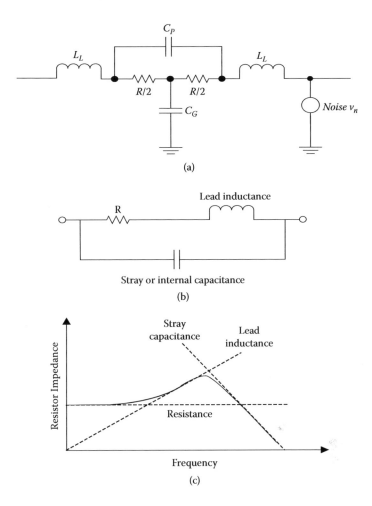

FIGURE 1.32
Resistors at higher frequencies: (a) an accurate model with stray effects and noise source; (b) simplified equivalent; (c) impedance value of a resistor at different frequencies.

1.10.2   Capacitors

There are many different types of capacitors. The common types are film, paper, ceramic, and electrolytic. Film capacitors can be subdivided into polyester, polycarbonate, polypropylene, and polystyrene; ceramic types can be subdivided into single layer or multilayer; and electrolytic types can be subdivided into nonsolid and solid aluminum and solid tantalum. Multilayer ceramics come in different types, such as COG, X5R, X7R, and Z5U, based on the dielectric. Details on the construction and characteristics of these can be found in Williams [18], Fowler [19], Bateman [20–22], van de Steeg [23], and Al-Abed and Gath [24].

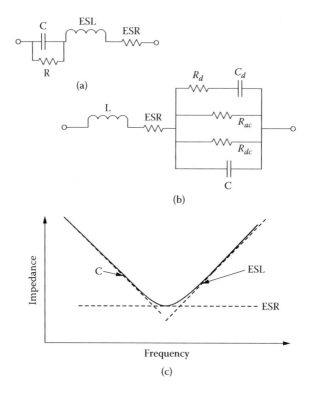

FIGURE 1.33
Capacitors in practical circumstances: (a) a simplified version; (b) a comprehensive equivalent circuit; (c) impedance characteristics based on the simplified model.

An ideal capacitor blocks DC, and its impedance is expected to drop with the inverse of the frequency based on the relationship $X_C = 1/2f\pi C$. However, in practical devices the device behaves as a complex combination of the ideal capacitor, with value C and other components as in Figure 1.33a. In this figure L_c indicates the equivalent series inductance (ESL) and R_C indicates the equivalent series resistance (ESR) composed of the lead resistances, electrodes, and terminating resistances, and these two values are important design parameters when the capacitor is expected to work at higher frequencies or in high-voltage situations, such as in power supply designs. R_{ac} and R_{dc} represent the equivalent resistance due to AC dielectric losses and leakage resistance due to DC dielectric losses, respectively. R_{ac} may vary nonlinearly with frequency and temperature, whereas R_{dc} may vary with temperature. Another annoying situation in some types of capacitors is the dielectric absorption property, in which the dielectric material does not become polarized instantly, causing a "memory effect." This is represented by R_d and C_d and can be problematic in sample and hold circuits if the wrong capacitor type is selected. In such a situation, if the capacitor is charged, discharged, and left open, it will recover some of its charge due to the parasitic effect of

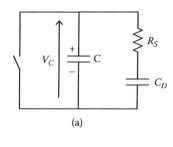

(a)

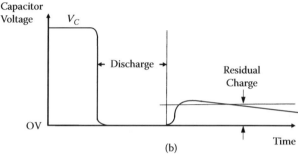

(b)

FIGURE 1.34
Effect of dielectric absorption: (a) equivalent circuit; (b) charge transfer.

R_d and C_d, known as the memory effect. Figure 1.33b shows a comprehensive equivalent circuit. Figure 1.33c shows the impedance versus frequency.

Figure 1.34 shows the effect of dielectric absorption when a capacitor is fully charged and then discharged by a short circuit, which should be seriously considered in designing sample and hold circuits [25].

1.10.3 Inductors

The ideal inductor is a simple magnetic element that should pass DC without any loss and at high frequencies creates an impedance given by $X_L = 2f\pi L$. However, the real-world situation is similar to the case for capacitors, with several other parasitic components, as in Figure 1.35, with frequency-dependent impedance characteristics. R_w is the resistance due to the winding and its terminations, which can vary with temperature. R_l is an equivalent parallel resistance due to core losses, which is dependent on frequency, temperature, and DC through the inductance. C_p is the self-capacitance of the winding.

1.10.4 Passive Component Tolerances and Worst-Case Design

Building reliable hardware requires accounting for all tolerances during the design stage. Whereas data sheets discuss active component errors due to parameter deviations, few designers value the need for considering the

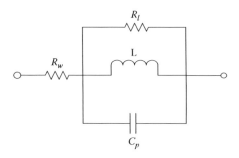

FIGURE 1.35
Inductor-equivalent circuit.

passive component tolerances in critical designs, particularly in worst-case designs. Worst-case design lets the components assume a wide range of values, which leads to a wide range of solutions; some of these solutions may be undesirable.

As an example, let's briefly discuss resistor tolerances. All resistors are specified with the purchase tolerances, Δ_p, which have common percentage values such as 0.5, 1, 2, 5, and 10. In many cases the values of the resistors may be almost at the maximum deviation, as some manufacturers select the better values to sell as tighter tolerance grades. In the manufacturing process, external stresses such as soldering can cause resistor values to change during assembly. Hence, resistor values may change beyond the purchase tolerance before the assembly leaves the factory. Also, component values change during their lifetimes because of external stresses such as temperature, humidity, and mounting. Component value changes during operation are called drift tolerances, or Δ_d, and are expressed as percentages. For a worst-case calculation, the designer should take the sum of the worst possibilities of Δ_p and Δ_d, which is $|\Delta_p| + |\Delta_d|$. Mancini [26] provides a practical overview of applying these in ratiometric circuits such as potential dividers, nonratiometric circuits such as voltage sources based on a simple resistor and a current source, and a difference amplifier. For example, a nonratiometric circuit designed with ±5% tolerance could result in an ultimate voltage source variation of ±10%. For a voltage divider, gain error could vary depending on the ratio of the two resistors, and for ±1% resistors the overall gain variation can be between ±1% and ±1.8% for resistor ratios from 0.5 to 0.1 [26].

Capacitor tolerances can be handled in the same way as resistors, bearing in mind that the variations are radically dependent on the manufacturing method. Electrolytic capacitors can have very large variations, from +80% to −20%, whereas some glass and ceramic capacitors have tolerances of only ±1%. In general, it is safe to triple the tolerances unless the data sheet gives specific values [26].

References

1. Kularatna, N., *Modern electronic test and measuring instruments*, Intitution of Electrical Engineers, London, 1996.
2. Beadle, E.R., Unifying overview of applied transform theory, *Electronic Design*, May, 107, 1995.
3. Lynn, P.A. and Fuerst, W., *Introducing Digital Signal Processing*, John Wiley & Sons, New York, 1990.
4. Kularatna, N., *Digital and Analogue Instrumentation: Testing and Measurement*, Institution of Electrical Engineers, London, 2003.
5. Sedra, A.S. and Smith, K.C., *Microelectronic Circuits*, 5th ed., Oxford University Press, New York, 2004.
6. Haykin, S.S., *Active Network Theory*, Addison Wesley, Reading, MA, 1970.
7. Ayers, J., *Analysis and Design of Digital Integrated Circuits*, CRC Press, Boca Raton, FL, 2004.
8. Gray, P.E. and Searle, C.L., *Electronic Principles*, John Wiley & Sons, New York, 1969.
9. Motchenbacher, C.D. and Connelly, J.A., *Low-Noise Electronic System Design*, John Wiley & Sons, 1993.
10. Gray, P.R., Hurst, P.J., Lewis, S.H., and Meyer, R.G., *Analysis and Design of Analog Integrated Circuits*, 4th ed., John Wiley & Sons, 2001.
11. Haitz, R.H., Controlled noise generation with avalanche diodes, *IEEE Transactions on Electron Devices*, 12, 198, 1965.
12. Peterson, D.G., Noise performance of transistors, *IEEE Transactions on Electron Devices*, 9, 296, 1962.
13. Abidi, A.A., High frequency noise measurements in FETs with small dimensions, *IEEE Transactions on Electron Devices*, 33, 1801, 1986.
14. van der Ziel, A., *Noise in Solid State Circuits*, John Wiley & Sons, New York, 1986.
15. Shaffer, D.K. and Lee, T.H., A 1.5-V, 1.5-GHz CMOS low noise amplifier, *IEEE Journal of Solid State Circuits*, 32, 745, 1997.
16. Israelsohn, J., Noise 101, *EDN*, January 8, 42, 2004.
17. Israelsohn, J., Noise 102, *EDN*, March 18, 47, 2004.
18. Williams, T., *The Circuit Designer's Companion*, 2nd ed., Newnes, London, 2004.
19. Fowler, K.R., *Electronic Instrument Design: Architecting for the Life Cycle*, Oxford University Press, New York, 1996.
20. Bateman, C., Understanding capacitors, *Electronics World*, April, 324, 1995.
21. Bateman, C., Understanding capacitors: aluminum and tantalum options, *Electronics World*, June, 495, 1998.
22. Bateman, C., Understanding capacitors, *Electronics World*, July, 594, 1998.
23. van de Steeg, T., Selecting optimum electrolytic capacitors requires an understanding of their characteristics and technology, *PCIM*, June, 38, 2000.
24. Al-Abed, B. and Gath, P.A., Internal construction boosts electrolytic capacitor operational life, *PCIM*, January, 64, 2001.
25. Kularatna, N., *Modern Electronic Components and Circuit Block Design*, Newnes, London, 2000.
26. Mancini, R., Worst-case circuit design includes component tolerances, *EDN*, April 15, 61, 2004.

2

Design Process

Shantha Fernando

CONTENTS

2.1 Introduction 58
2.2 Specifications for the Design 59
    2.2.1 Understanding Customer Requirements 59
    2.2.2 Marketing Requirements Specification 60
    2.2.3 Design Specifications 60
    2.2.4 Software Requirements 60
2.3 Internal Departments and Their Responsibilities in Product
    Design 61
    2.3.1 Design Engineering 61
    2.3.2 Component Engineering 62
        2.3.2.1 Reliability of Components 63
        2.3.2.2 Spares Requirement 63
        2.3.2.3 Preferred Component List 63
        2.3.2.4 Component Specifications 64
    2.3.3 Production Engineering 65
    2.3.4 Test Engineering 65
    2.3.5 Material Procurement 65
    2.3.6 Quality Assurance 66
        2.3.6.1 Quality Plan for Product Design 66
        2.3.6.2 Quality Plan Ensuring Performance and
                Reliability 66
        2.3.6.3 Product-Specific Quality Plan 66
    2.3.7 Engineering Services 66
    2.3.8 Marketing 67
2.4 Role of External Agencies 67
    2.4.1 Regulatory Agencies 67
        2.4.1.1 Electromagnetic Compliance 67
        2.4.1.2 Compliance Folder Management 68
    2.4.2 Test Houses 69
    2.4.3 Component Suppliers 69
        2.4.3.1 Component Order Codes 69

2.5   Reviews 69
       2.5.1   Specifications Review...70
       2.5.2   Review of the Test Plan..70
       2.5.3   Review of Design Release for Pilot Production..........................70
2.6   Product Documentation...70
       2.6.1   Production Documents 71
                2.6.1.1   Bill of Materials 71
                2.6.1.2   Factory Test Specifications...72
                2.6.1.3   Assembly Instructions 72
       2.6.2   User Manual 73
       2.6.3   Technical Manual..73
2.7   Design Management..73
       2.7.1   Getting Started..73
       2.7.2   Project Scope...73
       2.7.3   Project Strategy 74
       2.7.4   Managing Project Scope Changes..74
       2.7.5   Managing Project Time..74
       2.7.6   Implementing the Project Schedule 74
       2.7.7   Managing Project Costs 75
       2.7.8   Managing Project Quality 75
       2.7.9   Managing Project Personnel 75
       2.7.10   Project Closure 75
References 76

2.1   Introduction

The design process identifies important considerations in product development and ensures those considerations are implemented in the product design. Designers are constantly under pressure to have their designs production ready to meet "time-to-market" (TTM) deadlines. It is the responsibility of the engineering manager and design team leader to agree on a reasonable project timetable and ensure that the design team is not pressured to cut corners to release the design for production early. This chapter discusses the numerous activities of the design engineering department during product development in order to achieve a reliable end product. The chapter also discusses important reviews that will help to realize a production-friendly design. Such a design approach helps to achieve higher production yield and thus results in lower product cost. The role played by component engineering in the design process is another area discussed in this chapter.

The content of this chapter is based on the author's own working experience as a senior engineer in design and component engineering of large organizations during the 1990s. These companies practiced concurrent engineering concepts throughout the design phase of product development.

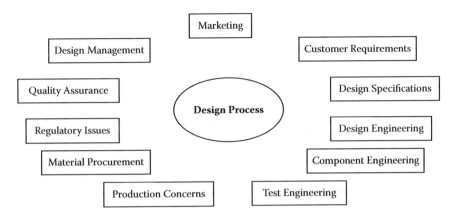

FIGURE 2.1
Key elements of the design process.

Some companies have now restructured their organizations to outsource production of their designs to specialized contract manufacturing firms. Such organizational changes created additional challenges for designers. However, the approach outlined in this chapter can be applied to the design process irrespective of where the production takes place.

2.2   Specifications for the Design

Translating customer expectations or marketing requirements to a design specification is the first and foremost challenge that design engineers encounter. This is a specialized area, and some companies employ functional or product analysts to perform this task.

2.2.1   Understanding Customer Requirements

Understanding what the customer expects from a product is of paramount importance to every department associated with the development of that product. If you are developing a product as the result of a successful bid in response to a tender, then customer requirements are embedded in the tender document. Furthermore, the engineering department may have been heavily involved in asserting the compliance statements for the bid specifically against the clauses in the technical requirements.

Alternatively, the marketing department may have made a successful unsolicited bid to a customer to develop a new product or to enhance the functionality of an existing product. In any event, obtaining a customer's required specifications is essential before detailed planning begins. This process must be followed whether the customer is an internal division, a small company, or a large agency. An accurately drafted customer requirements document can be the basis for a more elaborate marketing requirements specification. Figure 2.1 shows the key elements affecting the design process.

2.2.2   Marketing Requirements Specification

Companies will not spend money on developing products that do not have a sales potential. This development can include new product development or improvement of an existing product. Whatever the case may be, engineers will not develop a product without reference to the marketing requirements specification (MRS). It is the responsibility of the engineering manager, in consultation with experts, to clarify all design requirements outlined in the MRS. These requirements are then translated into proper design specifications, with reference to standard specifications wherever possible.

2.2.3   Design Specifications

Design specifications are developed per broadly set MRS guidelines. The design specifications define the operation of the product in much greater detail than what is stated in the MRS. Good design specifications detail how the product achieves its functional objectives. In this context the specifications must cover any design formulae, including calculations, input and output waveforms, cabling arrangements between subassemblies, and mechanical and physical properties. Table 2.1 presents a template for design specifications.

2.2.4   Software Requirements

Software design is defined as both "the process of defining the architecture, components, interfaces, and other characteristics of a system or component" and "the result of [that] process" [1]. In this chapter, emphasis will be placed on embedded software design and the procedures that are important to software designers in implementing embedded software for a product. Today, most electronic equipment operates under the control of embedded software. Domestic electrical appliances, from doorbells to dishwashers, now use microcontrollers in their operation. Software that is embedded in such hardware equipment is known as firmware. A definite advantage of having firmware is that firmware can be updated to improve a product's capability without changing its hardware. Companies try to avoid changing hardware, as such retrofits can be very costly.

The work flow of building software for a product is no different from any other product build. It must start with unambiguous requirement specifications. Requirements for a firmware design must be created with agreement between hardware and firmware engineers. Firmware design specifications are closely associated with hardware design specifications; therefore, firmware engineers must understand how the hardware of a product works. Firmware may interface with many hardware devices. These peripheral devices to firmware can range from a simple transistor circuit activating a relay to a complex field programmable gate array (FPGA) generating system timing.

TABLE 2.1

Sample Template for Design Specifications

Section	Subsection	Subsection Details
Introduction	Purpose	Describe the hardware
	Referenced documentation	Other applicable documentation, company and industry standards
	Abbreviations	Key to abbreviations
System overview	General description	
	Block schematics	Detail the function of each block
Design requirements	Printed circuit board	Dimensions of PCB, layering, component selections, technology, isolation requirements
	Mechanical/enclosure	Environmental [Ingress Protection (IP) rating]
	Circuit design	Design considerations for testing and production, component selections
	Physical design	Physical details (dimensions, color, weight)
	Connectors and cabling	Cabling diagrams, cable interconnection details between subassemblies
Power and thermal	Power and thermal requirements	Total power consumption, cooling needs
Regulatory compliance	EMC	Applicable standards and strategies to achieve compliance
	Safety	Applicable standards and design requirements, earthing arrangements
Environmental	Applicable standards	Temperature cycling
		Drop and vibration tests
Quality and reliability	Quality plan	Applicable generic quality plans or product-specific quality plan, reliability estimations
Documentation control	Version control	Change history
		Authors/approval authority

2.3 Internal Departments and Their Responsibilities in Product Design

2.3.1 Design Engineering

The design engineering department is the engine room of product design. It operates under the engineering manager, whose responsibility is to manage engineering resources effectively and efficiently and offer both product design and product support services to the organization. The engineering manager may also manage engineering services functions, such as component engineering, the drawing office, and the printed circuit board (PCB) design office.

Technology selection for a design is obviously an engineering responsibility. Selection of a technology can be influenced by various factors, such as cost, manufacturability, time-to-market constraints, availability of in-house design expertise and supporting tools, physical limitations, and compatibility with other designs. Technology selection for a design may influence the formation of the design team. Hardware, firmware, and software designers can be assigned or recruited by considering their related experiences. It may not always be possible to form a team that brings all necessary skills and experience to satisfy the design needs. In such circumstances, management should arrange training in the required disciplines.

Apart from having the desired skills in their respective design areas, whether they be in radio frequency engineering, analog circuit design, high-speed digital design, or embedded circuit design, today's hardware engineers are expected to use simulation software, possess knowledge of hardware description languages (HDLs), be competent in schematic capture, be able to produce PCB layout instructions to minimize electromagnetic compatibility (EMC) problems while maintaining signal integrity, and have the ability to write design documentation.

If the design involves firmware development, then the firmware engineer must be able to

- Capture embedded firmware requirements for the product
- Produce firmware architecture for review
- Design, develop, and write code for the application
- Develop test plans and verify the design using simulators and emulators and subsequently test the target product
- Produce documentation for firmware maintenance and factory programming of the microcontroller or associated memory chip

2.3.2   Component Engineering

Component engineering plays a vital role in any engineering organization. Component engineering is an engineering services role. This position liaises to the design, procurement, and manufacturing departments with respect to issues concerning the components proposed for the product.

Component engineering is responsible for the maintenance of the component database. The tasks related to the component database include supplier approvals, specifying order codes, creating part numbers, and assigning reliability values for parts. Component engineering also manages component obsolescence issues.

Circuit designers must work closely with component engineers during the initial phase of the design cycle. Generally designers should choose components from the preferred component list. They are also directed to choose components with multiple sourcing arrangements. These steps minimize the risk of production line disruptions in the event a supplier has difficulty delivering the ordered components for production. However, certain parts may have a single source. Often these suppliers are reputable manufacturers of components with a proven track record of managing customer relationship in case of discontinuity, substitution, or obsolescence of their products.

2.3.2.1 Reliability of Components

Every product must have a failure rate expressed in failures in time (FIT) and mean time between failures (MTBF). The FIT rate for a product is determined by summing the FIT rates for individual components used in the design.

Failures in time is expressed as failures per 10^9 hours. The MTBF is expressed as $(10^9/\text{FIT})$ hours. There are two widely used reliability prediction methods: standard MIL-HDBK-217 [2] and Bellcore Technical Reference TR-NWT-000332, which is the same as Telcordia SR-332.

Standard MIL-HDBK-217 is a widely used method. It presents two prediction methods:

> These methods vary in degree of information needed to apply them. The Part Stress Analysis Method requires a greater amount of detailed information and is applicable during the later design phase when actual hardware and circuits are being designed. The Parts Count Method requires less information, generally part quantities, quantity level, and the application environment. This method is applicable during the early design phase and during proposal formulation. In general, the Parts Count Method will usually result in a more conservative estimation (i.e., higher failure rate) of system reliability than the Parts Stress method [2].

As mentioned earlier, it is the responsibility of the component engineer to include the FIT rate for individual components in the component database. Moreover, designers can use commercially available software programs to predict accurate reliability figures for their designs. These programs allow the designers to set operating parameters for components and gauge more predictable reliability values for designs. Some larger companies have formulated company-specific reliability prediction methods, although such methods are fundamentally based on MIL-HDBK-217. They use the field return statistics and corresponding part failures to estimate more realistic or company-specific FIT rates for the components.

2.3.2.2 Spares Requirement

Reliability data of a product help determine recommended quantities of spare parts for the customers to stock.

2.3.2.3 Preferred Component List

Component engineering, in consultation with the procurement and manufacturing departments, creates and maintains a preferred component list.

Designers must select components from this list wherever possible in their designs. The advantages of choosing components from the preferred list are:

- The company may be buying these components in bulk quantities at a more competitive price.
- Design and testing times are reduced as preferred components are usually in stock for prototype assembly.
- Documentation pertaining to preferred components is already available (computer-aided design [CAD] symbols, inspection and test specifications)
- The components are procured from reliable suppliers.
- The components have been field tested.

2.3.2.4 Component Specifications

Every part in the component database relates to a specification. In most cases, the manufacturer's order code for the part and its data sheet define the electrical, mechanical, and packaging parameters for the part. Although this is true for the majority of the electronic components used in a design, there are other components in the bill of materials (BOM) where the construction details and expected performance characteristics must be supplied to vendors with the order. Such components are mainly electromechanical parts, such as cable assemblies, transformers and printed circuit boards. As an example, when ordering a cable assembly, the following can be supplied:

- Assembly drawing identifying the parts
- Specifications or manufacturer data for connectors and cables
- Termination details
- Labeling details

A draftsperson working for engineering services can create assembly drawings, but it is the responsibility of the design engineer to verify the accuracy of the drawing. The component engineer can then update the component database to include the documentation that is presented to the supplier of the cable assembly when ordering the part.

Other tasks of the component engineer are

- Advise engineers on technology trends
- Organize pilot production runs to qualify components
- Provide inspection and testing documentation for components when inward inspection is warranted
- Ensure multiple sources for parts

2.3.3 Production Engineering

The production or manufacturing engineering department must ensure that the product design includes the design for manufacturability (DFM) concept. Production engineering should examine the technology used, component packaging, and component layout and orientation on the PCB. Larger companies employ concurrent engineering concepts, so by the time the design is released for production, the majority of the issues related to production have been sorted out. This approach helps companies improve their time to market, ensure product quality, and achieve greater financial returns.

2.3.4 Test Engineering

In most organizations, test engineering is part of the manufacturing department. The two types of testing as far as production is concerned are:

- In-circuit testing. Once the PCB is populated with components, in-circuit testing identifies any faulty components, including PCB problems.
- Functional testing. Functional testing must be carried out on every product. Functional test specifications are produced and released by designers in consultation with test engineers.

Test engineering requires test point access to nodes on the circuit to perform tests on discrete components and component modules. This requirement must be factored in during PCB layout. Functional tests verify the product's operation at a higher level. Because every product is tested functionally prior to delivery to the customer, it is a challenge for test engineers to minimize the time in testing so that the cost of production is reduced while ensuring the product's performance to specifications.

2.3.5 Material Procurement

Material procurement is done by the purchasing arm of the company that builds the product. As more and more companies are outsourcing the assembly of products, this may be a department of a contract manufacturing company. Its major responsibilities include ensuring the material required for production is available as and when needed, and negotiating contract agreements with material suppliers. These contracts determine price, quantities for the contract period, delivery schedule, and lot sizes.

Involvement of the purchasing staff during design of a product is mutually beneficial to both departments. The purchasing department can source material for prototype production runs. It can help designers obtain sample components for design verification. Purchasing can also liaise with component engineering to have new parts qualified and approved through the MRP system.

2.3.6 Quality Assurance

Quality assurance (QA) is a key component of product development. There are company-specific and generic quality plans that must be followed at all stages of product development. The following plans are applicable to the design process.

2.3.6.1 Quality Plan for Product Design

This plan ensures that all cross-functional teams associated with the product's design evaluate and review the design, from the early design stage to design release for production. These cross-functional teams are design engineering (electrical and mechanical), production engineering (test and manufacturing), and material procurement.

2.3.6.2 Quality Plan Ensuring Performance and Reliability

In this plan the product is independently tested by QA engineers to verify its performance. This is carried out after design verification of the product is concluded by the engineering team. At this stage, all relevant documents pertaining to design, testing, and manufacturing have been reviewed and released. The QA department acquires a sample unit from a pilot production run to carry out the testing. The QA tests do not repeat the design verification tests carried out by the design team. At this stage, the QA team does not expect to uncover any design flaws. However, these tests may include environmental tests, such as temperature cycling, damp heat, vibration, and shock.

2.3.6.3 Product-Specific Quality Plan

The QA department is a major player in the design process. Its reviews, audits, and subsequent reports at various stages in the design process help uncover deficiencies in design methodology. Although it is not essential for all product designs, QA may produce a specific quality plan for the development of the product. Product marketing can use these specific quality plans to promote customer confidence and also to highlight the company's adherence to design and manufacturing standards.

2.3.7 Engineering Services

The engineering services department deals with the following:

- PCB design and liaising with PCB manufacturers
- Producing and maintaining assembly drawings prepared for products
- Preparing all necessary documentation (e.g., CAD files) for PCB manufacture

- Maintaining a document archive
- Library management and creation of physical and logical symbols for CAD services

2.3.8 Marketing

The marketing department initiates the product design. It justifies the cost of product design through a business case to company directors. Marketing personnel focus on customers and market the company's products as solutions to expand a customer's business capabilities. The MRS is produced by marketing and must be made available to engineering prior to initiating detailed design work.

2.4 Role of External Agencies

During the design process, internal departments deal with external agencies, including customers, regulatory agencies, component suppliers, and test houses.

2.4.1 Regulatory Agencies

Regulatory agencies enforce mandatory safety standards on products. They have the authority to identify and ban unsafe products in the market. In addition to enforcing safety standards, regulatory agencies can enforce mandatory information standards, such as the labeling that appears on products.

2.4.1.1 Electromagnetic Compliance

Electromagnetic compatibility (EMC) standards deal with two aspects of EMC: electromagnetic interference and electromagnetic susceptibility. In most countries, a product cannot be placed in the marketplace if that product does not comply with governing EMC standards and is labeled accordingly. As the radio frequency spectrum becomes more crowded and the electronic circuitry within products operates at increasingly higher frequencies, such standards have been justifiably imposed on products by regulatory authorities. EMC poses a challenge to designers, and companies must ensure that their design engineers understand EMC by providing training for them so they can engineer designs free from EMC problems. In this respect, companies may publish their own EMC design guidelines covering rules for technology, PCB layout, mechanical construction, and electrostatic discharge. Alternatively, designers can refer to application notes published by component manufacturers for guidance [3,4].

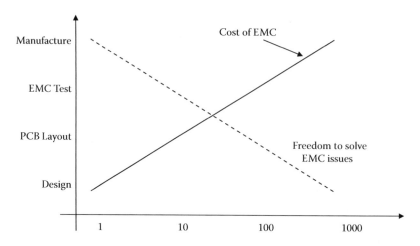

FIGURE 2.2
EMC cost diagram.

Solving EMC problems in later stages of the design cycle can be very costly. It may delay production, and companies may be penalized for these delays. Therefore EMC control work should start at the PCB layout level. Some companies employ engineers who specialize in EMC control to provide guidance and advice to the design team. Figure 2.2 shows the costs of making a product EMC compliant at various stages in the design cycle. It also shows the freedom of designers to take corrective actions at those stages. Note that at later stages designers have few corrective options available to overcome EMC problems, and these options could be prohibitively expensive. This means that the time designers spend addressing and solving EMC problems at the PCB layout level is justified to guarantee higher-level EMC problems are manageable and correctable.

2.4.1.2 Compliance Folder Management

The compliance folder is a folder containing the declaration of conformity [5] against governing standards, substantiated by the product's EMC test reports, list of compliant and labeled subassemblies used in the product, product description, and model number. The folder may also include sales literature, component placements, circuit diagrams, and software versions. Preparation of the compliance folder is a prerequisite to EMC labeling. The regulatory body checks the compliance folder to ascertain if all required testing, documentation, and declarations are included before approving the product's EMC compliance. Some companies obtain the services of external agencies that offer compliance folder management and support. These agencies may even retain the compliance folders and provide complete folder management for their clients. This is a popular service for companies such as importers that lack the technical knowledge and resources to manage the EMC compliance process.

TABLE 2.2

Order Codes for MC68HC05B6

Device Title	Package Type	Suffix			
		0 to 70°C	−40 to +85°C	−40 to +105°C	−40 to +125°C
MC68HC05B6	52-pin PLCC	FN	CFN	VFN	MFN
	64-pin QFP	FU	CFU	VFU	MFU
	56-pin SDIP	B	CB	VB	MB

2.4.2 Test Houses

Test houses offer testing services to verify a product's compliance with mandatory EMC and safety standards. They may also act as the agents for companies when dealing with regulatory issues.

2.4.3 Component Suppliers

Component suppliers solicit designers to use their components. They provide component samples, evaluation boards, and expert technical advice to designers. All reputable component manufacturers produce application notes for their components. Application notes provide tested and proven designs as recommended by the component manufacturer. Designers must be encouraged to refer to application notes during circuit design to ensure performance for the intended application. Furthermore, adherence to application notes offers a design advantage and may shorten design time by reducing debugging and test times.

2.4.3.1 Component Order Codes

The order code must describe the ordered component to the supplier without ambiguity. The order code specifies the component name, package type, and packing code. For example, an MC68HC05B6 in a 52-pin PLCC package at −40°C to +85°C would be ordered as MC68HC05B6CFN, as shown in Table 2.2. The packing code determines whether the components are packaged in tubes or in tape and reel with specified diameter. Production engineers must decide on the packing preferences for the component.

2.5 Reviews

Reviews of various specifications, test results, design release for PCB layouts, and final design release for pilot production are required during the design process. The QA plan for product design forces the design staff to perform such reviews at every milestone as the design progresses.

2.5.1   Specifications Review

Reviewing specifications is a necessary task that guarantees a quality outcome. For example, a design specification review must take place at an early stage in the design process prior to any detailed design work. Reviews empower the design team to embark on the detailed design confidently and with less ambiguity. Proper reviews require time, and reviews should not be rushed through the design process.

These reviews must be done as outlined by the QA plans developed for the design process. Depending on the type of specification being reviewed, the project manager or the team leader would convene a review meeting attended by stakeholders from the different functional areas.

2.5.2   Review of the Test Plan

The test plan is another name for the design verification specification. Test plans and subsequent test reports verify the performance of the product against the design specification. Test plans are reviewed by fellow designers in the design department. Test plans can be very descriptive and may describe testing methodologies in detail. The test report is the evidence that can be presented to internal reviewers (e.g., QA) and customers confirming the product's compliance with design objectives. Test reports must show the test configurations in detail, including the equipment used and their calibration status. Test reports must be kept with the project file for a designated period of time as required by company policies.

2.5.3   Review of Design Release for Pilot Production

Pilot production takes place prior to volume production. At this stage, all design-related problems have been rectified after thorough testing of the product. All necessary documentation required to manufacture the product has been released to the company's manufacturing resource planning (MRP) system. This means that purchasing can procure parts, production engineering can build and assemble the product, and test engineering can test the product without relying on support from those who engineered the design. This is the objective of the pilot production run. The pilot run includes product quantities for quality assurance and field trials. Even though a review has taken place prior to pilot production, the manufacturing process will find inadequacies in documentation. The pilot production report includes such shortcomings and the respective departments must attend to those discrepancies prior to volume production.

2.6   Product Documentation

Once the design is released for production, marketing and design specifications are of little use for ongoing manufacture of the product. The design process also generates production documents as well as user and technical manuals that will be used by production throughout the life cycle of the product.

2.6.1 Production Documents

Production documents related to a product are essential reference documents for engineers, testers, fitters, and assemblers working on the production floor. Documentation ensures the reproducibility and quality of the product. A bill of materials (BOM), assembly drawings, component inspection documents, and functional test specifications are some of the documents produced during the design process apart from CAD documentation and data files aiding computer-aided manufacture (CAM).

2.6.1.1 Bill of Materials

The BOM is produced by the engineering department and is used by the production and purchasing departments. Its ownership rests with engineering. Therefore, any changes to the BOM must be done with the consent of engineering. The BOM defines all necessary components required for assembly of the product. Basically, the BOM is a collection of part numbers; each part number identifies a component. These components can be electronic devices or mechanical items such as brackets, screws, washers, labels, and packaging material.

The part number must also refer to a circuit reference as described in the circuit schematics. Obviously, not all parts in the BOM have circuit references. They will also have a reference number, as depicted in the assembly drawing in Figure 2.3. Each line of the BOM has many fields. Essential fields

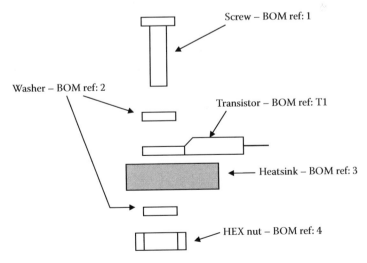

FIGURE 2.3
Example of an assembly drawing.

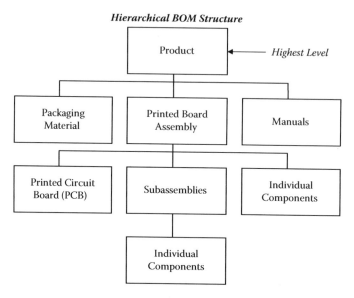

FIGURE 2.4
Typical BOM structure.

include a company-specific part number, a description of the component, and a circuit reference.

2.6.1.1.1 BOM Structure

A product will have a hierarchical BOM structure. The top of the BOM structure will be the part number assigned for the product for customer orders. Below this level, there are assemblies, packaging, manuals, and other documentation. Figure 2.4 shows a typical BOM structure.

In brief, when an order is placed in a company's MRP system for the highest-level part number, the system generates factory orders for the manufacturing department and material needs for the purchasing department. The MRP system will also generate a shortage list for parts in the BOM.

2.6.1.2 Factory Test Specifications

After assembly of the components, all products are tested for proper functioning according to the functional test specifications drafted and agreed to by the engineering and testing departments. Functional test specifications can be changed to reduce test time after analyzing failure statistics.

2.6.1.3 Assembly Instructions

Assembly instructions provide the information required to assemble the product. Assembly instructions are generated during the design phase and prepared by draftspersons. Assembly instructions are supported by diagrams and photographs, with references to the BOM.

2.6.2 User Manual

Every product must be sold with a user manual that should be written for the end user of the product. It should contain, among other things, safety precautions, product identification, parts supplied, installation procedures, operating instructions, warranty provisions, frequently asked questions, and product support contacts. User manuals are not technical manuals and should be written in plain language, avoiding technical jargon wherever possible.

2.6.3 Technical Manual

Technical manuals are written by technical writers. Technical writers gather information for the manual from all available design documents. They provide in-depth system-level technical details of the product that are useful for technical support personnel. Technical manuals may contain, depending on the complexity of the product, an installation guide, functional descriptions of the modules, circuit diagrams and bill of materials, recommended spares, user-configurable features, firmware version and fault diagnostics.

2.7 Design Management

All projects must have fixed start and end dates. The success of the project depends on how well it is managed by the project manager. Senior management may appoint a dedicated project manager to guide the design process for a complex project where there may be commercial implications for not delivering a desirable product. Alternatively, a team leader can manage a relatively simple product design and assume the role of project manager. Whatever the case may be, it is necessary to practice established project management principles wherever possible when managing product design.

2.7.1 Getting Started

No design work can begin without resources allocated to the project. The project manager must be able to justify the resources needed, whether they are monetary, equipment, or personnel. These justifications can be substantiated by providing evidence of resource allocations made to similar projects in the past.

2.7.2 Project Scope

It is necessary to clearly articulate and include in the scope definition as a project objective what the design is intended to achieve. The design objective must be a clear and unambiguous statement. It is essential that the client, product marketing, and all other stakeholders agree on the design objective. The scope of the project must remain current until the completion of the project.

2.7.3   Project Strategy

Once the design objectives are agreed to and understood by all stakeholders, it is necessary to develop a strategy to achieve these objectives. Project management can identify a range of options in consultation with design experts within design engineering that can lead to successful completion of the project. These options must be discussed and reviewed with senior management.

2.7.4   Managing Project Scope Changes

Project scope changes are not uncommon in product design and must be considered a challenge. These changes can originate from different sources, such as a customer's change request or changes to human resources. Changes can have a direct or indirect effect on the project; therefore, changes must be managed and controlled. They can impact resources, design milestones, cost, quality, and material.

2.7.5   Managing Project Time

Project management should determine the duration, effort, sequence, and dependencies of design work so that a project schedule can be created. Once the breakdown of tasks and activities is identified, project management can determine the proper sequence of events and their estimated durations and resource allocations. The durations and resource requirements for activities can be gauged from similar design tasks completed by designers in the past. Project management must also take contingencies into account when planning the schedule. The project schedule must mirror the objectives of the project. The schedule must be favorably reviewed by the stakeholders before design activities can begin.

2.7.6   Implementing the Project Schedule

Project progress can be monitored with the help of a Gantt chart. Project management must update the Gantt chart regularly to get a snapshot of progress related to the project. Progress must be reviewed at every project meeting to ensure that targeted outcomes and milestones do not slip within the schedule. The schedule is not a static document. It, too, will change due to a variety of events. Staff may become ill, components may not arrive on time, or testing of the prototypes may be prolonged; such events can have adverse effects on the schedule. If the changes happen to an event that is in the critical path, it should be examined thoroughly to determine the available options that may help minimize the effect on the schedule. If the change is due to an inadequacy of resources, project management should be able to negotiate for additional resources.

2.7.7 Managing Project Costs

Costs are estimated by analyzing resource requirements, salaries, and task duration as derived from the work breakdown structure. Project management can also access historical information from the organization's cost-tracking system for previous designs. Monitoring the costs of a project is done through cost control. The project team can charge material costs to a specific cost code assigned to the project by the finance department. Project management must alert higher authorities when cost increases occur. Summary reports sent to senior management at regular intervals must state whether the project is progressing within budget or not.

2.7.8 Managing Project Quality

The quality outcome of a project is said to be achieved if the project team has ensured the delivery of client requirements within the constraints identified. The project scope statement, overall quality policy of the company, project outcomes, and adherence to established product development processes are some of the inputs that can be taken into account when deciding on quality criteria for the project. During the life of the project, a representative from the QA department should work with the project team, performing scheduled and random audits to measure project performance against agreed-to quality criteria.

2.7.9 Managing Project Personnel

The management of designers is crucial to the success of a project. By monitoring the progress of the project, it is possible to assess individual performances against what was agreed to when the project schedule was formulated. By linking their performances to personal development reviews, designers can be encouraged to complete their tasks as planned. It is important that project management recognize achievements and establish a plan to reward good performance.

The morale of the team can be judged by the progress and enthusiasm displayed by team members throughout the project. If there is low morale, it may be necessary to consult with team members to ascertain the reasons. The reason may be a project-related issue, such as lack of training.

2.7.10 Project Closure

Project management can close the project when all design activities are concluded and the product is released for production. The project closure report outlines achievements against project objectives, lessons learned, and recommendations for future design work.

References

1. Standards Coordinating Committee of the IEEE Computer Society, IEEE 610.12-1990, *IEEE Standard Glossary of Software Engineering Terminology*, IEEE Computer Society, Washington, D.C., 1990.

2. Department of Defense, *Reliability Prediction of Electronic Equipment*, MIL-HDBK-217F, Department of Defense, Washington, D.C., 1995; available at http://www.sre.org/pubs/Mil-Hdbk-217F(2).pdf.

3. *ESG89001: Electromagnetic Compatibility and Printed Circuit Board (PCB) Constraints*, Application note, Philips Semiconductors, June 1989.

4. Lun, T.C., *AN2321: Designing for Board Level Electromagnetic Compatibility*, Rev. 1, AN2321, Freescale Semiconductor, Denver, CO, 2005.

5. Standards and Compliance, Ministry of Economic Development, Government of New Zealand; available at http://www.med.govt.nz/templates/Page____1270.aspx.

3

Design of DC Power Supply and Power Management

CONTENTS

3.1 Introduction 78
3.2 Design Approaches and Specifications 79
    3.2.1 Centralized Power Architecture versus Distributed Power Architecture 79
    3.2.2 Selection of DC-DC Converter Techniques 81
    3.2.3 Power Management Concepts..84
3.3 Specifying DC Power Supply Requirements 90
3.4 Loading Considerations...90
    3.4.1 Powering High-Power Processors and ASICs 94
        3.4.1.1 Advance Configuration and Power Interface (ACPI) Specification 98
    3.4.2 External Power Supplies and New Energy Standards.............99
3.5 Design of Off-the-Line Power Supplies 100
    3.5.1 Rectifier Section...100
        3.5.1.1 Fuses..102
        3.5.1.2 Inrush Current Limiting 103
    3.5.2 PFC in Off-the-Line Power Supplies 104
    3.5.3 Design of DC-DC Converters 106
        3.5.3.1 Forward Mode Converters..108
        3.5.3.2 Flyback Converters 109
        3.5.3.3 Two-Transistor Forward Mode Converters 129
        3.5.3.4 Bridge Converters 129
        3.5.3.5 SEPIC Converters 136
    3.5.4 Resonant Converters 137
        3.5.4.1 Comparison between Hard-Switching (PWM) and Resonant Techniques 137
        3.5.4.2 The Quasi-Resonant Principle.................................138
    3.5.5 Control of Switch Mode Converters..140
        3.5.5.1 Voltage Mode Control..140
        3.5.5.2 Current Mode Control..141
        3.5.5.3 Hysteretic Control 141
    3.5.6 Efficiency Improvements in Switch Mode Systems..................147
    3.5.7 EMI Reduction in Switch Mode Converters 148
    3.5.8 Control Loop Design 152

3.5.9 Low Dropout Regulators (LDO) 160
3.5.10 Charge Pump Converters 170
3.5.11 Achieving High Power Density 173
3.5.12 Postregulation Techniques 174
3.5.13 Digital Control 176
3.5.14 Power Supply Protection 184
    3.5.14.1 Thermal Design 184
    3.5.14.2 Overvoltage and Overcurrent Protection 185
    3.5.14.3 Protection against Input Transients 189
    3.5.14.4 Reliability of Input/Output Capacitors 189
    3.5.14.5 Age-Related Aspects 192
3.5.15 Testing of Power Supplies 193
Bibliography 197
References 197

3.1 Introduction

All electronic circuits require a clean and constant voltage DC power supply. However, the energy source available for the system may be a commercial AC supply, a battery pack, or a combination of the two. In some special cases, this energy source may be another DC bus within the system or the universal serial bus (USB) port of a laptop. In a successful total system design exercise, the power supply should not be considered as an afterthought or the final stage of the design process, because it is the most vital part of a system for good performance under worst-case circumstances. Another serious consideration in system design is the total weight and the volume, and this can be very much dependent on the power supply and the power management system. Also, it is important for design engineers to keep in mind that the power supply design may entail many analog design concepts.

Most power supply design issues are due to resource and component limitations within the power supply and the power management system. Non-ideal components—particularly, passives, commercial limitations to allocate sufficient backup energy storage within the battery pack, unexpected surges and transients from the commercial AC supply, and the fast load current transients—can create extreme and unexpected conditions within the system unless the power management system adequately addresses all the possible worst cases at an early design stage. Many product design experts choose to have the power supply and the power management system designed at an early stage with estimated parameters, with the actual system blocks powered from the system power supply. This approach may help minimize late-stage disasters in a large design project.

In the 1960s and early 1970s, power supplies were linear designs with efficiencies in the range of 30% to 50%. With the introduction of switching techniques in the 1980s, this rose to 60% to 80%. In the mid-1980s, power densities were about 50 W/in³. With the introduction of resonant converter techniques in the 1990s, this was increased to 100 W/in³. [1,2]. When high-speed and power hungry processors were introduced during the mid-1990s, much attention was focused on transient response, and industry trends were to mix linear and switching systems to obtain the best of both worlds. Low-dropout (LDO) regulators were introduced to power noise-sensitive and fast transient loads in many portable products. In the late 1990s, power management and digital control concepts and many advanced approaches were introduced into the power supply and overall power management [3].

In this chapter we consider design concepts and approaches, with a few design examples of how fundamental design concepts and practices can be applied to develop the power conversion stages and the power management system. Because of space constraints, for detailed theoretical aspects and deeper design considerations, the reader is referred to the many useful references cited herein.

3.2 Design Approaches and Specifications

A low-voltage power supply subsystem must fulfill four essential requirements: isolation from the mains, change of voltage level, conversion to a stable and precise DC value, and energy storage. In the modern world of power-hungry products with mixed power supply rail values where a battery pack or another limited capacity alternate energy source is used as the primary or the secondary source, a few additional requirements need to be considered. These are energy-saving aspects, quality of the output with respect to fast load transients, electromagnetic compatibility issues, protection and supervisory aspects, packaging aspects, and communication interfaces.

For the power supply, the design team has two basic choices: buy or build. The final decision must consider overall cost and the time to deliver. However, when a system becomes quite complex and requires multiple power rails and critical power management, design and build becomes the choice.

3.2.1 Centralized Power Architecture versus Distributed Power Architecture

Traditional power distribution techniques in a system have relied on a centralized architecture where all the required power rails are derived from a single high-power switch mode power supply (SMPS), as shown in Figure 3.1a. These conventional power supplies use an unregulated DC

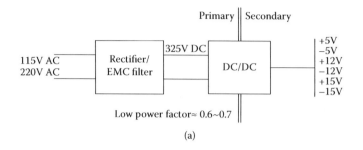

(a)

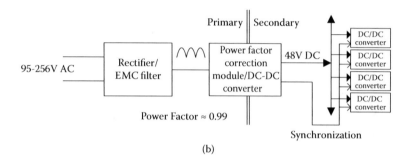

(b)

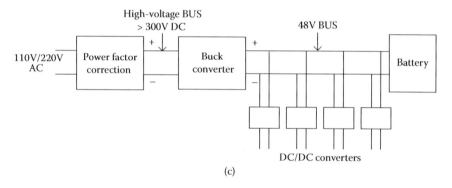

(c)

FIGURE 3.1
Comparison of the centralized power architecture and the distributed power architecture (DPA): (a) centralized power architecture; (b) distributed power architecture; (c) a DPA system with 48-V intermediate bus with battery backup.

power supply based on a rectifier and smoothing capacitors followed by a DC-DC converter with or without power factor correction (PFC) blocks to provide different DC power rails.

The distributed power architecture (DPA) approach is to distribute power at an intermediate "medium" voltage throughout the system and convert power locally. Distribution becomes more efficient with lower currents, and local power conversion occurs at lower power levels with the many different voltage levels required by individual circuit blocks. DPA systems are typically power factor corrected, have high efficiency (greater than 90%), and consist of a DC bus within the overall system. The selected DC bus voltage is determined by the safety requirements. Safe electrical low voltage (SELV) requires less than 60 V, and a common value for telecom systems is 48 V. Figure 3.1b shows an example of a universal AC input (95–265 V AC) power factor corrected DPA system with multiple DC-DC converter modules to derive individual power rail values. Davis [4] and Curatolo [5] provide and overview of DPA concepts, including system cost aspects, Cassani et al. [6] provide some design details of a system with a 36-V intermediate bus voltage for DC-DC converter inputs. Figure 3.1c shows a DPA system with a 48-V intermediate bus with a battery backup possibility [7]. Smith [8] provides key considerations for DPA systems, and Hemena and Malik [9] provide a case for a personal computer/server situation. Several important technical considerations in the DPA approach are partitioning of the load, determining the intermediate bus voltage value, end-to-end efficiency considerations (on the basis of AC line to load), and technology selection [5]. When the overall line-to-load efficiency (LTLE) is a serious design consideration, designers should carefully account for the efficiencies of individual power supply blocks [10].

Designing power solutions for line cards in information systems that handle multiple tasks at high speeds is complex. As these boards must process large amounts of data, they incorporate multigigahertz microprocessors, dual-logic application-specific integrated circuits (ASICs), and other high-performance devices. These new-generation devices can require two operating voltages per chip: one for the processor core (about 2.5 V or 1.8 V and rapidly moving downward) and the other for input/output (I/O) devices, which is higher (about 2.5 to 3.3 V). Until recently, these power supply requirements were addressed using multiple single-output isolated DC-DC converters. However, with rising currents and declining voltages, along with tighter regulation tolerances and faster slew rates, multiple isolated converter solutions are not as effective in such applications. These isolated converters are not space efficient and can cause higher ohmic losses along long interconnections. Under such conditions, maintaining high overall efficiency and tight point of load (POL) regulation becomes challenging. To alleviate these issues, now there are integrated building blocks that provide DPA architecture-based solutions for POL requirements. An example from International Rectifier (part number iP1201/iP1202), where an intermediate bus architecture (IBA) is based on an 8-V rail, is discussed in Bull and Smith [11]. Figure 3.2a illustrates the concept in this approach suitable for line cards up to about 150-W power consumption. Figure 3.2b illustrates implementation of a POL using iP1202 [11].

3.2.2 Selection of DC-DC Converter Techniques

In the modern world of electronics, there are three different basic approaches available for the process of DC-to-DC conversion. These are the linear approach, the switching approach, and the charge pump approach.

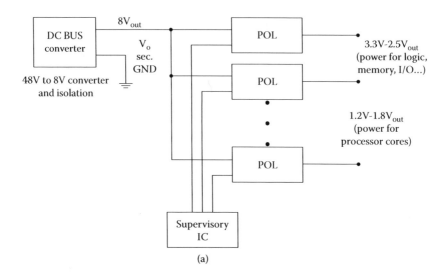

(a)

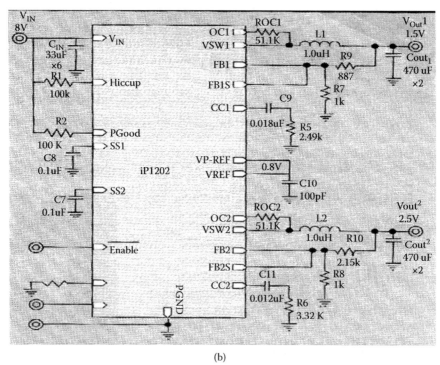

(b)

FIGURE 3.2
An 8-V intermediate bus architecture with 48-V to 8-V bulk converter and POL stages: (a) concept; (b) a complete schematic for a dual-output buck converter-based POL using iP1202. (Courtesy of *Power Electronics Technology* [11].)

In a practical system, one can mix these three techniques to provide a complex, but elegant, overall solution with energy efficiency, effective silicon or PCB area, and noise and transient performance to suit different parts of an electronic system. Switching-type DC-DC converters—once the clear choice for 5-V systems—suffer lower efficiency at lower voltages. In linear regulators where a series power semiconductor is connected between the unregulated DC rail (V_{in}) and the regulated output (V_{out}), if the control circuits are designed with a low-power approach, efficiency is approximately given by (V_{out}/V_{in}) × 100. This situation has led to LDO voltage regulators based on linear designs. Compared to a high-frequency switching technique-based SMPS solution, LDO regulator ICs (or simply LDOs) are faster in responding to load current changes, produce less noise, and are more compact on a PCB or as an integrated solution for a complete silicon solution. Often, combining a switcher and an LDO makes more sense in electronic systems where the DC rail voltages are less than 5 V or 3.3 V and a combination of many different low voltages are within the same system. This is illustrated in Figure 3.3, which is derived from the characteristic curves for the Power Trends PT6305 integrated switching regulators [12]. The graph indicates that the efficiency drops from approximately 79% for a 3.3-V output model to 56% for a 1.2-V output device. This is mostly due to the rectification losses at very low output voltages.

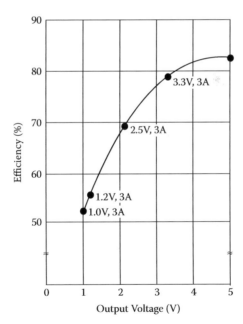

FIGURE 3.3
Efficiency versus input voltage for a compact switching regulator family—a typical example. (Courtesy of *EDN*.)

TABLE 3.1

Comparison of Popular DC-DC Converter Techniques

Feature	LDOs (and VLDOs)	Charge Pump Converters	Switching Regulators
Design complexity	Low	Moderate	Moderate to high
Cost	Low	Moderate	Moderate
Noise	Lowest	Low	Low to moderate
Efficiency	Low to moderate	Moderate to high	High
Thermal management	Poor to moderate	Good	Best
Output current capability	Moderate	Low	High
Requirement of magnetic parts	No	No	Yes
Limitations	Cannot step up	V_{in}/V_{out} ratio	Layout considerations

Charge pumps, switched capacitors, flying capacitors, and inductorless converters are all different names for DC-DC converters that use a set of capacitors rather than an inductor or transformer for energy storage and conversion. For many years, designers have used charge pumps for DC-DC conversion in applications for which the regulation tolerance, current conversion efficiency, and noise specifications are not very stringent. As discussed later, these circuits use capacitors combined with switches to boost or invert the input voltage, and they do not occupy more PCB or silicon area to implement as a single-chip converter [13]. Recent generations of charge pumps have become viable DC-DC conversion methods for cellular phones, portable wireless equipment, notebook computers, and PDAs, where high-density DC-DC conversion is necessary and circuit area is at a premium [13].

In a practical system, the power supply designer has the possibility of combining these techniques. For an effective overall solution taking all specifications and cost into account, combining a large-capacity bulk SMPS in tandem with LDOs and charge pumps becomes a very effective approach. Table 3.1 provides a comparison of the three popular DC-DC converter techniques. In portable application design, the designer should be careful when selecting techniques, and in many situations the method can be a mix of the three techniques discussed.

3.2.3 Power Management Concepts

Modern electronic product design requires dealing with different low-voltage DC rails for the longest battery life, thermal management, EMC compliance, and PCB area optimization. Modern gigahertz-order processors and peripherals generate fast load current transients on their multiple DC rails, and users expect longer run times from batteries. Therefore, in designing the DC power supply subsystem, one should consider the following:

- Battery pack or energy source–related aspects;
- Load partitioning aspects to minimize the number of DC rails to be used;

- Effects of fast transient loading at the output rails and the stability of the converter blocks;
- Dealing with power factor, harmonics, and other EMC issues;
- Packaging and thermal issues;
- Transient protection of the power supply and the product;
- Issues related to swapping of modules;
- Electrical isolation requirements.

To achieve the above, designers have access to a wide variety of technologies and power management IC families, battery management technologies, architectures, and standards. The important consideration is to deal with all of the above in an integrated manner and in a cost-effective way for a "power management solution" [14]. If the overall system consumes more than 50 W with several DC rail requirements, one has to first carefully analyze the load and have an overall view of its DC rail voltages and the transient behavior of the load. Let's look at a few examples, such as a digital still camera and a cellular phone.

Figure 3.4 depicts the system-level block diagram of a digital still camera, based on a TMS320DM270 programmable DSP-based media processor

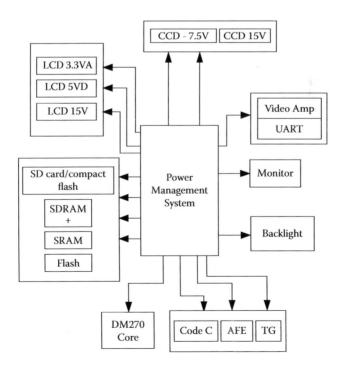

FIGURE 3.4
Digital still camera system components. (Courtesy of Power Electronics Technology [15].)

("DM 270") [15]. The input power source for a case like this will be a (Li)-ion or Li-polymer rechargeable battery pack. If this particular system operates from a Li-ion battery, the input operating voltage will be from 3.6 to 4.2 V. This operating range is a critical factor in selecting the power supply topologies in the design process. In this kind of product, there are different mixes of blocks, such as the processor, memory, analog front ends, video amplifier, motor, liquid crystal display (LCD), CCD module, and backlight for the LCD. Table 3.2 shows each system component and its power requirements. A few concepts used in such a system include:

- DC rails for low-noise analog circuits need be separated from digital block DC rails.

- System components frequently turned on and off need to be separately powered from the continually running circuit blocks.

- 3.3-V rails (a total of eight) can be grouped into three separate sets of 3.3-V rails.

- Secure digital (SD) card memory, synchronous dynamic random access memory (SDRAM), static random access memory (SRAM), and flash memory on the DM 270 can all be powered by the same bus.

- The codec, analog front end (AFE), and video timing generator require low-noise LDOs.

- The 3.3-V rail for the LCD needs to be separated because it goes on and off during use.

Appropriate supply topology selection depends on the input voltage, output voltage, noise, efficiency, cost, and space. The last three items usually compete with each other. The first two restrict the choice of the topology. If the output voltage is always lower than the input voltage, a buck converter or an LDO will work. Otherwise, a buck-boost or a single-ended primary inductance converter (SEPIC) (discussed later) will work. A typical Li-ion battery's voltage profile varies from a high value of 4.2 V (at full charge) to 3.0 V when fully discharged. Between these limits, the battery maintains approximately 3.7 V. All selections need to be based on these values. In terms of efficiency, it is important to consider the overall efficiency for prolonged battery life, and for this reason a special approach is to use the weighted efficiency for each power rail, as indicated in Table 3.2. Weighted efficiency is calculated by multiplying each efficiency value by the typical bus power divided by total estimated power ratio. In real-world operation, the example here can have different modes of operation, and for each mode one has to develop a table and analyze and estimate the best options. Then it is necessary to have an estimate of the percentage of the time the camera spends in each mode. For lower-noise considerations, even if an LDO solution is considered (in lieu of a switcher), weighted efficiency indicates otherwise for items such as the

TABLE 3.2

Partitioning of the System Loads

System Component	Indication in Figure 3.4	Bus Voltage (V)	Typical Bus Current (mA)	Typical Bus Power (mW)	Efficiency of an LDO-Based DC Supply (%)	Efficiency of a Switcher (%)	Weighted Efficiency of an LDO (%)	Weighted Efficiency of a Switcher (%)
SD card/compact flash	3.3 V I/O	3.3	30					
SDRAM		3.3	125					
SRAM		3.3	140					
Flash		3.3	30					
Total I/O (3.3 V)		3.3	325	1072.5	89.2	93.0	34.03	35.48
DM 270 core	1.5 V_Core	1.5	185	277.5	40.5	91.0	4.00	8.98
DM 270 Codec	3.3 VA	3.3	5					
Timing generator		3.3	80					
Analog front end		3.3	15					
Total analog circuits (3.3 V)		3.3	100	330	89.2	93.0	10.47	10.92
LCD	3.3 VA_LCD	3.3	20	66	89.2	93.0	2.09	2.18
LCD	5 V LCD	5	20	100	n/a	94.0	n/a	3.34
LCD	15 V_LCD	15	6	90	n/a	85.0	n/a	2.72
CCD	15 V_CCD	15	6	90	n/a	85.0	n/a	2.72
CCD	7.5 VN_CCD	−7.5	−6	45	n/a	80.0	n/a	1.28
Video amp	5 VA	5	3					
UART		5	10					
Total analog (5-V rail-based)		5	13	65	n/a	94.0	n/a	2.17
Lens motor	<2.7 V	2.7	250	675	73	95.0	17.52	22.81

Source: Power Electronics Technology [15].

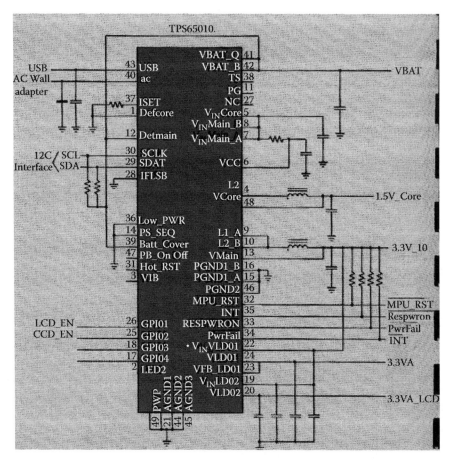

FIGURE 3.5
Part of a power management (PM) solution for the DSC from Texas Instruments.

DM270_Core. For the 5-V analog bus for the video amp, low-noise operation may be mandatory, but at lower battery voltages such as 3.5 V or lower, an LDO will cease to operate. In this situation, a switcher is to be selected and then followed up with an LDO. After defining the power supply requirements, the designer can start choosing individual ICs required for the system. Figure 3.5 indicates part of the solution using a Texas Instruments TPS 65010 power management IC, which has several switcher outputs and two LDOs. In addition, a TPS 61120 IC will be required. For details, see Day [15] and the data sheets of the relevant ICs [16,17].

Let's briefly discuss the case of powering a modern cellular phone. By 2007, most cellular phones allowed many features in addition to voice communication. A trend is for cellular phones to act as MP3 players or to add a micro hard disk or a very large amount of silicon memory. Figure 3.6 shows

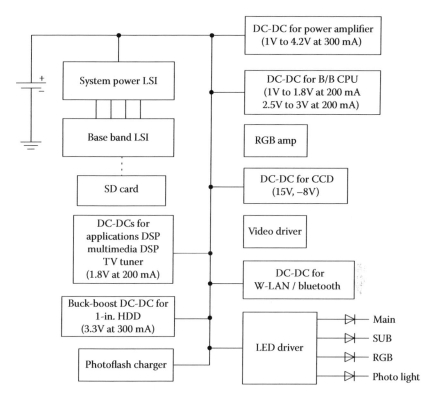

FIGURE 3.6

Block diagram of a cellular phone in 2005 shows many low-voltage DC supply requirements. (Courtesy of *Power Electronics Technology.*)

a generic block diagram of a 2005-generation cellular phone. In the older generation phones the main power rail was 3.3 V, but the newer-generation chipsets use a 1.5-V main power rail because the majority of large scale integrated circuits (LSIs) operate with voltage rails of 1.5 V or less. Examples are baseband chipsets running from 1.375-V rails and video-processing DSP chips running from 1.2-V rails. In these situations it is possible to mix very low dropout (VLDO) linear regulators with conversion efficiencies from 80% to 90% [18]. With the common acceptance of 600-mAh Li-based cell phone battery packs, and dealing with packaging problems, thermal management, and noise issues, power management of the product becomes quite critical and the designer has to make a well-informed and critically analyzed approach [18]. A simple example is the case of lower than 50% efficiency of LDO solutions for converting 3.3 V to 1.8 V compared to the situation of modern-generation power rail requirements from 1.5 V to 1.2 V, which can be supported by an LDO solution with efficiencies of about 80%. For more discussion on the practical design considerations for portable wireless products, see Armstrong [18] and Maxim Integrated Products [19].

3.3   Specifying DC Power Supply Requirements

Let's start with the simple fact that the DC supply subsystem is expected to provide a constant set of output voltages at a maximum set of load currents. Given this requirement to be derived from an AC input rail such as 230 V/50 Hz or 120 V/60 Hz or an energy source such as a battery or a fuel cell, the designer has to start the list of specifications with input voltage, output voltage, and load current. Then we can add as much secondary information as possible. The more requirement specifications we list, the easier it is to narrow down the available options.

Design specifications act as the performance goal that the ultimate power supply must meet in order for the product to meet its overall performance specification. When developing the specification, the power supply designer must keep in mind what is a reasonable requirement and what is an idealistic requirement. Most specification-related parameters are measurable using common test setups under different environmental conditions. These specifications can be grouped into several subsets, as shown in Table 3.3.

In developing these specifications, the designer should have a clear idea of the load requirements and steady-state and transient behaviors. In a very simple case where load consumes a few watts to about 50 W in a single- or dual-rail requirement, the load can be supplied by a simple linear or switching supply where only a few of the above items need to be specified. Many complex loads require multiple rails, power management and green design concepts, transient loading conditions, tight space or weight requirements, and cost restrictions, and the designer may have to start with the generalized concept of a power supply, as in Figure 3.7. In the case of an AC-powered situation, concepts in Figure 3.7a apply, whereas for battery-powered products the concepts in Figure 3.7b apply. Brown [20] and Rubadue [21] provide useful details on specifications and design concepts. A discussion of battery management for longest run time and standby time is beyond the scope of this chapter.

3.4   Loading Considerations

Load connected to a power supply can be as simple as a single-rail requirement, which can be easily met by a simple linear or switching power supply. In extreme cases, the load may consist of several complex processors or other mixed signal loads that may require multiple power rails, specialized power management aspects, and ultra-low-voltage DC rails that consume 100 A or more. Some processor loads may demand digitally controlled adjustable power rails for effective power management. In communication subsystems, the load may demand extra low-noise and low-voltage power rails. Designers

TABLE 3.3

Power Supply Specifications

Subset	Item	Remarks
Input specifications	Nominal input voltage	Product is expected to work around this voltage most of the time.
	Range of input voltages	Product is expected to withstand this range of fluctuations.
	Frequency (for AC input systems) or total energy available from a battery pack (in mAh or Ah)	In the case of a battery pack, an off-the-line charger may be designed for the input frequency.
	In-rush current	Important for the start-up conditions.
	Voltage transients	Important for the reliability of the power supply and the load for reliable operation.
	Permissible harmonics or power factor	Governed by various standardization bodies.
	Fusing	Speed of fusing is based on the I^2t rating of the device.
Output specifications	Nominal output voltage	The load is expected to operate at this voltage.
	Average and peak currents	RMS values to be used.
	Turn-on delay	• Capacitor/inductor energy storage dependent at the time of initial switch-on. • In multiple rail output situations, carefully timed sequencing may be necessary.
	Stability over a specified period	Based on the age of the components.
Regulation specifications	Load regulation	• Variation of the output voltage versus current. • Specified as a percentage or graphically shown for different input voltages.
	Line regulation	• Variation of the output voltage in response to changes in the input line. • Specified as a percentage or graphically shown for different load currents.
	Hold-up time	Amount of time the output remains within usable limits when the input source is disconnected temporarily.
	Output voltage temperature coefficient	The stability of the output voltage rails. Dependent primarily on the reference source stability and the temperature tolerances of the output sampling chain resistors.

(continued on next page)

TABLE 3.3 (continued)

Power Supply Specifications

Subset	Item	Remarks
Regulation specifications (continued)	Transient specifications • Overshoot, undershoot, and settling time • Step response of the output for sudden changes of load current	• Overshoot, undershoot, etc. are dependent on the control loop behavior. • Step response is very important in dealing with complex high-current processor loads.
	Output impedance	Represents the Thevenin equivalent resistance of the power supply output.
	Ripple and noise limits	These specifications may be very significant for the reliable operation and accuracy of analog- and mixed-signal circuitry.
Protection conditions	Over- and undervoltage limits at the output	These specifications indicate the safe operation limits of the load.
	Output current limit	Maximum load current expected.
	Thermal limits	To avoid excessive temperature conditions within the product or the power supply.
Power conversion specifications	Overall efficiency	In battery-powered products and green designs, this is a critical specification.
	Thermal dissipation	Determines the need for cooling and packaging limitations.
Safety and regulatory agency specifications	Isolation requirements • Dielectric withstand voltage • Insulation resistance	• For the safety of the user, a power supply should have galvanic isolation between the AC power input side and the load side. • Insulation resistance is usually specified for the transformers involved in the design.
	RFI/EMI requirements • Conducted EMI • Radiated RFI	• Conducted EMI specifies the line filtering requirement. • Radiated RFI affects the physical layout and enclosures.
Power management requirements	Energy conservation and green design aspects	Particularly important in high-power loads.
	Sequencing and resetting of the output rails	Critically important specification in multiple-rail situations.

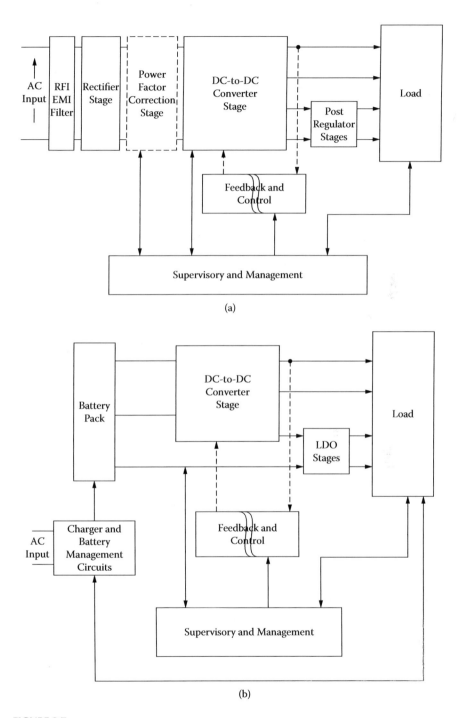

FIGURE 3.7
Overall design approach to a complete power supply subsystem (a) based on an AC input source, (b) based on a battery pack with a charging subsystem.

should have an initial estimate of the load requirements and an idea of the nature of the load in general. The key considerations include:

- The number of voltage rails and regulation aspects,
- The nature of the load transients and stability considerations,
- Efficiency and power management aspects,
- Protection requirements, and
- Noise and EMC considerations and regulatory requirements.

For the simple cases of single- or dual-rail power requirements, there is a choice of three-terminal linear regulator chips; these are low cost and easy to implement, with excellent noise and drift characteristics. The most useful property is their speed of response to transient loads. The only major disadvantage of these solutions is their low efficiencies, which in general range between 30% and 50%. There are ways to improve the efficiencies of linear regulators by manipulating the rectifier circuits in the input stages using silicon-controlled rectifiers [22]. Many common loads can tolerate slower responses and greater amounts of high-frequency noise. For such simple requirements, there are switching regulator solutions where the equivalent of a three-terminal linear IC solution is provided by integrated switching regulator (ISR) techniques by companies such as Power Trends. ISRs are able to provide buck, boost, or inverting voltage values from a single DC bus supply such as 5 V [23]. Figure 3.8 indicates a DPA solution based on an intermediate bus architecture (IBA) of 5 V. Another fully packaged switching solution for high-current-capability DC rails is the "brick converter," where a wide range of voltages (0.9 V to 48 V) is possible at currents up to a few tens of amperes [24,25]. Figure 3.9 shows the relative sizes of quarter, eighth, and sixteenth brick sizes.

Further to examples given in Section 3.2, powering portable devices such as Palm computers, pose different issues. Some of these are powered by a few AA cells from which different voltages need to be generated. The typical power source for a Palm computer is a disposable alkaline cell, and using such a cell creates other design challenges, such as generating higher voltages (such as a 5-V rail from two alkaline cells of 3 V) efficiently, generating LCD bias generators (typically –24 V), and generating miscellaneous lower voltages such as 3.6 V, 2.4 V, and other values [26]. In such cases, boost converters, such as the LT 1173 from Linear Technology, with charge pump configurations (to invert the +24 V to –24 V) can be used [26]. See Figure 3.10.

3.4.1 Powering High-Power Processors and ASICs

Advanced microprocessors and ASICs are power hungry and can consume as much as 100 A from power rails of 1.0 V or less. When the Pentium range processors entered the market in the mid-1990s (with only a few hundred megahertz clock speeds), their power consumption was a few amps to more

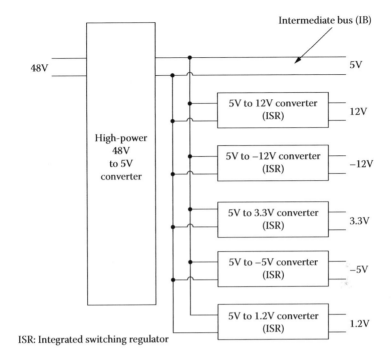

FIGURE 3.8
ISR devices from Power Trends, Inc. in a DPA solution. (Adapted from Narveson [23].)

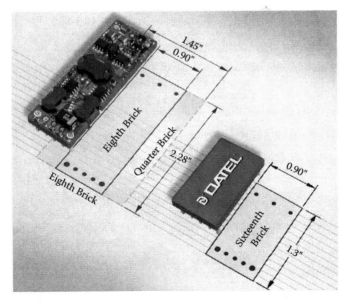

FIGURE 3.9
Brick converter examples. (Courtesy of *Power Electronics Technology*.)

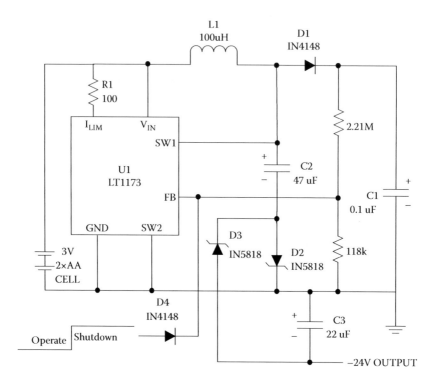

FIGURE 3.10
Use of a boost converter IC and a charge pump to convert 3 V from two alkaline cells to generate a +24-V and −24-V LCD bias supply. (Reprinted by permission of *EDN Magazine*, © 1993. Reed Business Information, a division of Reed Elsevier. All rights reserved.)

than 10 A from voltage rails of 1.8 V to 3.3 V. In CMOS digital circuits, the power consumption is proportional to V^2, and this fact encourages chip designers to develop processors that operate with lower rail voltages [27]. Two other important facts about high-end digital components are that they require multiple rail voltages for efficiency and speed, and the load currents can have slew rates easily up to 100 A/μs. To achieve these rates, most modern high-power processors have digital command signals based on four- to five-bit code to command a voltage regulator module (VRM) to output different voltages to power the processor. This concept is shown in Figure 3.11, where the processor outputs a four- or five-bit code via a special set of pins (called voltage identification [VID] pins) that command a VRM to change the voltage output from about 0.8 V to about 3.5 V in steps of 100 mV or 50 mV. For more details, see Mannion [28]. The VRM is capable of adjusting its own output voltage under the command code bits from the processor within the range of values specified by the processor. An early example of this is the LXM1700 from LinFinity [28]. The example in Figure 3.11 is for a five-bit VID pin code from the processor. With the load demands of extremely high current slew rates, the VRM should be capable of responding quickly with low-ESR capacitors at the output. More of these design aspects are discussed

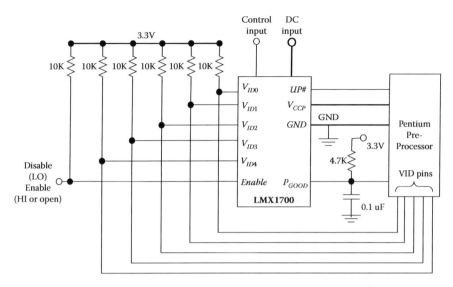

FIGURE 3.11
A voltage regulator module under the VID command signals (four or five bits) from the processor.

in later sections. The PCB track inductances can jeopardize the required slew rate; thus, designers must pay special attention to PCB layout.

To power Intel, AMD, and other high-end processors, the concept of VRMs has created a special set of power modules coming under the VRM specifications series. For details, see Brown [29], Gentchev [30], and Wong et al. [31]. Most of the VRMs used on processor boards are powered by the 5-V rail of the PC power supply ("silver box").

In most processor-based systems, power rails have typical value combinations such as 1.8 V, 2.5 V, or 3.3 V, and these rails need to be properly sequenced for the reliable operation of a system. There are different possibilities of sequencing depending on the application, as shown in Figure 3.12. Common themes are sequential power-up, ratiometric method, and output tracking (or simultaneous or coincidental power-up).

In sequential power-up (Figure 3.12a), the system turns on the core voltage, and when it reaches the voltage set point, it turns on the second I/O rail. This technique can be used to delay the start-up of the second rail at some predetermined time after the first rail is turned on. In this technique, interlocks can be introduced, as shown in Figure 3.12d. Switching off can be based on the same principle in the reverse order [32].

In Figure 3.12b, the ratiometric technique is shown. The two rails are turned on simultaneously, reaching regulation at their respective set points at the same time. In this method, the two rails are controlled with different slew rates. Figure 3.12c shows the output tracking or simultaneous power-up scheme where the two rails reach their output voltages at the same rate, and once the core voltage is reached, it remains constant, while the I/O rails reach the 3.3-V value [33].

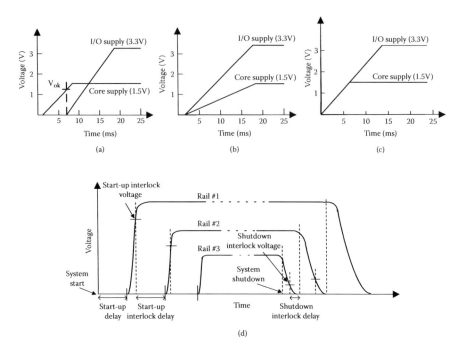

FIGURE 3.12
Sequencing of voltage rails: (a) sequential power-up; (b) ratiometric method; (c) output tracking (simultaneous or coincidental); (d) sequential tracking with interlocks.

3.4.1.1   Advance Configuration and Power Interface (ACPI) Specification

In the early days of personal computing, there was no power management. Around 1989, Intel shipped processors with technology to allow the central processing unit (CPU) to slow down, suspend, or shut down part or all of the system platform, or even the CPU itself, to preserve and extend battery life. In 1992, an early version of PC power management techniques based on the Advanced Power Management (APM) specification was introduced. With the need for energy saving simultaneously with the instantly available PC (IAPC), an advance version of the APM was introduced, called the Advanced Configuration and Power Interface (ACPI). The history and technical details related to this process are available in Kolinski et al. [34].

The ACPI specification provides a platform-independent, industry-standard approach to operating system–based power management. The ACPI is the key constituent in operating system power management (OSPM). OSPM and ACPI apply to all classes of computers, including handheld, notebook, desktop, and server machines. In ACPI-enabled systems, the basic input/output system (BIOS), hardware, and power architecture must use a standard approach that enables the operating system to manage the entire system in all operational situations. From a computer power system designer's viewpoint, ACPI power management means generating and managing a multitude of voltages on the motherboard and riser cards with no user intervention to enable the processing of audio, video, and data streams. ACPI-compliant computers require the generation of these multiple voltages at various current ratings as the system transitions between sleep states. The ACPI defines six possible discrete system operating states, which are referred to as S0 to S5, in order of highest to lowest power consumption. For more details on the ACPI and ACPI power controllers, see Kolinski et al. [34] and Lakkas and Duduman [35].

3.4.2 External Power Supplies and New Energy Standards

For certain types of consumer products, by using an external or a wall plug-in supply, designers can gain board space, eliminate a major source of heat, and reduce EMI noise. In addition, this may even provide the option of freeing the designer from the burden of getting safety approvals for the product. Until the late 1980s, wall plug-in supplies were limited to less than 25 W, but as of 2005 this capability has grown to about 250 W. Most low-power, older versions were linear regulator based, but recent energy-saving and related standards worldwide have pushed these products to adopt switch mode designs. It is important for designers to note the recent standards released by the California Energy Commission (CEC), Environmental Protection Agency (EPA), and the European Union (EU) regarding efficiency requirements and the nameplate output power ratings, as summarized in Figure 3.13. For more

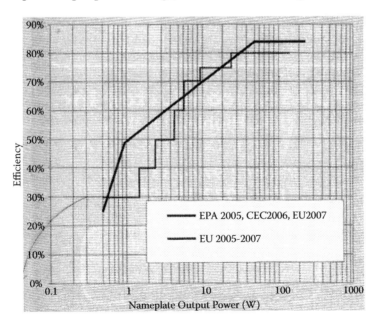

FIGURE 3.13
Efficiency requirements for external power supply curves based on CEC, EPA, and EU recommendations. (Courtesy of *Power Electronics Technology*.)

details on external power supplies, international power converter requirements, and recent energy saving standards, see Everett [36], Forrester [37], and Jovalusky [38].

3.5 Design of Off-the-Line Power Supplies

In this section we discuss the important aspects of designing a complete power supply using AC mains as the energy source. Figure 3.7a shows the blocks of such a system; some of the blocks, such as PFC and postregulation stages, may not be used in all cases, especially in lower-power versions.

This block diagram depicts the AC line RFI filter block, the supervisory and power management blocks, and the requirement of I/O isolation in the overall system. (Note the isolation indicated in the feedback and control block, which is essential for this purpose.). The EMI/RFI filter can be either part of the power supply or external to it, and it is generally designed to comply with national or international specifications, such as the FCC class A or class B and VDE-0871. Within the past two decades, because of the emphasis placed on power quality issues, PFC as applicable to the nonlinear behavior of the input current waveform of a rectifier has become an important issue. For power supplies with output capacities greater than 600 VA, PFC is necessary.

Input/output isolation is essential to off-the-line switchers. The isolation used within different stages may be optical or magnetic, and it should be designed to comply with Underwriters Laboratories (UL)/Canadian Standards Association (CSA) or Verband Deutscher Electronotechniker (VDE)/International Electrotechnical Commission (IEC) safety standards. The UL and CSA require 1000-V AC isolation voltage, whereas VDE and IEC require 3750-V AC. Consequently, any step-down power transformer or high-frequency switching transformer within the DC-DC converter stage has to be designed to the same safety isolation requirements.

3.5.1 Rectifier Section

In older linear power supplies, a step-down transformer was used with low-voltage rectifier circuits. In off-the-line switching power supplies, a high-voltage rectifier set is used directly off the AC line without any low-frequency line isolation transformer between the AC mains and the rectifiers. Because in most of today's electronic equipment the manufacturers are generally addressing an international market, power supply designers must use an input circuit capable of accepting many different line voltages, normally 90- to 130-V AC or 180- to 260-V AC. Figure 3.14 shows such a circuit using a voltage doubler technique. When the switch S_1 is closed, the circuit may be operated at a nominal line of 110-V AC. During the positive half-cycle of the AC, capacitor C_1 is charged to the corresponding peak voltage,

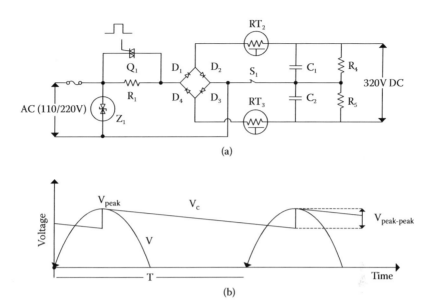

FIGURE 3.14
Input section of an off-the-line power supply: (a) overall arrangement; (b) ripple voltage calculations.

approximately 160-V DC, through diode D_1. During the negative half-cycle, capacitor C_2 is charged to 160-V DC through diode D_4. Thus, the resulting DC output is the sum of the voltages across $C_1 + C_2$, or 320-V DC. When the switch is open, D_1 to D_4 form a full-bridge rectifier capable of rectifying a nominal 220-V AC line and producing the same 320-V DC output voltage. In many universal input systems, there is an automatic changeover capability built into the design.

Two important aspects of the representative circuit in Figure 3.14a should be noted. One is the energy storage in capacitors C_1 and C_2. The other is the resistor R_1 and special components such as negative temperature coefficient (NTC) thermistors for the purpose of inrush current limiting. Figure 3.14b illustrates the simplified case of a half-wave rectifier where the charging and discharging of the capacitor is simplified by a sawtooth assumption for the discharge during the negative half of the AC cycle. During this process, approximate average DC output voltages (V_{DC}) for the half-wave or full-wave cases are given by

$$V_{DC} = \sqrt{2}V_{rms} - \frac{I_{load}}{2fC} \qquad \text{(3.1a) [half wave]}$$

$$V_{DC} = \sqrt{2}V_{rms} - \frac{I_{load}}{4fC}, \qquad \text{(3.1b) [full wave]}$$

where V_{RMS} is the rms AC input voltage, I_{load} is the average load current, f is the line frequency, and C is the value of the effective smoothing capacitor. The rms ripple voltage is given by

$$V_{ripple(rms)} = \frac{I_{load}}{2\sqrt{3}\,fC} \qquad \text{(3.2a) [half wave]}$$

$$V_{ripple(rms)} = \frac{I_{load}}{4\sqrt{3}\,fC}. \qquad \text{(3.2b) [full wave]}$$

For more details, see Smith [39,40].

For diodes, maximum forward rectification current capability, peak inverse voltage (PIV) capability, and the surge current capability (to withstand the peak current associated with turn-on) are the most important specifications.

Proper calculation and selection of the input rectifier filter capacitors are very important, because this will influence performance parameters such as low-frequency AC ripple at the output power supply and the holdover time. Normally, high-grade electrolytic capacitors with high ripple current capacity and low ESR need to be used with a minimum working voltage of 200-V DC. Resistors R_4 and R_5 (Figure 3.14b) provide a discharge path when the AC supply is switched off.

3.5.1.1   Fuses

Even the selection of the fuse for the input section needs to be done based on proper specifications. Fuses are categorized by three major parameters: current rating, voltage rating, and, most important, "let-through" current, or I^2t rating. The current rating of a fuse is the RMS value or the maximum DC value that it must exceed before blowing. The voltage rating of a fuse is not necessarily linked to the supply voltage. Rather, the fuse voltage rating is an indication of the fuse's ability to extinguish the arc that is generated as the fuse element melts under fault conditions. The voltage across the fuse element under these conditions depends on the supply voltage and the type of circuit. For example, a fuse in series with an inductive circuit may see voltages several times greater than the supply voltage during the clearance transient.

The I^2t rating of a fuse is defined by the amount of energy that must be dissipated in the fuse element to cause it to melt. This is sometimes referred to as the prearcing let-through current. To melt the fuse element, heat energy must be dissipated in the element more rapidly than it can be conducted away. This requires a defined current and time product. The heat energy dissipated in the fuse element is estimated in watt-seconds (or joules), or I^2Rt for a particular fuse with an internal resistance of R. As the fuse resistance is a constant, this is proportional to I^2t, normally referred to as the I^2t rating for a particular fuse or the prearcing energy. The I^2t rating categorizes fuses into the more familiar slow-blow, normal, and fast-blow types. It should be noted that the I^2t energy can be as much as 20 times greater in a slow-blow fuse of the same DC current rating. For example, a 10-A fuse can have an I^2t rating ranging from 5 A^{2}s for a fast fuse to 3000 A^{2}s for a slow fuse. The selection of fuse ratings for off-line SMPSs is discussed by Billings [41]. For high-power semiconductors such as power diodes and transistors, manufacturers indicate a value of I^2t, from 10 ms (for 50 Hz) or 8.3 ms (for 60 Hz), that should not be exceeded. Comparing this value with the fuse I^2t permits us to verify the protection [42].

3.5.1.2 Inrush Current Limiting

An off-the-line switching power supply may develop extremely high peak inrush currents during turn-on unless it incorporates some form of current limiting in the input section. These currents are caused by the charging of the filter capacitors, which at turn-on present a low impedance to the AC lines, generally limited by the ESR plus the total input resistance within the charging path. If no protection is employed, these surge currents may approach very large values to blow the input fuses.

Several methods are widely employed in introducing a high impedance to the AC line at turn-on. Figure 3.15 illustrates a few common methods. In Figure 3.15a, an NTC thermistor limits the inrush. Initially, the thermistor resistance is high, which limits the inrush. As the inrush current flows

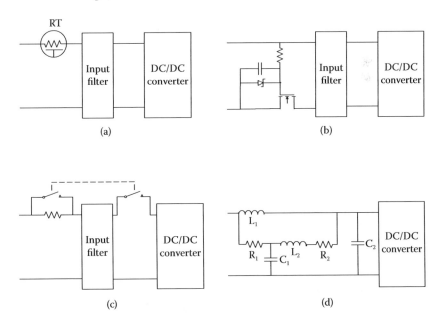

FIGURE 3.15
Power supply inrush current limiter techniques: (a) thermistor technique; (b) MOSFET-based approach; (c) resistor relay; (d) inductor based. (Courtesy of *Power Electronics Technology*; Bell [43].)

through the thermistor, it heats up and the resistance decreases for normal operation. Figure 3.15b indicates a power semiconductor (MOSFET)–based technique. The FET limits the inrush by turning on slowly as its gate capacitor charges. Figure 3.15c shows an approach using a resistor and a relay. The input filter charges through the resistor until the relay is commanded to connect the filter to the converter and short the resistor. Each of these has its advantages and drawbacks [43]. Figure 3.14a illustrates the combination of thermistor and series power semiconductor (such as a triac)–based techniques, which is very popular in computer power supplies.

One of the self-limiting techniques is to use an input inductance to limit inrush current, as in Figure 3.15d. This filter implements the required damping using inductor L_2 and resistors R_1/R_2 in parallel to the main DC-carrying inductor L_1. The value of L_1 is typically 5 to 10 times larger than L_2. More details of this technique are discussed by Bell [43], giving attention to the design of the magnetics.

3.5.2   PFC in Off-the-Line Power Supplies

Power factor correction aligns the current waveform with the input voltage waveform. If the waveforms are not aligned, the power factor (PF) is less than 1; if they are aligned, the PF is greater than 1. Figure 3.16 illustrates this. Figure 3.16a illustrates a typical non-PFC case where the nonsinusoidal current waveform does not align with the voltage waveform. Figure 3.16b illustrates the case of a half-sinusoidal input current waveform with PF = 1. In most cases, PF-corrected designs can achieve PF values of 0.95 to 0.98, and supplies that are not corrected can have a PF that is significantly less, usually less than 0.65. Given this simple explanation, it is also necessary to appreciate the case of nonsinusoidal rectifier currents, which can generate many harmonics of the line frequency waveform. When a repetitive waveform is not sinusoidal, it generates many harmonics, and the total harmonic distortion is given by

$$V_{THD} = \frac{\sqrt{V_2^2 + V_3^2 + \ldots + V_n^2}}{V_1}. \tag{3.3a}$$

Only current and voltage components with the same frequency can produce nonzero active power. In practice, sometimes it is reasonable to assume that the input voltage waveform is purely sinusoidal, despite any distortions in the current waveform. Under such an assumption, V, the RMS voltage of the input waveform, is approximately equal to the RMS of the fundamental component, V_1, and we can get the approximate relationship

$$PF = Cos\Phi_1 \cong \frac{VI_1 \cos\Phi_1}{VI} = \frac{I_1 \cos\Phi_1}{I}, \tag{3.3b}$$

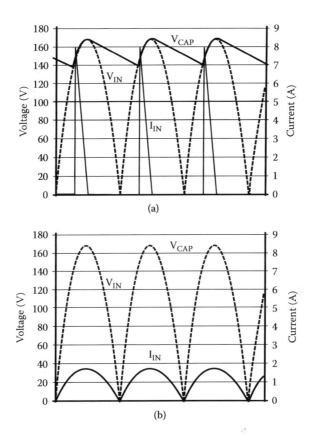

FIGURE 3.16
Input current and voltage relationships of a rectifier stage such as in Figure 3.14: (a) case of a typical rectifier without PFC; (b) ideal full-wave rectified PFC case with $PF = 1$.

where I_1 is the in-phase fundamental RMS current and I is the total RMS current. With this discussion, we can appreciate the regulatory bodies defining the limits of harmonics generated by electrical systems connected to the AC utility grid. The EU put into effect EN 61000-3-2 to establish limits on harmonics up to the 40th harmonic of the AC line–powered equipment's input current. Amendments in 2001 clearly state that PCs, PC monitors, and TV receivers with power ratings from 75 to 600 W must have PFC power supplies.

In practical cases of off-line SMPS systems, the fundamental idea is to use a power MOSFET switch and an inductor in the series path of the charging capacitor to artificially align the voltage and charging current waveforms. For the case of a boost converter configuration, this can be achieved as shown in Figure 3.17a, where the smoothing capacitor is now shifted toward the DC-DC converter stage. A power factor control IC that switches the MOSFET at higher frequency smoothes out the current waveform and aligns its fundamental component with the input voltage waveform.

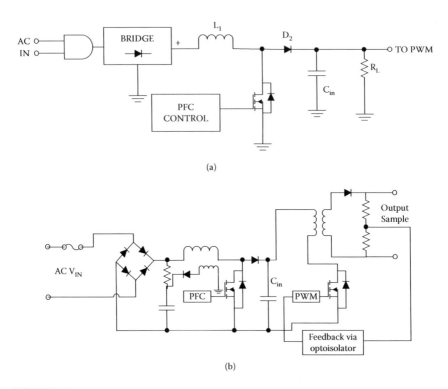

(a)

(b)

FIGURE 3.17
Practical approach to power factor correction: (a) basic concept of implementation; (b) use of a stand-alone PFC controller IC; (c) use of a PFC/PWM combo IC in a boost topology; (d) waveforms in discontinuous mode. (Courtesy of *EDN*; Zuk [44].)

Several IC manufacturers, such as Fairchild, Microlinear, and International Rectifier, supply various controller ICs for PFC or PFC/PWM combo operations. Figure 3.17b and Figure 3.17c are typical configurations in a boost converter topology [44] based on Fairchild parts FAN 752B and FAN 4803. It is important to note that the main smoothing capacitor C_{in} has now moved further toward the DC-DC converter stage. For the case of discontinuous conduction, such as the case of using a FAN 752B-type controller for outputs up to 200 W, the boost converter's MOSFET turns on at zero inductor current and turns off when the current meets the desired input reference voltage. Figure 3.17d indicates the waveforms in discontinuous mode. For details of this application case, see Zuk [44]; Sandler et al. [45], Valentine [181] and Chapter 9 of Kularatna [46] provide a practical overview of PFC techniques and application information.

3.5.3   Design of DC-DC Converters

The heart of any power supply is the DC-DC converter section, shown in Figure 3.7. In most state-of-the-art off-the-line power supplies as well as in

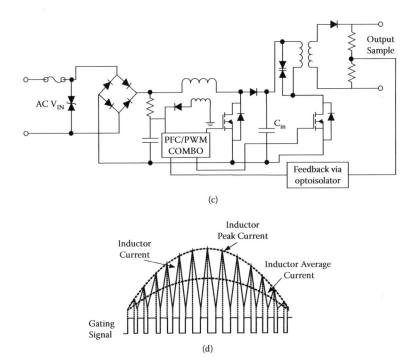

(c)

(d)

FIGURE 3.17 (continued)

battery-powered systems, this block is designed based on switch mode techniques for reasons of efficiency, PCB area, and multiple rail requirements. However, when low-noise requirements and fast transient response at the output side are required, LDOs are considered as follow-on blocks or postregulator circuit blocks. The following section provides an overview of techniques available and important design considerations. For complete theory and analysis of switch mode and linear converters, the reader should refer to the bibliography at the end of the chapter.

As discussed in Section 3.2.2, there are three possible approaches to DC-DC converter design. Let's discuss the switcher techniques first. The switching converter topology has a major bearing on the conditions in which the power supply can operate safely and on the amount of power it can deliver. Cost versus performance tradeoffs are also needed in selecting a suitable converter topology for an application. The primary factors that determine the choice of topology are whether DC isolation is needed, the peak currents and voltages the power switches are subjected to, the voltages applied to transformer primaries, cost, and reliability.

Some relative merits and demerits of the converter topologies and typical applications are summarized in Table 3.4, including essential mathematical expressions for important design relationships. The industry has settled on several primary topologies for a majority of applications. Figure 3.18 illustrates the approximate range of usage for these topologies. The boundaries

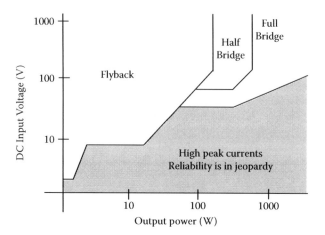

FIGURE 3.18
Industry-favorite configurations and areas of usage.

to these areas are determined primarily by the amount of stress the power switches must endure and still provide reliable performance. The boundaries delineated in Figure 3.18 represent approximately 20 A of peak current in power switches.

Nonisolated basic converters (buck, boost, and buck-boost types) are generally used for lower-power PCB-level converter circuits and are not so popular for higher-power applications. Isolated versions such as forward mode, flyback, and bridge types are generally used for applications where higher power, galvanic isolation, and multiple output rails are required. Because of the advantages of operation in buck or boost modes without inversion, SEPIC converters are gaining popularity in battery-powered applications. The following sections provide an overview of important design aspects of these topologies.

3.5.3.1 Forward Mode Converters

The forward converter is derived from the buck topology family, generally employing a single switch. The power switch in the forward topology is ground referenced (also called a low-side switch), whereas in buck topology the switch source terminal floats on the switching node. The main advantage of the forward topology is that it provides isolation and the capability to provide step-up or step-down function. Relevant figures in Table 3.4 indicate the similarities, and the only difference is the inclusion of the transformer turns ratio (N_2/N_1) in the transfer function. One important design consideration in this topology is the magnetizing inductance and the need to reset the transformer core. Figure 3.19 shows a simplified transformer model including the magnetizing inductance (L_M) and the leakage inductance (L_L). The value of L_M can be measured at the primary terminals with the secondary winding open-circuited. The peak current in L_M is proportional to the maximum flux density within the core, and a given core can handle only a limited flux density before saturation occurs. At saturation, a rapid reduction of inductance occurs. The other element added to the transformer model is L_L, and this can be measured at the primary terminals with the secondary winding(s) short-circuited. This term represents the stray value, which does not couple primary to secondary. With careful design, this value can be kept small and the effect on the converter is limited to voltage spikes on the power switch.

An important consideration in forward mode converter design is the core-resetting requirement to avoid core saturation (the effect of L_M in Figure 3.19). A few techniques are available for this purpose. The common method is to have an additional winding and a diode for core resetting. A few advanced techniques used for solving the same problem are active clamp reset and resonant reset forward converters. Figure 3.20 compares these techniques. Mappus [47], King and Gehrke [48], and Hariharan and Schie [49] provide analysis and design aspects of the active clamp reset techniques. Hariharan and Schie [49] and Khasiev [50] detail the important design aspects of resonant reset forward converters.

3.5.3.2 *Flyback Converters*

Low cost and simplicity are the major advantages of the flyback topology. In multiple output applications, the addition of a secondary winding, a diode, and an output capacitor is all that is required for an additional output. Flyback converter operation can lead to confusion if the designer approaches the design of its magnetics as if it were a transformer. Except for the case of multiple output windings, the magnetics in a flyback converter are not a transformer. An easy way to view this is as an energy bucket that is alternately filled (when the switch is on) and dumped (when the switch is off). In other words, a flyback magnetic (sometimes called a transformer choke) is an energy-in, energy-out power transfer device where input and output windings do not conduct current simultaneously. A gapped core is used in general to have adequate leakage inductance at the input side for energy storage during the switch-on period. The primary specifications for an off-line flyback power supply design, such as a power adapter for a notebook computer or a PDA, are based on the following:

- Nominal AC input voltage (V_{ACnom}),
- Minimum and maximum AC input voltage (V_{ACmin} and V_{ACmax}),
- Output voltage (V_{out}) (a typical value is about 16 V),
- Maximum output overshoot, full load to no load (ΔV_o),
- Maximum output power (P_o),
- Target efficiency at full load (η),
- Holdup time at nominal AC input voltage and full load at output (T_{hold}).

TABLE 3.4

A Summary of Power Supply Topologies

Topology and Relative Cost	Config- uration	Wave- forms	Ideal Transfer Function	Peak Drain Current (ID_{MAX})	Peak Drain Voltage (V_{DS})
Buck (1.0)	See Table Figure 3.4a		D	$I_{DMAX} = I_{RL} + \dfrac{\Delta I_{L1}}{2}$	$V_{DS} = V_{IN} + V_D$
Boost (1.0)	See Table Figure 3.4b		$\dfrac{1}{1-D}$	$I_{DMAX} = I_{RL}\left(\dfrac{1}{1-D}\right) + \dfrac{\Delta I_{L1}}{2}$	$V_{DS} = V_O + V_D$
Buck-boost (1.0)	See Table Figure 3.4c		$\dfrac{-D}{1-D}$	$I_{DMAX} = I_{RL}\left(\dfrac{1}{1-D}\right) + \dfrac{\Delta I_{L1}}{2}$	$V_{DS} = V_{IN} + V_O + V_D$
SEPIC	See Table Figure 3.4d		$\dfrac{D}{1-D}$	$I_{DMAX} = I_1 + I_{RL} + \dfrac{\Delta I_{L1} + \Delta I_{L1}}{2}$	$V_{DS} = V_{IN} + V_O + V_D$
Forward	See Table Figure 3.4e		$\dfrac{N_2}{N_1}D$	$I_{DMAX} = \dfrac{N_2}{N_1}\left(I_{RL} + \dfrac{\Delta I_{L1}}{2}\right) + \hat{I}_{MAG}$ $\hat{I}_{MAG}$....Peak magnetizing current	$V_{DS} = V_{IN}\left(1 + \dfrac{N_1}{N_3}\right)$

Average Diode Currents	Diode Reverse Voltage (V_{RM})	Advantages	Disadvantages	Typical Efficiency
$I_{CR1} = I_{RL}(1-D)$	$V_{RM} = V_{IN}$	Simple High efficiency No transformer Low switch stress Small output filter Low ripple	No isolation Potential overvoltage if switch shorts High-side switch drive required High input ripple current	78%
$I_{CR1} = I_{RL}$	$V_{RM} = V_o$	High efficiency Simple No transformer Low input ripple	No isolation High peak drain current Regulator loop hard to stabilize High output ripple Unable to control short-circuit current	80%
$I_{CR1} = I_{RL}$	$V_{RM} = V_o + V_{IN}$	Voltage inversion Simple	No isolation High-side switch required Regulator loop hard to stabilize High output ripple High input ripple current	80%
$I_{CR1} = I_{RL}$	$V_{RM} = V_o + V_{IN}$	Low ripple input current Buck and boost with no inversion No transformer Capacitive isolation against a switch failure	No isolation Switch has high RMS/peak currents (limits power) Capacitors have high ripple currents (low ESR needed) High output ripple Loop stabilization difficult	
$I_{CR1} = \hat{I}_{MAG}\left(\dfrac{D}{2}\right)$ $I_{CR2} = I_{RL}(D)$ $I_{CR3} = I_{RL}(1-D)$	$V_{CR1} = V_{IN}\left(1 + \dfrac{N_3}{N_1}\right)$ $V_{CR2} = V_{IN}\left(\dfrac{N_2}{N_3}\right)$ $V_{CR3} = V_{IN}\left(\dfrac{N_2}{N_1}\right)$	Drain current reduced by ratio N_2/N_1 Low output ripple	Poor transformer utilization Poor transient response Transformer design critical Transformer reset limits D Switch requires high voltage capability High input current ripple	

(continued on next page)

TABLE 3.4 (continued)

A Summary of Power Supply Topologies

Topology and Relative Cost	Configuration	Waveforms	Ideal Transfer Function	Peak Drain Current (ID_{MAX})	Peak Drain Voltage (V_{DS})
Flyback (1.2)	See Table Figure 3.4f		$\dfrac{N_2}{N_1}\left(\dfrac{D}{1-D}\right)$	$I_{DMAX} = I_{RL}\left(\dfrac{N_2}{N_1}\right)\left(\dfrac{1}{1-D}\right) + \dfrac{\Delta I_{L1}}{2}$	$V_{DS} =$ $V_{IN} + \left(\dfrac{N_1}{N_2}\right)(V_{OUT} + V_D)$
Push-pull (2.0)	See Table Figure 3.4g		$2\dfrac{N_2}{N_1}D$	$I_{DMAX} = \dfrac{N_2}{N_1}\left(I_{RL} + \dfrac{\Delta I_{L1}}{2}\right) + \hat{I}_{MAG}$	$V_{DS} = 2V_{IN}$
Two-switch forward	See Table Figure 3.4h		$\dfrac{N_2}{N_1}D$	$I_{DMAX} = \dfrac{N_2}{N_1}\left(I_{RL} + \dfrac{\Delta I_{L1}}{2}\right) + \hat{I}_{MAG}$	$V_{DS} = V_{IN} + V_{D1}$ (for both transistors)
Half bridge (2.2)	See Table Figure 3.4i		$\dfrac{N_2}{N_1}D$	$I_{DMAX} = \dfrac{N_2}{N_1}\left(I_{RL} + \dfrac{\Delta I_{L1}}{2}\right) + \hat{I}_{MAG}$	$V_{DS} = V_{IN}$
Full bridge (2.5)	See Table Figure 3.4j		$2\dfrac{N_2}{N_1}D$	$I_{DMAX} = \dfrac{N_2}{N_1}\left(I_{RL} + \dfrac{\Delta I_{L1}}{2}\right) + \hat{I}_{MAG}$	$V_{DS} = V_{IN}$

Average Diode Currents	Diode Reverse Voltage (V_{RM})	Advantages	Disadvantages	Typical Efficiency
$I_{CR1} = I_{RL}$	$V_{CR1} = V_{IN}\left(\dfrac{N_2}{N_1}\right)$	Drain current reduced by ratio N_2/N_1 Low parts count Isolated output No secondary inductors	Poor transformer utilization Transformer stores energy High output ripple CR_1 requires fast recovery	80%
$I_{CR1} = \dfrac{I_{RL}}{2}$ $I_{CR2} = \dfrac{I_{RL}}{2}$	$V_{CR1} = V_{CR2} = 2V_{IN}\left(\dfrac{N_2}{N_1}\right)$	Good transformer utilization Drain current reduced as function of N_2/N_1 Good at low V_{IN} values Low output ripple	Cross-conduction of Q_1/Q_2 possible High parts count Transformer design critical Q_1/Q_2 should be high-voltage capable High input current ripple	75%
$I_{CR1} = I_{CR2} = \hat{I}_{MAG}\left(\dfrac{D}{2}\right)$ $I_{CR3} = I_{RL}(D)$ $I_{CR4} = I_{RL}(1-D)$	$V_{CR1} = V_{CR2} = V_{IN}$ $V_{CR3} = V_{CR4} = V_{IN}\left(\dfrac{N_2}{N_1}\right)$	Drain current reduced by turns ratio Lossless snubber recovers energy Drain voltage is half that of single-switch forward converter Low output ripple	Poor transformer utilization High parts count High-side switch required Transformer reset limits D High input current ripple	
$I_{CR3} = \dfrac{I_{RL}}{2}$ $I_{CR4} = \dfrac{I_{RL}}{2}$	$V_{CR3} = V_{CR4} = V_{IN}\left(\dfrac{N_2}{N_1}\right)$	Good transformer utilization Transistors rated at V_{IN} Isolated output I_D reduced as a ratio of N_2/N_1 High power output Low output ripple Zero voltage switching possible near $D = 1$	Poor transient response High parts count High side switch required High input current ripple C_1/C_2 has high ripple current Cross-conduction of Q_1/Q_2 possible	75%
$I_{CR5} = I_{RL}$ $I_{CR6} = I_{RL}$	$V_{CR1} = V_{CR2} = V_{IN}$ $V_{CR5} = V_{CR6} = 2V_{IN}\left(\dfrac{N_2}{N_1}\right)$	Nearly same as half bridge	High parts count High-side switch required High input current ripple C_1 has high ripple current Cross-conduction of Q_1/Q_2 or Q_3/Q_4 possible	73%

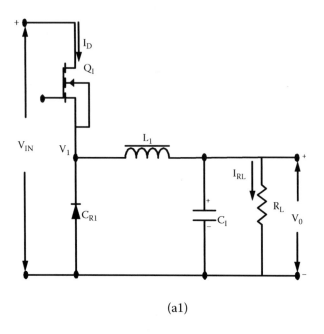

(a1)

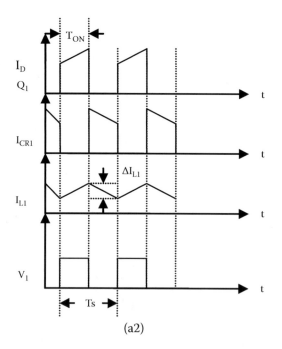

(a2)

TABLE FIGURE 3.4a
Buck converter: (a1) configuration; (a2) waveforms.

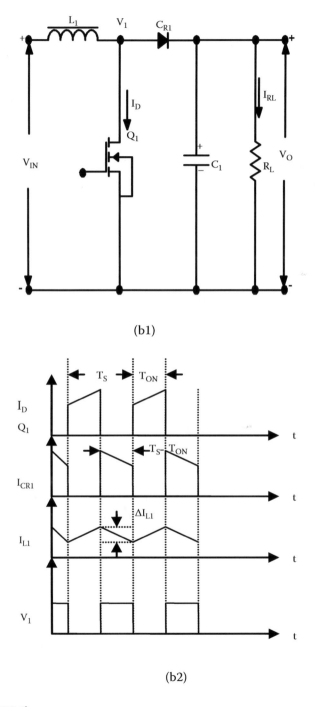

(b1)

(b2)

TABLE FIGURE 3.4b
Boost converter: (b1) configuration; (b2) waveforms.

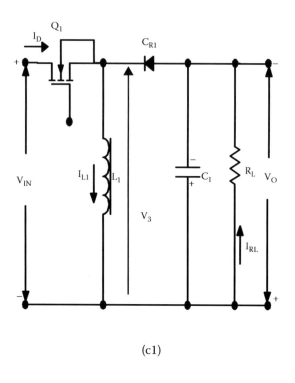

(c1)

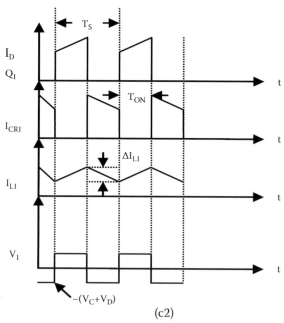

(c2)

TABLE FIGURE 3.4c

Buck-boost converter: (c1) configuration; (c2) waveforms.

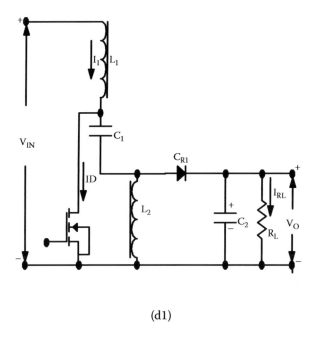

(d1)

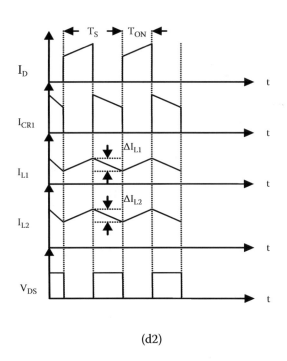

(d2)

TABLE FIGURE 3.4d
SEPIC converter: (d1) configuration; (d2) waveforms.

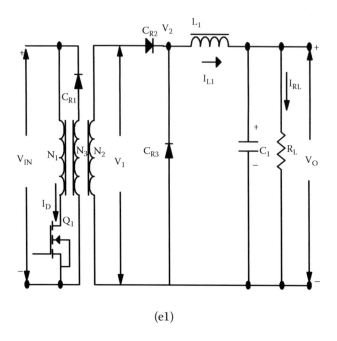

(e1)

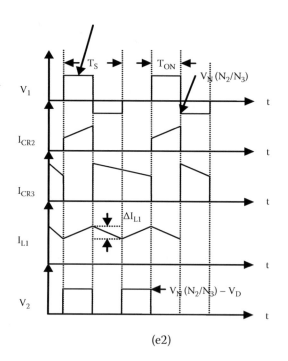

(e2)

TABLE FIGURE 3.4e
Forward converter: (e1) configuration; (e2) waveforms.

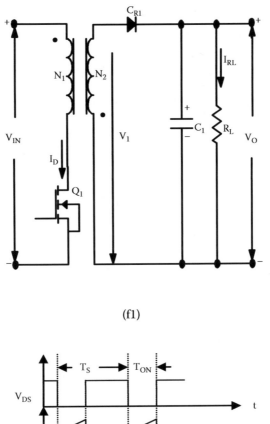

(f1)

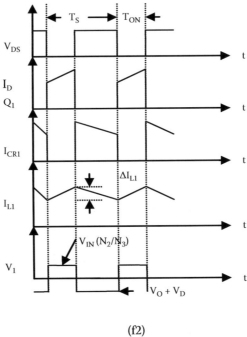

(f2)

TABLE FIGURE 3.4f
Flyback converter: (f1) configuration; (f2) waveforms.

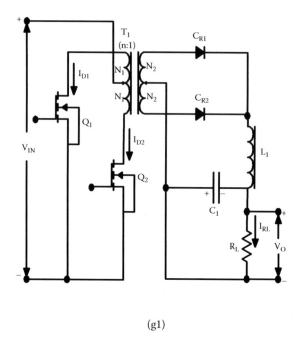

(g1)

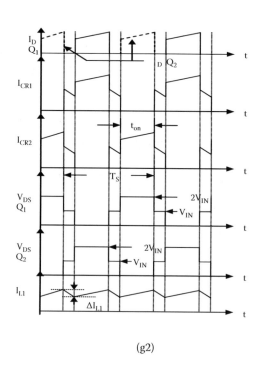

(g2)

TABLE FIGURE 3.4g
Push-pull converter: (g1) configuration; (g2) waveforms.

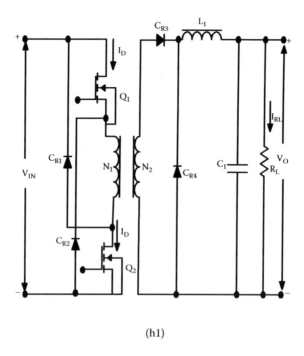

(h1)

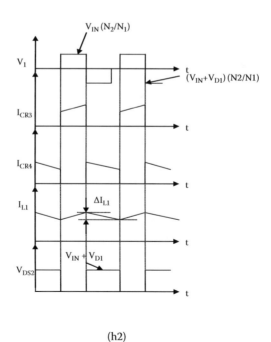

(h2)

TABLE FIGURE 3.4h
Two-switch forward converter: (h1) configuration; (h2) waveforms.

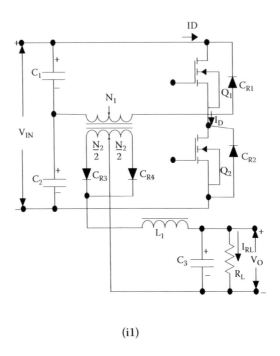

(i1)

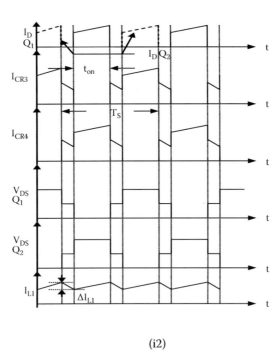

(i2)

TABLE FIGURE 3.4i
Half bridge converter: (i1) configuration; (i2) waveforms.

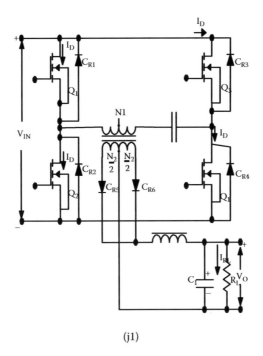

(j1)

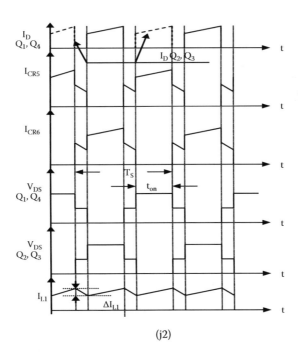

(j2)

TABLE FIGURE 3.4j
Full bridge converter: (j1) configuration; (j2) waveforms.

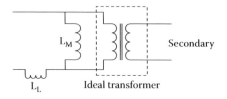

FIGURE 3.19
Equivalent circuit for the forward mode transformer.

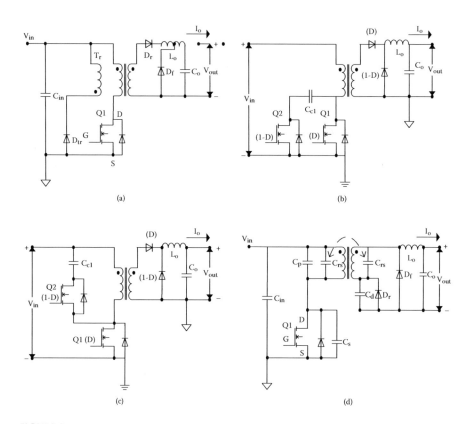

FIGURE 3.20
Transformer core reset techniques: (a) additional winding and a diode; (b) low-side active clamp technique; (c) high-side active clamp technique; (d) single-switch resonant reset technique. (Source: *Power Electronics Technology* [45,47,48].)

The above data give the designer the necessary maximum input power, $P_{in\text{-}max} = P_o/\eta$. To design for a low-input line situation with cycle skip holdup time requirements (when there is a short-duration AC voltage failure), a minimum DC bus regulation voltage target must be selected and DC bus filter capacitance, C_3 in Figure 3.21, must be calculated. Based on an approximate DC bus

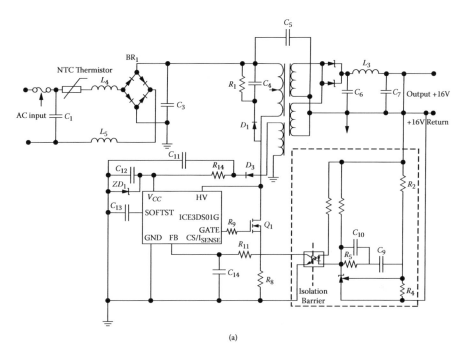

(a)

FIGURE 3.21
A representative flyback converter and transformer current waveforms related to different modes: (a) basic circuit arrangement; (b) continuous mode current waveforms; (c) discontinuous mode current waveforms. (Adapted from [51], courtesy of Power Electronics Technology.)

typical voltage of $V_{DCtyp(pk)} = \sqrt{2}\,V_{ACnom}$, the nominal value for DC bulk capacitor C_3 can be calculated as

$$C_{Bulk(nom)} = \frac{2P_o \times T_{hold}}{\left(V^2_{DCtyp(pk)} - V^2_{DCmin}\right)\eta}. \tag{3.4a}$$

From this DC rail, the circuit can operate in two different modes: the continuous conduction mode (CCM), with a large primary inductance of the transformer choke, or the discontinuous conduction mode (DCM), where the primary current is shown in Figure 3.21b and Figure 3.21c. For a typical case of DCM with a duty cycle limit of $D_{max} = 0.5$, the peak primary current, $I_{PRI(pk)}$, will be

$$I_{PRI(pk)} = \frac{2P_{in(max)}}{V_{DCmin}D_{max}}. \tag{3.4b}$$

The primary RMS current, $I_{PRI(rms)}$ (which is useful in determining the MOSFET capability and the primary conduction losses), is given by

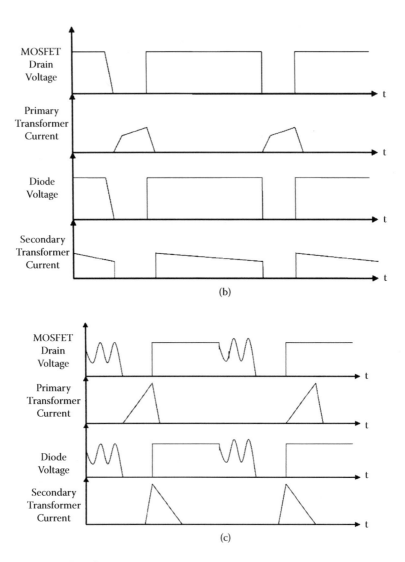

(b)

(c)

FIGURE 3.21 (continued)

$$I_{PRI(rms)} = I_{PRI(pk)}\sqrt{\frac{D_{max}}{3}}\ . \tag{3.4c}$$

From the peak primary current and target switching frequency, f_s (determined by the controller in Figure 3.21), the primary inductance of the flyback transformer can be calculated:

$$L_{PRI} = \frac{D_{max}V_{DC\,min}}{I_{PRI(pk)}f_s}\ . \tag{3.4d}$$

Assuming a maximum flux density value for the core (typically within 0.12T to 0.3T for a core such as an ETD series), a core can be selected from a magnetics manufacturer's data sheet. The core set and the gap must be chosen for an A_L product that supports a reasonable number of turns in such a way that it meets the other requirements as well [51]. The number of primary turns can be calculated as

$$N_p = \sqrt{\frac{L_{PRI}}{A_L}} \qquad (3.4e)$$

and rounded down to the nearest integer value. Then the secondary turns are calculated from the following:

$$N_s = \frac{N_P(V_O + V_{Diode})}{V_{R(max)}}, \qquad (3.4f)$$

where V_{diode} is the estimated peak diode forward voltage and $V_{R(max)}$ is the reflected voltage on the primary. The reverse voltage for the rectifier diode, $V_{R(Diode)}$, is given by

$$V_{R(Diode)} = V_{OUT} + \left(V_{DC\,max} \frac{N_s}{N_p}\right). \qquad (3.4g)$$

(In practice, the value required may be much higher due to overvoltages related to parasitic inductances, etc.) For more details related to the DCM-type flyback converter, see Hancock [51] and Ruble and Clarke [52,53].

Figure 3.21a indicates the essential circuit elements of a flyback converter based on a modern SMPS controller chip, such as the ICE3DSO1 from Infineon Technologies [51], with a power MOSFET driving the primary side winding. Another approach for flyback converters based on a complete power IC (an SMPS controller and a power MOSFET) is shown in Figure 3.22 from Power Integrations. Figure 3.22a to Figure 3.22d indicate different levels of feedback circuit arrangements, where output regulation performance can vary from average (lowest cost) to extra-high accuracy. For more details related to these design approaches, see Leman [54] and Power Integrations, Inc. [55]. In this design approach, current waveform parameter K_P simplifies calculations for both continuous and discontinuous modes [54]. For critical mode control–based design approaches, see Basso [56]. Flyback topology design using a MATHCAD-based approach is discussed in Huber and Jovanovic [57,58]. In this kind of design, for the best performance in charging a battery, the constant voltage mode (CVM), constant power mode (CPM), and constant current mode (CIM) are combined. It is also possible to use either an active clamp or RCD clamp approach for transformer demagnetizing, and critical mode conduction is used on the boundary of the CVM and CPM regions [58].

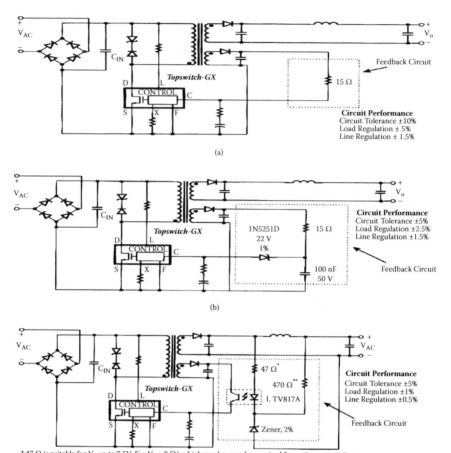

* 47 Ω is suitable for V$_o$ up to 7.5V. For V$_o$ > 7.5V, a higher value may be required for optimum transient response
** 470 Ω is good for Zeners with I$_{ZT}$ = 5 MA. Lower values are needed for Zeners with higher I$_{ZT}$. (E.g., 150 Ω fore I$_{ZT}$ = 20 mA).

(c)

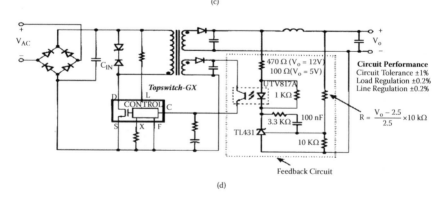

(d)

FIGURE 3.22

Reduced component designs using power ICs: low-cost version with (a) simple feedback circuit; (b) enhanced feedback; (c) opto/zener feedback with tighter regulation; (d) opto/TL431-based feedback with excellent load and line regulation performance. (Courtesy of Power Integrations, Inc.)

3.5.3.3 Two-Transistor Forward Mode Converters

A single-transistor flyback converter is an almost uncontested choice for off-line converters delivering fewer than 150 W. They are inexpensive because the transformer (which really works as a coupled inductor) is part of the output filter and generating multiple outputs merely requires the addition of another secondary winding along with diodes and output filter capacitors. However, at power levels greater than 150 W, because of excessive peak currents in the switching transistor and excessive voltages across the switches, this topology reaches its limitations. In these situations the two-transistor forward converter approach is a solution. Design aspects and calculation guidelines for the two-transistor forward converter are available in Gauen [59,60].

3.5.3.4 Bridge Converters

As shown in Figure 3.18, bridge converters are generally used for higher-power and higher-voltage converters because there are several power switches to share the dissipation and the voltage stress. In general, full-bridge topology is used for very high-power applications, and it is quite important to consider the losses in the circuits and the design complications due to their high-side switches operating with their source terminals (in the case of MOSFETs) or emitters (in IGBTs or power transistors) at floating levels. As indicated in the topology diagrams in Table 3.4, the transistors on the upper parts of the bridge (high-side transistors) require special circuitry to drive floating gate terminals. Gate driver ICs help solve this problem.

In a high-power DC-DC converter design, to achieve adequate efficiency, the designer should develop an "efficiency budget" or a loss calculation. In general, losses are contributed by many different sources, the important ones being:

- Rectification losses (low-frequency rectifiers on the input side and high-frequency rectification circuits on the output side),
- Switching losses in power semis (static and dynamic dissipation),
- Core losses in magnetic components,
- Losses due to control and supervisory circuits, and
- PCB loses associated with high-current tracks of the PCB.

Using a simple calculation based on an Excel sheet, the designer can determine where optimization can be achieved. Figure 3.23 indicates the losses associated with a switching power supply with an output capacity of about 10 W. In a larger-capacity power supply, the percentage values may be different.

3.5.3.4.1 Use of Gate Driver ICs

The essential idea of a gate driver IC is to achieve two important design requirements: to provide correct voltage drive levels required by the MOSFET

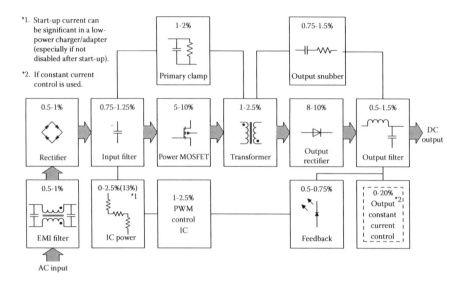

FIGURE 3.23
Losses associated with a switching supply. (Courtesy of *Power Electronics Technology*; Jovalusky [38].)

or IGBT gates where floating voltages are required, and to provide fast charge/discharge gate capacitances for MOSFETs or IGBTs. For example, in half- or full-bridge circuits based on MOSFETs, low-side (n channel) transistors need to be driven by a positive gate voltage with respect to the ground plane, but the high-side transistor gate needs to be driven by a positive voltage with respect to its source terminals, which will be at floating voltage values. Table 3.5 shows the different techniques used for gate driver circuits and their key features [61]. Gate driver circuits are useful in any switching system topology where two switches operate at high and low sides. To justify the use of these for efficient power circuit designs, the designer should understand and pay adequate attention to the parasitic capacitances at the gate input [62]. For IGBT-based bridge topologies, there are hybrid ICs available as gate drivers [63]. In some of these, optoisolators are used for electrical isolation between the drive side and the power stage.

3.5.3.4.2 An Overview of a Design Example: A 1-kW Full-Bridge DC-DC Converter

This section discusses a practical design process of a full-bridge configuration. Several years ago the author was asked to develop a DC-DC converter based on the following specifications:

- Input voltage: 20–30 V DC (nominal value of 24 V DC)
- Output voltage: 220 V DC at 1-kW output
- Topology: full bridge

TABLE 3.5

Comparison of Gate Driver Techniques

Technique	Basic Circuit Configuration	Features
Pulse transformer	 (a)	• Simple and cost effective • Size increases with lower frequencies • Operation over wide duty cycles need complex techniques • At higher switching frequencies, parasitics come into play
Bootstrap technique	 (b)	• Simple and inexpensive • Duty cycle and on time are constrained by the need to charge the bootstrap capacitor • At higher voltages, charging bootstrap capacitor may make up significant losses • A level shifter is required
Floating gate drive supply	 (c)	• Full gate control over wide range • Level shifting can demand complex circuitry • Cost due to isolated power supply for each high-side switch • Optoisolator use can be relatively expensive

(continued on next page)

TABLE 3.5 (continued)

Comparison of Gate Driver Techniques

Technique	Basic Circuit Configuration	Features
Charge pump based	 (d)	• Level shifting problems need to be tackled • Useful to generate a gate drive voltage above the rail voltage • Turn-on times can be too long • Inefficiencies of voltage multipliers can require more than two stage capacitor circuits
Carrier drive	 (e)	• Provides full gate control • Limited in switching performance • Could be improved by adding complex circuits

Source: Adapted from Clemente, S. and Dubhashi, A., HV floating MOS gate driver IC, Application Note 978A, International Rectifier, El Segundo, CA, 1990.

- Frequency of operation: 150–250 kHz
- Transformer configuration: planar
- Regulation load and line: ±2%
- Ripple: 0.5 V_{pp}
- Protection: Overload, overvoltage, and overtemperature, inhibit control

Several options were available for a project of this nature, and after considering hard-switching PWM to resonant converters, the ultimate decision was for a Intersil HIP 408X-based full-bridge configuration, as shown in Figure 3.24a. With the requirement for an extremely compact version, with a percentage-efficiency target in the high 70s, an estimate of losses was

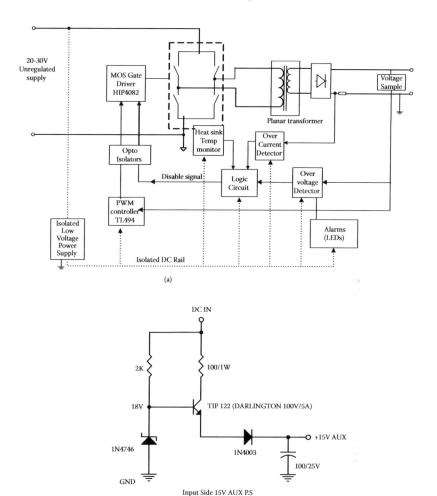

(a)

(b)

FIGURE 3.24
Design approach to a 1-kW, 24-V input, 220-V DC output full bridge with supervisory circuits and auxiliary power supplies: (a) overall design approach; (b) kick-start power supply based on a simple circuit; (c) load regulation; (d) efficiency.

done based on a set of MOSFETs with 70 V, 180 A, and 6 mΩ (or better) as the four switches. The overall unit was expected to have an isolated low-voltage power supply for the control and supervisory circuits, as shown in Figure 3.24a. For initial startup requirements, a simple auxiliary power supply of 15 V was proposed. Once the 220-V DC output appeared, an auxiliary winding in the planar transformer would handle overpowering of the control and supervisory circuits. Figure 3.24b shows this auxiliary (kick start) supply. With the decision to hard-switch the PWM, a TL494 was chosen as the PWM controller. As indicated in the power stage of Figure 3.24 Appendix Intersil 4081 gate driver [64] was used to simplify the design and to achieve

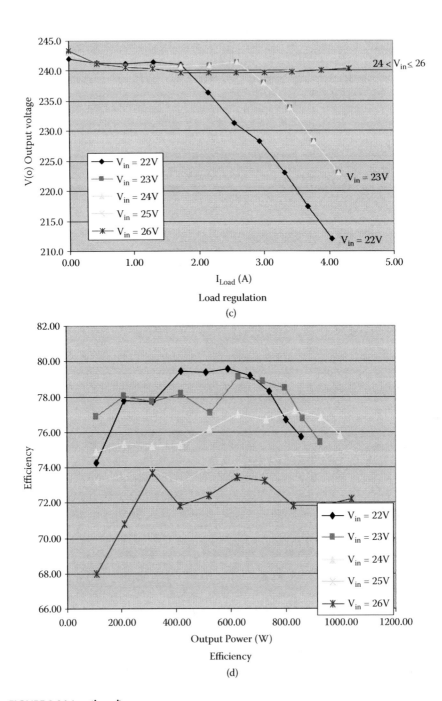

Load regulation

(c)

Efficiency

(d)

FIGURE 3.24 (continued)

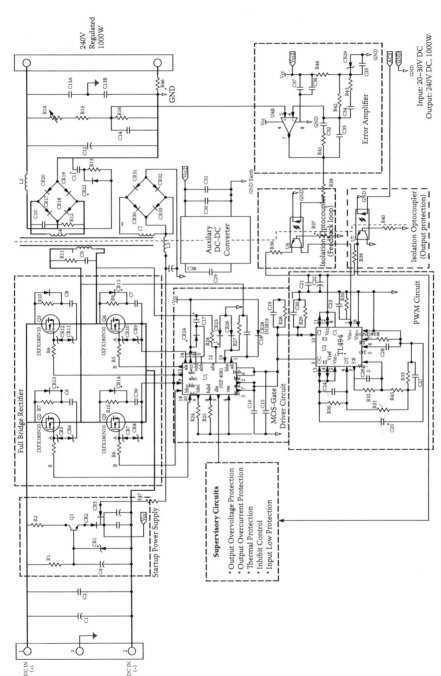

FIGURE 3.24 APPENDIX.

a smaller PCB area after carefully considering the simplicity achievable by the pulse transformers.

To achieve an extremely flat profile with a very small PCB area, a planar transformer from Payton America, Inc. was used [65]. In achieving a low component count and associated high reliability, it was necessary to drop the temptation to use standard logic IC-based supervisory circuits and use (a component count optimized) simple comparator circuit-based subcircuit. A kick-start supply was used to power up the TL494 PWM controller [66], gate drivers, etc. during startup, and once the system started running, a single-turn auxiliary winding in the transformer would take over, increasing efficiency. The efficiency achieved was about 75% at full load. The performance of the circuit is shown in Figure 3.24c and Figure 3.24d.

3.5.3.5 SEPIC Converters

Theoretical concepts related to the SEPIC topology have been of interest since its development in the mid-1970s. However, practical use of the technique was limited until battery-powered applications proliferated, particularly Li-ion types, where the battery pack's useful voltage can range from about 4.2 V to about 2.7 V. The SEPIC is definitely worth considering for a typical portable system, in which 3-V circuitry is powered by a Li-ion cell. Although SEPIC circuits require more components than buck or boost converters, they allow operation with fewer cells in the battery, where the cost of extra components is usually offset by the savings in the battery. An important use of the SEPIC is in PFC [67,68]. SEPIC topologies possess the following advantages:

- Single switch
- Continuous input current (similar to boost)
- Any output voltage (like in the buck-boost case)
- Ripple current can be steered away from the input, reducing the need for input noise filtering
- Inrush/overload current limiting capability
- Switch location is a simple low-side case, hence easier gate drive circuits
- Outer loop control scheme is similar to a boost converter's case

Disadvantages of the SEPIC are

- Higher switch/diode peak voltages compared to boost topology
- Bulk capacitor size and cost will be greater if operated lower than boost

A SEPIC converter can have six operating modes. A more detailed analysis [69,70] with design approach can be found in Dixon [67,68,71], Nuefeld [69], and Rahban [70]. Figure 3.25 shows the concept and two SEPIC application

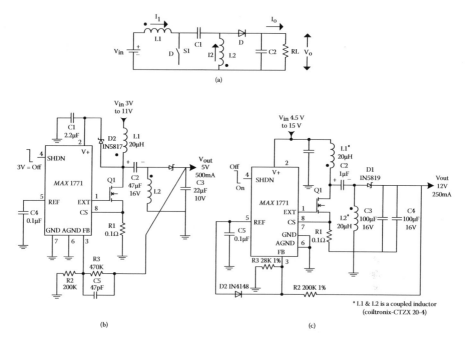

FIGURE 3.25

SEPIC concept and application circuits: (a) basic topology; (b) a circuit for 5-V or 3.3-V output from a 3-V to 11-V input; (c) circuit for 12-V output from 4.5-V to 15-V input. (Courtesy of *Power Electronics Technology*; Nuefeld [69].)

circuits. Achievable efficiencies are about 85% [69]. Another useful practical consideration for easy construction and lower cost in SEPIC circuits is to have the two (nearly equal) inductors coupled [69,71].

3.5.4 Resonant Converters

Resonant converters, which use the principle of an LC tank circuit that resonates, are those which process power in a sinusoidal form and have long been used with high-power systems. However, due to its circuit complexity, it had not found application in low-power DC-DC converters until about the early 1990s. The thrust toward resonant supplies has been fueled by the industry's demand for miniaturization, together with increasing power densities and overall efficiency, and low EMI. All resonant control circuits keep the pulse width constant and vary the frequency, whereas all PWM control circuits keep the frequency constant and vary the pulse width.

3.5.4.1 Comparison between Hard-Switching (PWM) and Resonant Techniques

The PWM or hard-switching converters discussed so far process power in pulse form. With available devices and circuit techniques, PWM converters have been designed to operate with switching frequencies generally in the range of 50 kHz to more than 500 kHz. In the 1990s, advances in power MOSFETs enabled the switching frequencies to be increased to several megahertz. However, increasing the switching frequency, although allowing for miniaturization, leads to increased switching stresses and losses. This leads to a reduction in efficiency. The detrimental effects of the parasitic elements also become more pronounced as the switching frequency is increased.

Resonant circuits in power supplies can also operate in two modes: continuous and discontinuous. In the continuous mode, the circuit operates either above or below resonance. The controller shifts the frequency either toward or away from resonance, using the slope of the resonant circuit's impedance curve to vary the output voltage. This is a truly resonant technique, but it is not commonly used in power supplies because of its high peak currents and voltages.

In the discontinuous mode, the control circuit generates pulses having a fixed on-time but at varying frequencies determined by the load requirements. This mode of operation does not generate continuous current flow in the tuned circuit. This is the common mode of operation in a majority of resonant converters and is called the quasi-resonant mode of operation.

3.5.4.2   The Quasi-Resonant Principle

The quasi-resonant principle in power converters incorporates a resonant LC circuit with the power switch. The power switch is turned on and off in the same manner as in PWM converters, but the tank circuit forces the current through the switch into a sinusoidal form. The actual conduction period of the switch is governed by the resonant frequency, f_r, of the tank circuit. This basic principle is illustrated in Figure 3.26.

The main advantages of the quasi-resonance arise from the near-sinusoidal switching currents and voltages. The switching losses are reduced,

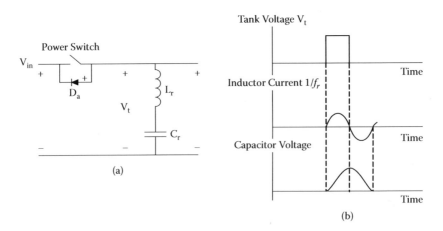

FIGURE 3.26
The resonant principle: (a) the basic circuit; (b) associated waveforms.

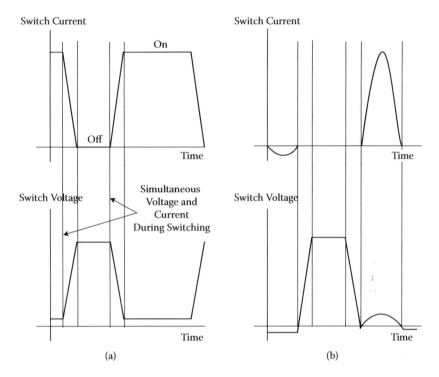

FIGURE 3.27
Comparison of waveforms in PWM and resonant converters: (a) PWM switching; (b) resonant switching.

leading to higher efficiency, and EMI is greatly reduced. The operation of a quasi-resonant switching power supply is analogous to a PWM supply of the same topology. The difference lies in the fact that the switching waveform in quasi-resonant supplies has been preshaped into a sinusoidal form. Figure 3.27 provides a comparison. With PWM converters, there is simultaneous conduction of current and voltage during part of the switching period. In resonant conversion, switching can be achieved at either the zero current point or the zero voltage point of the sinusoidal switching waveform, thus minimizing the switching losses.

However, increasing the switching frequency is accompanied by increasing switching stresses and detrimental effects due to parasitics. The advantages and disadvantages of PWM and quasi-resonant conversion techniques are summarized in Table 3.6.

Because resonant circuits generate sinusoids, designers can operate the power switches either at zero current or at zero voltage points in the resonant waveform. Based on this, there are two types of resonant switches: zero current switches (ZCSs) and zero voltage switches (ZVSs). The two types of switches are opposites of each other. Table 3.7 compares the circuits, waveforms, and four different operating states of ZCS and ZVS quasi-resonant systems, taking a buck topology as an example. Figure 3.28 compares the

TABLE 3.6

Comparison of Quasi-Resonant Converters (QRCs) and PWM Converters

Advantages	Disadvantages
Lower switching losses, hence higher efficiency.	More complex circuit.
Lower EMI.	Requires a longer design time and a higher level of expertise from the designer.
Higher maximum operating frequencies.	Parasitics must be taken into account.
Smaller size of components.	As current waveforms in QRCs are sinusoidal, peak values are higher than in PWM converters. Increased device stress.

operation of the power switches in PWM and resonant versions, indicating the advantage of resonant techniques. Fichera [72] provides the selection guidelines for quasi-resonant converters. Calculating the operating frequency of a quasi-resonant converter is not very straightforward due to various parasitic effects; some practical guidelines are given in Basso [73] with an example of a quasi-resonant flyback supply. The design aspects of phase-shifted full-bridge converters are discussed in Shennai and Trivedi [74] and Andreycak [75,76].

3.5.5 Control of Switch Mode Converters

In developing the control circuits for a DC-DC converter, there are two important practical steps: selection of the control IC and design of the feedback loop. In this section, a brief introduction to control ICs is provided. More design aspects of the feedback loop will be discussed later.

The primary function of the control IC in a switch mode power supply is to sense any change in the DC output voltage and adjust the duty cycle of the power switches to maintain the average DC output voltage constant. In general, an oscillator within the IC allows the designer to set the basic frequency of operation. A stable, temperature-compensated reference is also provided within the IC. There are two basic modes of control used in PWM converters: voltage mode and current mode.

3.5.5.1 Voltage Mode Control

This is the traditional mode of control in PWM converters. It is also called single-loop control, as only the output is sensed and used in the control circuit. A simplified diagram of a voltage mode control circuit is shown in Figure 3.29. The main components of this circuit are an oscillator, an error amplifier, and a comparator. The output voltage is sensed and compared to a reference. The error voltage is amplified in a high-gain amplifier. This is followed by a comparator, which compares the amplified error signal with a sawtooth waveform generated across a timing capacitor.

The comparator output is a pulse-width modulated signal that serves to correct any drift in the output voltage. As the error signal increases in the positive direction, the duty cycle is decreased, and as the error signal increases in the negative direction, the duty cycle is increased. The voltage mode control technique works well when the loads are constant. If the load or the input changes quickly, the delayed response of the output is one drawback of the control circuit, as it senses only the output voltage. Also, the control circuit cannot protect against instantaneous overcurrent conditions on the power switch. These drawbacks are overcome in current mode control.

3.5.5.2 Current Mode Control

The current control mode is a multiloop control technique that has a current feedback loop in addition to the voltage feedback loop. This second loop directly controls the peak inductor current with the error signal rather than controlling the duty cycle of the switching waveform. Figure 3.30 shows a block diagram of a basic current mode control circuit. The error amplifier compares the output to a fixed reference. The resulting error signal is then compared with a feedback signal representing the switch current in the current-sensing comparator. This comparator output resets a flip-flop that is set by the oscillator. Therefore, switch conduction is initiated by the oscillator and terminated when the peak inductor current reaches the threshold level established by the error amplifier output. Thus, the error signal controls the peak inductor current on a cycle-by-cycle basis. The level of the error voltage dictates the maximum level of peak switch current. If the load increases, the voltage error amplifier allows higher peak currents. The inductor current is sensed through a ground-referenced sense resistor in series with the switch.

The disadvantages of this mode of control are loop instability above 50% duty cycle, less than ideal loop response due to peak instead of average current sensing, and a tendency toward subharmonic oscillation and noise sensitivity, particularly at very small ripple currents. However, with careful design using slope compensation techniques [77,78], these disadvantages can be overcome. Therefore, current mode control becomes an attractive option for high-frequency switching power supplies. Some special problems, such as pulse-skipping oscillations, and solutions are discussed in Dobrenko [79].

3.5.5.3 Hysteretic Control

For processors such as Pentiums, current requirements are on the order of 20 to 30 A in desktops and similar systems. Other requirements are extremely low-output ripple voltages (on the order of only 50 to 100 mV, at most) with step load current transients on the order of 30 to 50 A/μs. In such situations the speed of the controller becomes very critical. Table 3.8 indicates the core voltage requirements of an old 300-MHz Pentium processor. A relatively newer technique using a hysteretic controller or ripple regulator has become prominent. Ripple regulation combines the advantages of voltage mode regulation and current mode solutions to power supply regulation. Voltage mode regulation is noted for reliable operation within a specified window, but it

TABLE 3.7

Comparison of ZCS-QR and ZVS-QR Operation In Buck Topology

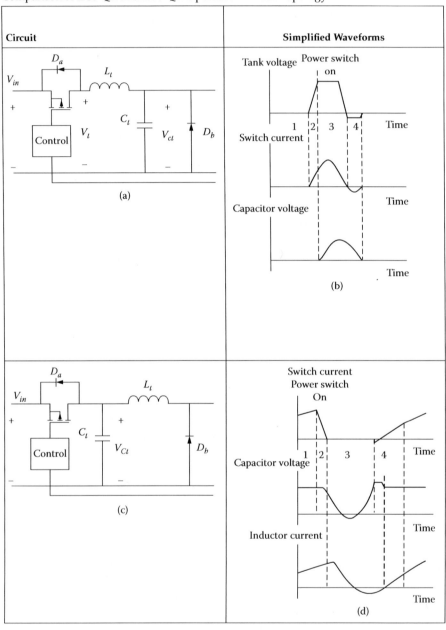

Circuit	Simplified Waveforms

Operating States			
Period 1	**Period 2**	**Period 3**	**Period 4**
The power switch is off, and the diode D_b is conducting the load current.	The power switch turns on. The voltage across the switch makes a step change. The resonant capacitor appears to be short-circuited at this time because of the conducting diode, and the power switch sees only the inductor turning on. Therefore, the switch current increases linearly from zero. This continues until all the load current is taken up by the current through the switch and the resonant inductor, displacing the current through the diode.	As the diode current is displaced, it turns off in a zero-current fashion, and the resonant capacitor is released into the circuit. Now the current waveform assumes a sinusoidal shape as the circuit resonates. During this period, the capacitor voltage lags the current waveform by 90°. The switch current proceeds over its crest and passes through zero. The resonant inductor's current then starts to flow in the opposite direction through the antiparallel diode D_a.	When the inductor current passes through zero, the resonant capacitor begins to dump its charge into the load, thus reducing its voltage in a linear ramp. The diode begins to conduct. When the capacitor voltage reaches zero, the diode takes up the entire current and the circuit awaits the next conduction period of the power switch.
The power switch is on. The switch current is determined by the converter stage configuration. The resonant inductor is saturated and is effectively short-circuited. The input voltage appears across the resonant capacitor, and the diode D_b is off.	The resonant period is initiated by the power switch turning off. As the voltage across the capacitor cannot change instantaneously, the power switch voltage remains constant while the current reduces to zero.	The capacitor voltage starts falling, together with the inductor current. The diode starts to conduct, taking over the load current from the resonant inductor, which gradually falls out of saturation. The tank circuit begins to resonate. The capacitor voltage rings back above the input voltage, at which point the current is conducted by the antiparallel diode.	The power switch turns on. The diode D_b is also on at this time, and the capacitor is shunted out of the circuit. Therefore, the switch current increases linearly through the resonant inductor. When this current exceeds the load current being conducted through the diode, the diode turns off. Then the resonant inductor can enter saturation and await the next cycle.

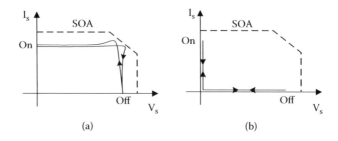

FIGURE 3.28
Comparison of load loci (a) for PWM (hard-switching) and (b) resonant (soft-switching) techniques.

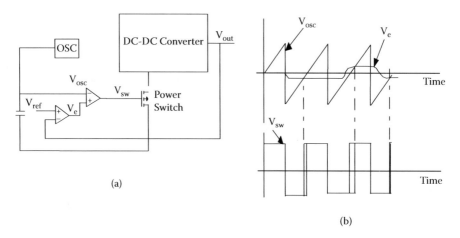

FIGURE 3.29
Voltage mode control: (a) block diagram; (b) associated waveforms.

is slow in responding to transient current demands, and loop compensation (discussed later) is difficult to implement. Current mode regulation offers better transient response than voltage mode regulation but at the expense of additional losses due to current monitoring resistors in the circuit.

A ripple regulator responds quickly to step current demands, and it is power efficient. In addition, a well-designed hysteretic controller keeps the ripple-regulated V_O within the specified window, maintaining general conditions required by a power-hungry CPU. Figure 3.31a shows the basic concept of a hysteretic-controlled buck converter that compares the actual output voltage to a reference signal corresponding to a desired output. Within the V_{hys} margins set by the Schmitt trigger/comparator, the output voltage will ramp up and down. One important aspect of this approach is that the frequency of operation is not constant and depends on the input–output voltage differential, inductor value, and ESR of the output capacitor. The typical range of frequencies is from about 150 to 700 kHz. The graph in Figure 3.31a shows the shape of the output ripple current flowing through the output capacitor. The

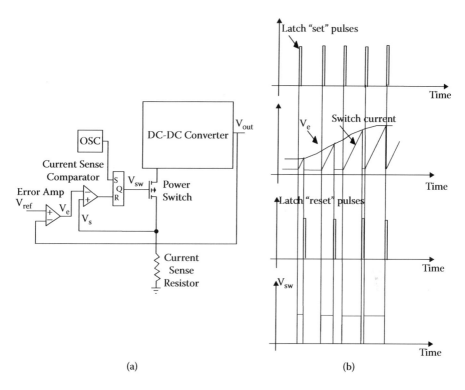

(a)   (b)

FIGURE 3.30
Current mode control: (a) block diagram; (b) associated waveforms.

TABLE 3.8

Design Constraints for a 300-MHz Pentium II Processor

Parameter	Typical Value
Core voltage (V_{core})	2.8 V
Static voltage tolerance	+100 to -60 mV
Dynamic voltage tolerance	±140 mV
Maximum processor core current ($I_{CC(max)}$)	14.2 A
Typical processor core current ($I_{CC(typ)}$)	8.7 A
Step current slew rate	30 A/μs
Input current di/dt	<0.1 A/μs

Source: Vosicher, E., Hysteretic controller fits processor needs, *PCIM*, January, 28, 2000.

hysteretic control concept is easy to implement, but EMI control may be difficult due to the unpredictable noise spectrum from the variable operational frequency. Figure 3.31b shows voltage tolerance budgeting waveforms at the application of a typical load (upper waveform with no droop compensation;

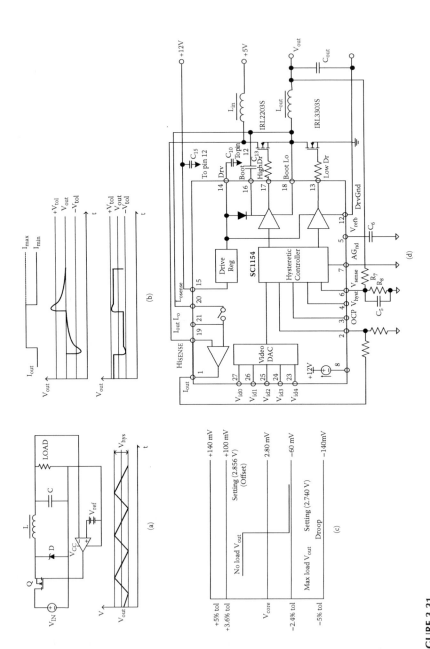

FIGURE 3.31

Hysteretic control and a typical circuit: (a) basic concept of hysteretic control; (b) voltage tolerance budgeting waveforms; (c) actual voltage tolerancing for the case of the Intel VRM 8.3 specification; (d) a typical circuit. (Courtesy of *Power Electronics Technology*; Vosicher [80].)

lower with droop compensation.). Figure 3.31c shows the actual voltage tolerances of a hysteretic control–based power supply where some offset voltages are applied above and below the nominal core voltage. In this design approach using adjustable values of offset and droop values for no-load and maximum-load conditions, respectively, variation of the output voltage is maintained within the specified limits of the processor. For example, SC1154 (from Semtech Corporation) and TPS 5211(from Texas Instruments) provide the design capabilities for the Intel VRM Spec 8.3. These controllers allow the output voltage to be adjusted from 1.3 V to 3.5 V, depending on a five-bit DAC output, which is a common requirement of VRM specifications. A guide to design calculations is available in Vosicher [80] and Nowakowski and Hodson [81]. Figure 3.21d provides a typical power supply schematic based on the SC1154 [80].

3.5.6 Efficiency Improvements in Switch Mode Systems

In Section 3.5.3.4 we briefly discussed the losses associated with a switching power supply. In improving the efficiency of a power supply, every item must be carefully evaluated and optimized. In low-output voltage switching power supplies, losses in the output rectifiers and the switching losses in the transistors contribute a significant share. High-frequency rectifiers usable in the output stage may be of three types: high-efficiency fast-recovery types, high-efficiency very fast rectifiers, and Schottky barrier rectifiers. A discussion of these can be found in Kularatna [46]. Recently, gallium arsenide (GaAs) and silicon carbide (SiC) devices have been introduced, and some of these could help improve the efficiency of the design.

For the lower DC rail voltage requirements, such as 1.2 to 3.3 V, of high-performance digital circuits, the high-efficiency requirement comes at very low-output voltages, and the general design approach using a Schottky diode becomes inadequate. In such circumstances, synchronous rectifiers (SRs) configured using low-$R_{DS(on)}$ power MOSFETs can provide much better efficiencies. In all switching topologies, the output rectifiers can be conceptually replaced by power MOSFETs. This basically replaces the DC loss component because of the combination forward voltage ($V_F I_{RMS}$) and the $I_{rms}^2 r_D$ (the resistive loss component in the diode) by a single element ($I_{rms}^2 R_{Ds(on)}$ of the MOSFET), providing significant efficiency improvements. Figure 3.32 indicates a comparison of a typical Schottky diode and a MOSFET usable in an SR. For more information, see Sherman and Walters [82], Moore [83,84], and Christiansen [85].

There are two basic types of SRs: self-driven (SDSR) and control-driven (CDSR). Figure 3.33 shows these types, and Table 3.9 compares their advantages and disadvantages. More design details can be found in How [86]. There are many controllers suitable for SR systems, and details on these advanced techniques can be found in Khasiev [50], Bindra [87,89], Yee [88], Elbanhawy [90], and Mappus [91].

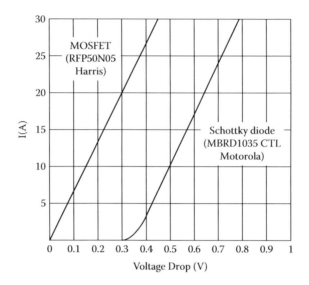

FIGURE 3.32
Voltage drop of a MOSFET compared with a Schottky diode.

In off-the line SMPS similar to the silver box in PCs, there is much room for efficiency improvement. Easily identifiable areas of improvement can be classified into three major categories:

- An appropriate harmonic reduction front end with active PFC.
- Architectural-level improvements to eliminate losses.
- Component-level improvements and upgrades to reduce losses.

More detailed discussion and guidelines can be found in Dalal [92]. Another important design consideration is the reduction of startup current-related losses [93]. Another aspect is to reduce the losses due to stray and leakage inductances [94]. Current-sensing resistors in switchers can also add significant losses. In situations where an inductor current is to be sensed, there are special techniques using a parallel RC network to sense the inductor current and minimize the loss across the series resistor inserted with the inductor [95].

3.5.7 EMI Reduction in Switch Mode Converters

Because of high-frequency nonsinusoidal waveforms within the circuitry, a switching power supply generates noise that can emerge through conduction or radiation. This noise can be conducted into the load or input source, and radiated components can create annoying situations in portable products such as cellular phones, PDAs, and laptops. United States and international standards for EMI-RFI have been established that require manufacturers to minimize the radiated and conducted interference of

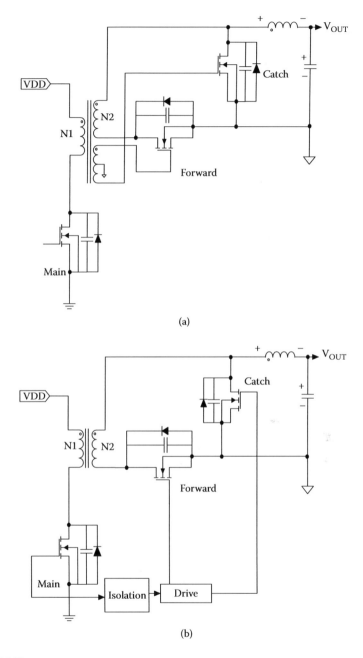

(a)

(b)

FIGURE 3.33
Different types of synchronous rectifiers: (a) self-driven SR (SDSR); (b) control-driven SR (CDSR).

their equipment to acceptable levels. In the United States, the guiding doc-ument is FCC Docket 20780, and internationally the West German VDE-08XX series is a well-accepted example. These standards generally exclude

TABLE 3.9

Comparison of Synchronous Rectification Techniques

Technique	Advantages	Disadvantages
Self-driven SR (SDSR)	Simple	For most topologies, drive signal amplitude varies with the line voltage When driven by transformer windings, there is no gate driving voltage during dead time Problems when parallel operated
Control-driven SR (CDSR)	Constant gate signal, irrespective of line voltage and load changes Constant $R_{Ds(on)}$ Suitable for wide input voltage ranges Applicable for all topologies Dead time can be kept to a minimum	Complex circuitry Needs accurate timing to prevent cross-conduction

Source: Sherman, J. and Walters, M.M., Synchronous rectification: improving the efficiency of buck converters, *EDN*, March 14, 111, 1996.

subassemblies from compliance, but the overall system should strictly adhere to the specifications.

The main sources of high-frequency noise are the switching transistors, input and output rectifier stages, protective diodes, and the control ICs. The RFI noise level can vary with the topology used. Flyback converters, because of their near-triangular input current waveforms, generate less conducted RFI than topologies with rectangular input current waveforms. EMI noise reduction is generally accomplished by three means: suppression of the noise source, isolation of the noise coupling path, and filtering and shielding. Some advanced techniques are frequency modulation techniques and slew rate control. Because the total spectrum of high frequencies should be minimized, measurement of conducted EMI noise using a spectrum analyzer is generally carried out, with adequate attention to the resolution bandwidth (RBW) of the spectrum analyzer or a similar test setup [96].

The most common method of conducted noise suppression at the input of an off-line SMPS is the utilization of an LC filter for differential-mode and common-mode RFI suppression. Normally, a coupled inductor is inserted in series with each of the AC input lines, and capacitors are placed between lines (*X* capacitors) as well as between the lines and the earth terminal (*Y* capacitors). Figure 3.34 indicates different line filter schemes used. The resistor *R* is for the discharge of the *X* capacitors and the values are recommended by the relevant safety specifications under the VDE or IEC series.

Proper component layout and selection are important in controlling EMI. In a typical offline power supply with common mode filters, the main source of common mode noise is the MOSFET. With the requirement of a heat

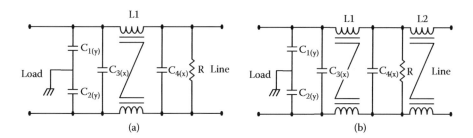

FIGURE 3.34
Input line filters for SMPS systems: (a) a basic version; (b) an improved version with two filter chokes.

sink, for example, in the case of a TO-220 package, the capacitance formed between the drain and the ground plane (CP_1 in Figure 3.35a) can conduct some common mode noise. In a typical power supply, such as in Figure 3.35, the transformer can also conduct some high-frequency current through the parasitic capacitances formed between the windings. One technique to reduce the effects of these capacitances, such as C_{P2A} and C_{P2B}, is to use a Faraday shield for the windings. The situation is compared in Figure 3.35b and Figure 3.35c.

Another possible source of EMI is gapped cores, such as in flyback transformers. Although the gap increases the stored energy, it can lead to increased EMI problems, and for this reason experienced designers avoid bobbin cores. Details on the selection of transformer cores can be found in Schindler [97].

Another important approach is to pay adequate attention to the layout of the circuit. The most important consideration is the power stage, because it creates the highest circulating currents and acts as the main source of EMI. Next is the drive stage, where currents can be a few hundred milliamps to 10 A or higher. Because of the relatively high currents possible in the drive stage, it should be placed very close to the power stage. In the MOSFET drive stages, if the gate connections are longer than about 5 cm, a rule of thumb is to place a series resistor of 10 Ω near the power MOSFET. Another important consideration is to place the power traces and the returns close to each other to minimize the loop area between the traces and to increase coupling capacitance. An ideal layout design for a multilayer board would have these two traces on adjacent layers, one directly above or below the other. Another important consideration is to have the ground of the controller IC close to the feedback circuit's ground to minimize feedback voltage errors. More details can be found in Rogers [98] and Scolio [99].

Another more recent advancement is the spread spectrum–based controller ICs, whereby modulating the PWM frequency, the noise gets spread across the band [100]. Another advanced approach is to introduce dither to the system clock of the DC-DC converter. This approach, with its resulting spread spectrum operation, allows the switching frequency to be modulated

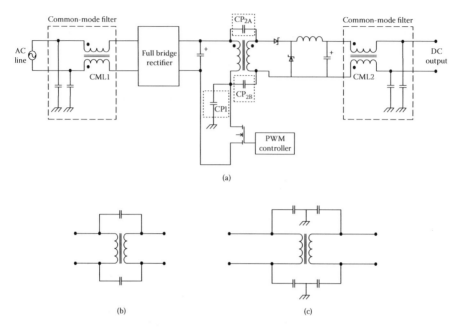

FIGURE 3.35
A simplified example of a typical power supply with common mode filters: (a) a general case indicating parasitic capacitances; (b) parasitic capacitances across a winding without a Faraday shield; (c) Faraday shield reducing the stray capacitance coupling. (Courtesy of *EDN Magazine,* © 2006, Reed Business Information, a division of Reed Elsevier. All rights reserved. [97]).

by a pseudo-random number (PRN) sequence to eliminate narrowband harmonics [101]. In this approach, a charge pump technique is used. For analysis and spectral characteristics of these techniques, see Tse et al. [102].

In some controller ICs, such as the LT 1533 from Linear Technology Corporation, the slew rate of the switcher voltage and current waveforms are controlled to reduce noise, at the expense of the efficiency [103]. Wittenbreder [104–107] provides a guideline for designing converters for lower EMI. Reducing the ground bounce problems in DC-DC converters is discussed in Barrow [109].

3.5.8   Control Loop Design

For most designers, feedback control loop stability is shrouded by a cloud of mystery. Although most designers understand the problem of unwanted oscillations of a switching supply, many use trial-and-error procedures or fancy mathematical models that require computing resources. In this section we discuss feedback loop stability, adding to the basic concepts discussed in Chapter 1 with some practical insights into the theory and suggesting some useful practical procedures for refining the process.

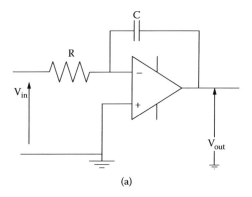

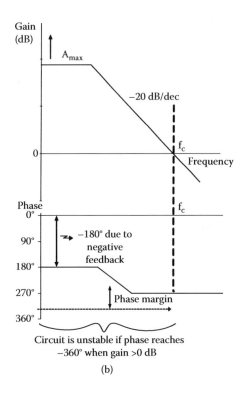

FIGURE 3.36
Simple RC network and its gain and phase plots: (a) RC circuit; (b) gain and phase plot.

A good place to start is the simple RC integrator, as in Figure 3.36a, and its gain and phase plot. For this simple single-pole circuit, the transfer function is given by

$$G(s) = \frac{V_{out}(s)}{V_{in}(s)} = \frac{1}{1 + sRC} \qquad (3.5a)$$

and the pole is given by

$$s = -\frac{1}{2\pi RC}.$$ (3.5b)

Equation 3.5b shows an important result, that a pole will cause the transition of the gain plot from 0 to –1 at a frequency of $f_c = 1/2\pi RC$. At this corner frequency f_c, the asymptote breaks and the slope is –6 dB/octave or –20 dB/decade. In general, a pole in a transfer function will cause a transition from +1 to a 0 slope, or 0 to –1, or –1 to –2, or –2 to –3, etc., with a gain change of –6 dB/octave (or –20 db/decade). This is associated with a phase shift of –90°/octave or –45°/decade. Zeros are the points where the Bode plot breaks upward, causing an opposite change of slopes with leading phase shifts.

For a closed loop of a switching supply, as shown in Figure 3.37a, the loop consists of two typical blocks: the modulator and the error or feedback amplifier. The case shown is a simple buck converter only, but a similar simplified block diagram can be developed for any other configuration. Figure 3.37b indicates typical transfer function characteristics of the LC filter and the modulator in a circuit similar to Figure 3.37a. An error amp based on the op amp can be designed to have any pole-zero combination to change the Bode plot to attain unconditionally stable characteristics.

Once we have the overall system transfer function estimated, the stability of the circuit can be estimated using the phase margin and the gain margin discussed in Chapter 1 (Figure 1.14). In this situation, the combination of the modulator and the op amp (error amp) provides the overall closed-loop gain, similar to the fundamental block diagram discussed in Figure 1.14. Figure 3.37b shows the effect of the LC filter with two poles, where the gain function has a slope of –40 dB/decade, and in a typical practical circuit the output filter capacitor (with a finite ESR) makes the gain plot return toward –20 dB/decade.

The op amp circuit can have different configurations, and in general there are three common types used in practical switching power supply environments. Table 3.10 shows the simplified circuit configurations and the associated transfer functions. In error amplifier configurations, such as in Table 3.10, break or corner frequencies are predetermined by the design objectives. Type 1 is a simple RC low-pass filter with a single pole; type 2 is with a pole-zero pair; and type 3 is with two pole-zero pairs. In type 3, loop crossover should occur between f_2 and f_3 for better stability. More details can be found in Chryssis [110]. Another mathematical technique useful in this process is called the K-factor method, which is a mathematical tool for defining the shape and characteristics of the transfer function; details can be found in Chryssis [110].

In practice, the overall loop can be complicated, and in the final stage of a design project, the design team can make use of some tests to measure the loop transfer function. One test is to inject a signal into the loop and then

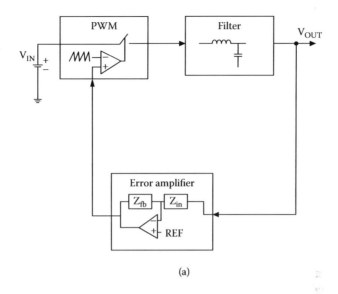

(a)

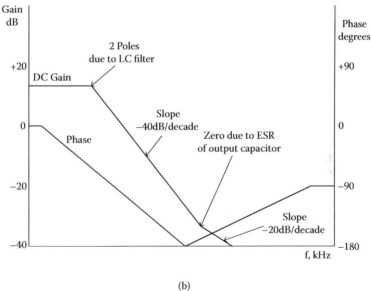

(b)

FIGURE 3.37
Control loop of a buck converter: (a) modulator and the feedback amplifier; (b) transfer function of the LC filter.

measure the loop transfer function. However, due to monolithic ICs that do not allow injection of the signal in the best location, one may have to compromise with an "achievable method." Figure 3.38 shows the case from Venable [111], where the desired versus achievable methods are indicated for a simplified case of a computer power supply. In Figure 3.38a, for the case of a common PWM control IC type UC 3844, two cases of feedback paths exist (via

TABLE 3.10

Different Error Amp Types Usable in Switching Power Supplies

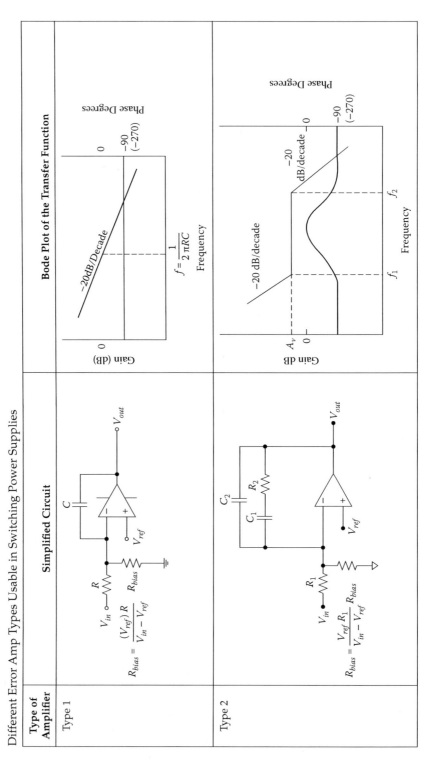

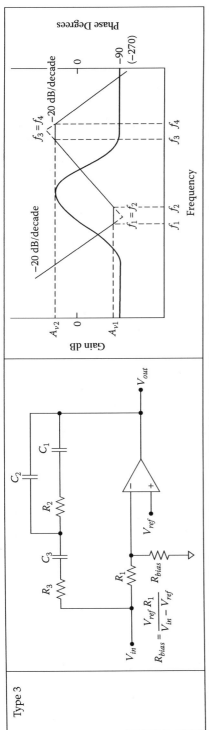

Source: Adapted from Hsiu, L., Goldman, M., Caristen, D., Witulski, A.F., and Kerwin, W., Characterization and comparison of noise generation for quasi-resonant and pulsewidth modulated converters, *IEEE Transactions on Power Electronics*, 9, 425, 1994.

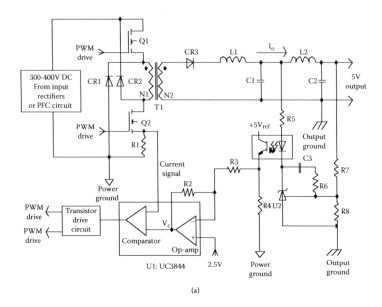

(a)

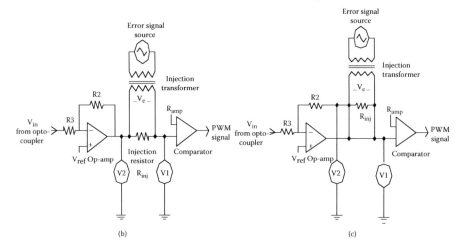

(b)   (c)

FIGURE 3.38

Loop measurements and typical response: (a)simplified schematic; (b) desired injection method; (c) achievable injection method; (d) transfer function of slower loop (measurement in series with R_7); (e) transfer function of fast loop (measurement in series with R_5); (f) transfer function of entire loop (measured in series with R_2). (Courtesy of *PCIM*. Source: [111]).

resistor R_5, a fast loop, and a slow loop via R_7). Figure 3.38d to Figure 3.38f show these measurements in this typical example [111].

     One practical difficulty with such measurements is to select an injection transformer with wide frequency response. Although some companies sell such transformers, they can be expensive. One solution proposed is to modify

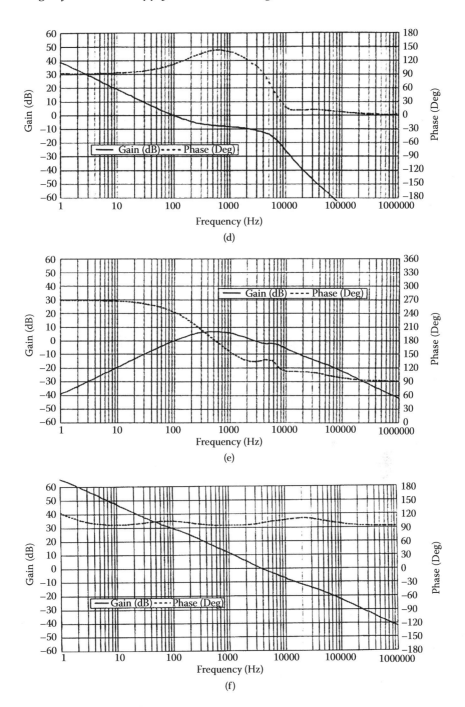

FIGURE 3.38 (continued)

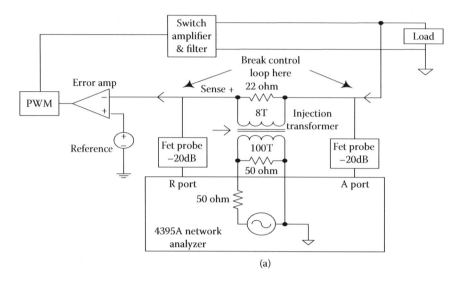

(a)

FIGURE 3.39
Typical injection transformer setup for loop-gain measurements: (a) setup; (b) frequency response of the PE-51687 transformer with eight-turn secondary; (c) frequency response for the same with 22 Ω on the secondary as in (a). (Courtesy of *Power Electronics Technology*. Source: [112].)

an off-the-shelf current transformer and use it with a typical network analyzer-based measurement setup. Figure 3.39 shows a typical injection transformer setup with a network analyzer [112]. Figure 3.39b and Figure 3.39c show a typical frequency response curve for a commercial current transformer, such as the PE-51687 from Pulse Engineering [112]. Williams [113] discusses practical guidelines for an iterative procedure for easy frequency compensation using a test setup. Hesse [114] provides some analytical aspects of a battery-powered buck converter example. Gain equalization aspects of flyback converters are discussed in Sandler [115].

3.5.9 Low Dropout Regulators (LDO)

With the unprecedented development of switch mode systems, one tends to think that the linear regulator is totally obsolete and is suitable only for low-end applications or applications where low efficiency is acceptable. However, higher noise and lower load current transient response of switch modes have helped in developing low dropout regulators (LDOs). The demand for LDOs is increasing due to the growing demand for portable products, and typically they are cascaded onto switching regulators to suppress noise and to respond rapidly to high slew rate load currents. In general, LDOs are commonly used to provide power to low-voltage digital circuits, where point of load (POL) regulation is important. As discussed in Section 3.2.2, the designer should use informed judgment when selecting and mixing different techniques to get the best out of a power management system design.

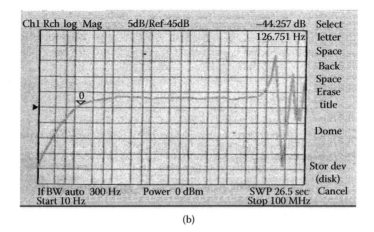

(b)

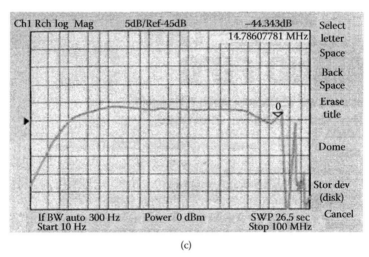

(c)

FIGURE 3.39 (continued)

Figure 3.40 shows the circuit elements of a typical LDO application. The main components are the pass element, precision reference, feedback network, and error amplifier. The input and output capacitors are the only key components of an LDO solution that are not contained within the monolithic LDO. Table 3.11 compares different options available for the pass transistor and the advantages and disadvantages of the approaches. LDOs are available in a wide variety of output voltages and current capacities. Many LDOs are tailored to applications where a good response to a fast step current transient is important. Figure 3.40b shows a block diagram of an LDO with a secondary loop for fast transient response. Figure 3.40c and Figure 3.40d show the response of a typical LDO, such as the TPS75433 from Texas Instruments, for low currents and high currents, indicating the effect of the fast transient loop [116].

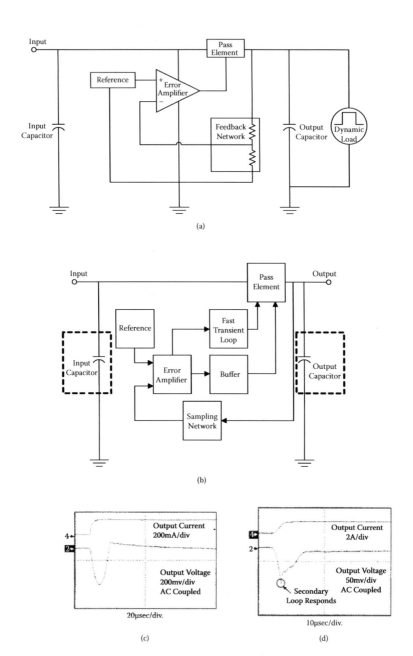

FIGURE 3.40
LDO block diagrams: (a) basic version; (b) version with a secondary loop for fast transient response; (c) response to a low current; (d) response to a high current. (Courtesy of PCIM. Source: [116].)

TABLE 3.11

LDO Pass Transistor Configurations and Their Characteristics

	Single NPN	Darlington NPN	Single PNP	PNP/NPN Combination	p-MOSFET
Configuration	V_{in} — Control	V_{in} — Control	V_o — Control	V_{in} — Control	V_{in} — V_o — Control
Minimum dropout voltage (minimum pass transistor drop)	$\approx 1V$	$\approx 2V$	$\approx 0.1V$	$\approx 1.5V$	$\approx R_{DS}(on) \times I_L$
Load current capability	<1A	>1A	<1A	>1A	>1A
Output impedance (Z_{out})	Low	Low	High	High	High
Bandwidth	Wide	Wide	Narrow	Narrow	Narrow
Effect of load capacitance (C_L)	Immune	Immune	Sensitive	Sensitive	Sensitive

In the majority of LDOs and quasi-LDOs (where a composite NPN-PNP pair is used as the pass device), the pass device or the driver is a lateral PNP. Even though a PNP is better at providing a lower dropout voltage than an NPN [117], a lateral PNP is a low-frequency cutoff device with a poor transient response. For this reason, proper selection of the external output capacitor is important for the stability of the loop and adequate transient response. The compensation capacitor determines three key characteristics of an LDO: startup delay, load transient response, and loop stability.

The startup time is approximately given by

$$T_{\text{startup}} = CV_{\text{o}}/I_{\text{limit}}, \tag{3.6a}$$

where C is the value of the output capacitor and I_{limit} is the current limit of the regulator. If C is fully discharged before the regulator is powered up, the regulator will limit current during startup, and the time to reach the nominal V_{o} will be delayed. Conversely, if C is too small, the output voltage will overshoot the nominal V_{o} during startup. Because it is impossible to investigate all three characteristics at once, the designer should concentrate on first achieving a stable loop design and then check the startup delay and load transient response. Figure 3.36b shows the case of an ideal system where we can achieve a single-pole response that determines the system's crossover frequency. Its crossover frequency, f_c, should be selected to ensure that the system can quickly respond to load transients without undue ringing at the output. For stability, the phase margin should be at least 30° away from 360°.

Unfortunately, an LDO has three dominant poles, and two are set by the regulator IC and the third is a function of the load and the output capacitor. The first pole, determined by the error amplifier, generally occurs between 10 and 300 Hz; the second pole, due to the pass device (or the PNP bias device of a compound regulator), is usually between 100 and 300 kHz. The third pole, set by the load and the output capacitor, occurs within the same range as the error amplifier or even slightly lower at light loads. Figure 3.41a shows the simplified case of a load and output capacitor combination.

It can be shown that the pole and the zero created by the load are given by

$$f_{pL} = \frac{1}{2\pi(R_L + ESR)C} \tag{3.7a}$$

$$f_{zL} = \frac{1}{2\pi(ESR)C}. \tag{3.7b}$$

Based on the discussion in Section 3.5.8, where added poles and zeros change the Bode plot, it is apparent that the zero due to capacitor ESR modifies the total response of the circuit. Figure 3.41b1 shows the case where the

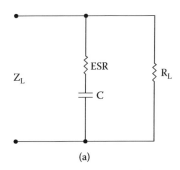

(a)

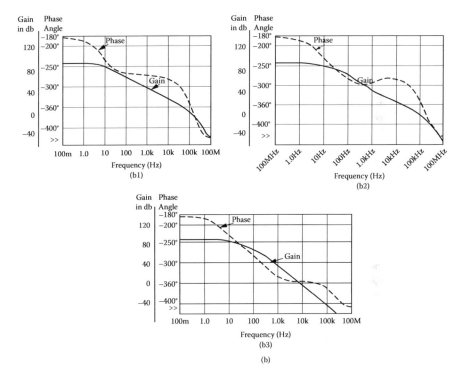

FIGURE 3.41
Effect of the load on stability: (a) simplified load and output capacitor combination; (b) Bode plot of the output for a practical LDO based on a regulator IC such as CS 8156 from ON Semiconductor for different cases of ESR values: (b1) $R_O = 120 \ \Omega$ and $C = 22 \ \mu F$ with ESR = 3 Ω; (b2) $R_O = 120 \ \Omega$ and $C = 22 \ \mu F$ with ESR = 1 Ω; (b3) $R_O = 120 \ \Omega$ and $C = 22 \ \mu F$ with ESR = 0.01 Ω. (Courtesy of ON Semiconductor; O'Malley [118].)

output is marginally stable for ESR = 3.0 Ω. As depicted in Figure 3.41b2, when the ESR is reduced to 1.0 Ω, the system's phase margin increases and the system becomes stable. When the ESR is lowered further, the system can become unstable, as in Figure 3.41b3. The capacitor used at the output should have some stability within the operational temperature ranges. Figure 3.42 shows typical aluminum electrolytic capacitor characteristics over frequency

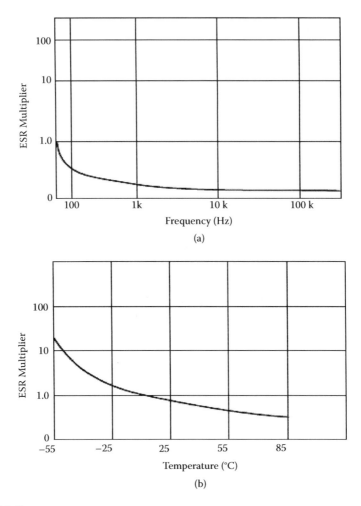

FIGURE 3.42
Behavior of typical aluminum electrolytic capacitors at different frequencies and temperatures: (a) frequency behavior; (b) ESR change with temperature; (c) capacitance change with temperature. (Source: ON Semiconductor, Application Note SR003AN/D.)

and temperature. Based on the discussion related to Figure 3.41, it is important for designers to carefully examine the parameter changes of capacitors over frequency and temperature to achieve a stable design. Details can be found in King [116], O'Malley [118], Simpson [119,120], and Goodenough [121].

LDOs find applications in automotive environments because of the rapid voltage changes of the 12-V rail during cold startup [122]. Most LDOs are used in powering high-power processors where load current changes in step mode. Schiff [123] and Rincon-Mora and Allen [124] provide design guidelines to deal with these conditions. Details on frequency compensation of

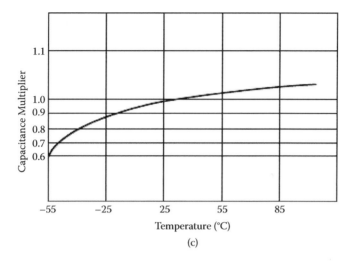

(c)

FIGURE 3.42 (continued)

LDOs are available in Kwok and Mok [125] and Chava and Silva-Martinez [126]. For applications with extra LDO voltages, ultra-low-dropout (ULDO) linear regulators based on bipolar CMOS-DMOS (BCD) technologies are available [127].

Some of the LDO regulators are specially designed for low-noise requirements [128] within cellular handsets and other portable applications, because most switching techniques are too noisy for these applications. The noise performance of these components sometimes needs to be quantified, and special measurement setups may be necessary. In this process one should ensure that the LDO meets the system's noise requirement within the entire bandwidth of interest, typically in the range of 10 Hz to 100 kHz. Figure 3.43 indicates a suitable filter structure for testing the noise performance of LDOs in this frequency band. In measurement, special consideration should be given to ground loop elimination; hence, all power supplies should be battery based, and thermally responding RMS meters should be used for measurements [129].

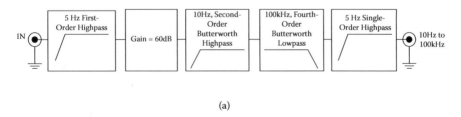

(a)

FIGURE 3.43
Noise measurements of LDOs: (a) block diagram of the filter arrangement; (b) a typical circuit configuration. (Courtesy of *EDN*; Williams and Owen [129].)

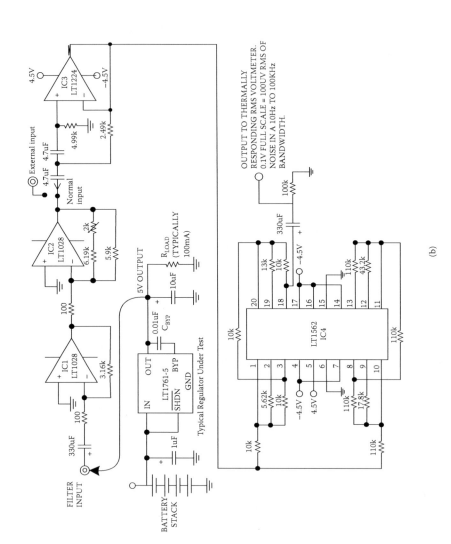

(b)

FIGURE 3.43 (continued)

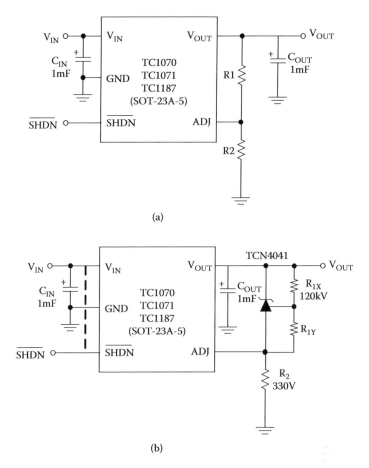

FIGURE 3.44
Adjustable LDO circuits: (a) a simple circuit with two resistors to adjust the output; (b) use of an adjustable reference source for improving accuracy (Courtesy of *EDN*; Paglia [130].)

For applications where nonstandard voltages are required, an adjustable LDO is a good choice, but getting the highest accuracy from such an IC may require a few circuit tricks. Figure 3.44 shows a few examples, including the use of an adjustable reference for improving accuracy [130]. For applications with hot-swap requirements, LDO ICs can be used with special current limiting arrangements [131]. Performance verification of LDOs is discussed in Williams and Owen [132].

For battery-powered applications, PMOS-based LDOs provide acceptable solutions. The factors to be considered include dropout voltage, ground current, noise, input voltage, and thermal response. Typical ground current components in an LDO are shown in Figure 3.45a. Figure 3.45b and Figure 3.45c show the comparative performance of typical PNP LDOs and PMOS-based LDOs. For details, see Christ [133].

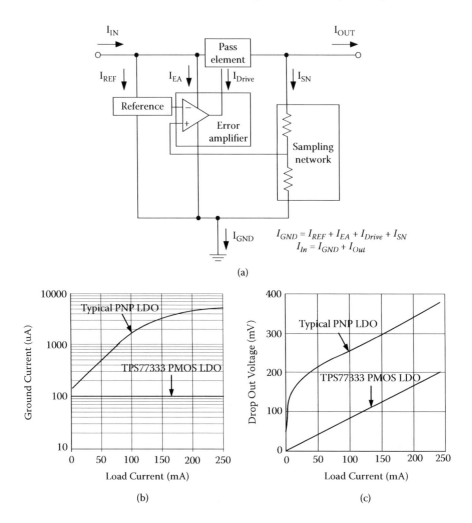

$$I_{GND} = I_{REF} + I_{EA} + I_{Drive} + I_{SN}$$
$$I_{In} = I_{GND} + I_{Out}$$

(a)

(b)　　　　　　　　　　　　(c)

FIGURE 3.45
Ground currents in an LDO and comparison of PNP and PMOS types: (a) ground currents; (b) comparison of ground currents in PNP and PMOS types; (c) comparison of dropout voltage in PNP and PMOS types. (Adapted from Christ [133].)

3.5.10　Charge Pump Converters

Switched capacitor (charge pump) converters use capacitors rather than inductors or transformers to store and transfer energy. The most compelling advantage is the absence of inductors, which have greater component size, more EMI, greater layout sensitivity, and higher cost. Compared with other types of voltage converters, the switched capacitor converter can provide superior performance in applications that process low-level signals or require low-noise operation. These converters offer extremely low operating current—a useful feature in systems where the load current is either uniformly low or low most of the time. Thus, for small handheld products, light-load

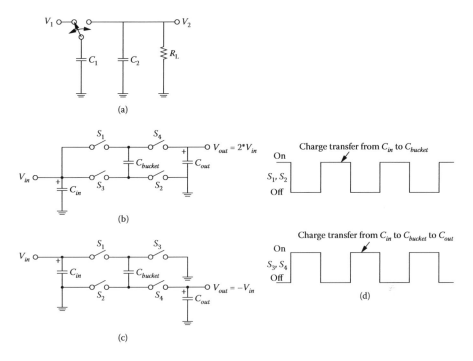

FIGURE 3.46
Switched capacitor converters: (a) basic principle of operation; (b) doubler; (c) inverter; (d) timing.

operating currents can be much more important than full-load efficiency in determining battery life. The basic operation of switched capacitor voltage converters is shown in Figure 3.46 [134].

When the switch is in the left position, C_1 charges to V_1 (Figure 3.42c). The total charge on C_1 is given by $q_1 = C_1 V_1$. When the switch moves to the right position, C_1 discharges to V_2. The total charge on C_1 is now given by $q_2 = C_1 V_2$. The total charge transfer is given by

$$q = q_1 - q_2 = C_1(V_1 - V_2).$$

If the switch is cycled at a frequency f, the charge transfer per second, or the current, is given by

$$I = fC_1(V_1 - V_2) = (V_1 - V_2)/R_{eq}, \qquad (3.8)$$

where R_{eq} is given by $1/fC_1$. The reservoir capacitor C_2 holds the output constant. A basic charge pump can work as a doubler or an inverter, as shown in Figure 3.46b and Figure 3.46c, respectively. Figure 3.46d shows the switch drive waveforms for the two cases. Some variations of the basic doubler exist for which the output voltage is about 1.33 to 1.5 times the V_{in} value [13]. In some cases, output voltage can be programmed with an external resistor divider. Charge pumps can be either regulated or unregulated.

Recent generations of charge pumps offer improved specifications and have become a viable DC-DC conversion method for many portable appliances where high-density converters are necessary and circuit area is limited. Two common charge pump types are hysteretic and fixed frequency. Figure 3.47a shows the concept of hysteretic control in charge pumps. With

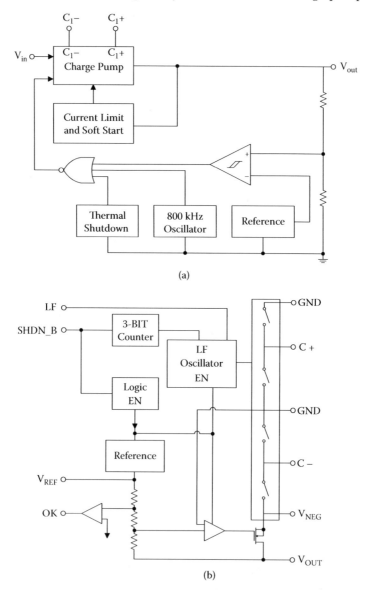

(a)

(b)

FIGURE 3.47

Different types of charge pumps: (a) hysteretic control based; (b) fixed-frequency type. (Courtesy of *EDN Magazine*, © 2000, Reed Business Information, a division of Reed Elsevier. All rights reserved. [13].)

this technique, an output voltage that falls below the reference voltage enables the oscillator. During the first clock cycle, the bucket capacitor charges to the input voltage. During the next cycle, the total charge, consisting of C_{bucket} and C_{in}, transfers to the output capacitor. This cycle repeats until the output voltage reaches the upper hysteretic threshold, at which point the comparator disables the oscillator. The internal comparator continues to enable/disable the charge pump switches based on the output level [13]. The example shown is for an SC1517-5.

In a fixed-frequency type with linear regulation, shown in Figure 3.47b, the internal oscillator runs at a fixed frequency when the device is not shut down. The charge pump provides an unregulated voltage to an internal linear regulator that adjusts this voltage to a fixed output. The device achieves regulation by using an internal comparator that senses the output voltage and compares it with an internal reference while adjusting the gate drive to the internal pass MOSFET for fixed output voltage. The oscillator frequency that controls the charge pump is usually outside the sensitive frequency spectrum of cellular communication bandwidths. Unlike the hysteretic types, this type can restrict any generated switching noise to noncritical bandwidths [13]. For details, see Khorshid [13], Arimoto [135,136], and Vitchev [137].

3.5.11   Achieving High Power Density

Consider the requirements of a 1-V, 100-A DC-DC converter, such as a VRM or a processor power supply. At this power level, the converter is driving an effective 10-mΩ load. With a 10-mΩ load, practically anything could contribute to the detriment in efficiency. Such causes include, for example, PCB traces, series resistance in inductors, and $R_{DS(on)}$ of MOSFETs. Each milliohm in the path to the load represents an additional 10% loss. Efficiency management is only one of the important issues. Another is the transient response for load current changes.

One common approach to design converters for such requirements is to have a multiphase supply. A multiphase supply interleaves the clock signals of the paralleled output stages and thereby reduces input and output ripple current without increasing the switching frequency. Figure 3.48 indicates the operating principle of a two-phase converter. In general, these converters are based on synchronous rectifiers to achieve higher efficiencies. Most multiphase buck converters operate at about 250 kHz to get the best compromise of switching losses and the values of capacitors and inductors. To achieve high-power capability, sometimes interleaving two separate forward or flyback converters with appropriate clock phases is a common approach. Details are available in Travis [138,139], Wei [140], Benport [141], Peterson [142], Harriman [143], O'Loughlin [144], Bindra [145], and Betten and Kollman [146]. Use of coupled inductors in a multiphase buck converter could further reduce the ripple in each phase [147].

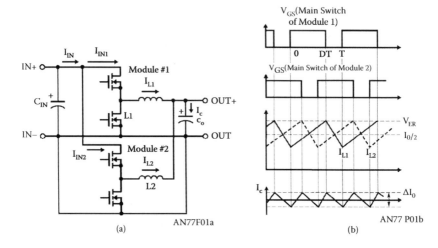

FIGURE 3.48
Multiphase converter approach: (a) a two-phase converter; (b) waveforms in a two-phase converter. (Courtesy of Linear Technology Corp., [140].)

3.5.12 Postregulation Techniques

In many common processor-based load environments, the load may require very tightly regulated outputs together with excellent transient response at the power supply output for fast-changing load currents. The output ripple of transformer-isolated multiple-rail DC-DC converters is typically between 0.5% and 1% of the nominal output, whereas a linear regulator can perform within a millivolt- or even a microvolt-order ripple. The transient response of a 50% to 100% (or vice versa) load step can range from 100 to 300 μs for an isolated DC-DC converter, whereas a linear supply can respond within 1 to 5 μs for a similar load current step. In demanding situations of loads where tight regulation and fast transient response are required, postregulation techniques can be used. This is particularly the case when multiple voltage rails are available in a power supply system. The basic principle behind all postregulation techniques is to use some extra regulator circuits at the output side of the switching power supply (Figure 3.7a).

Some of the popular postregulation techniques include linear regulators, added secondary side DC-DC converters, coupled inductors, magnetic amplifiers (mag amps; sometimes called saturable reactors), and secondary side postregulators (SSPR). Figure 3.49 shows these concepts. A linear regulator block used as a secondary side controller (Figure 3.49a) or a postregulator is quite common and provides excellent transient response and minimum ripple. However, for efficiency reasons, these linear regulators should be in the form of LDOs. Adding a second stage DC-DC converter (Figure 3.49b) seems simple but can present design difficulties. Coupled inductors (Figure 3.49c) are

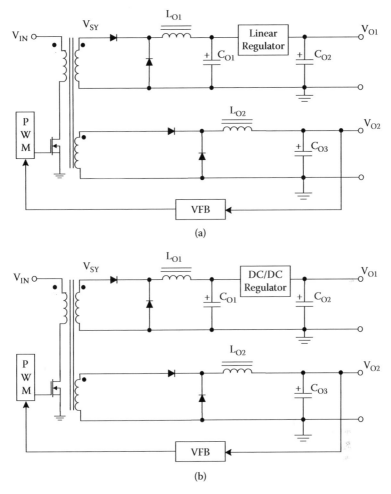

FIGURE 3.49

Postregulation techniques: (a) linear regulator based; (b) switch mode controller based; (c) coupled inductor based; (d) magnetic amplifier based; (e) secondary side postregulator (SSPR) based.

suitable for secondary rail situations, where lower regulation is acceptable, in a typical tolerance range of ±5% to ±8%. Tight coupling can create unwanted interactions between coupled outputs. Magnetic amplifier techniques are based on a saturable reactor, which can act as a magnetic switch (Figure 3.49d), exhibiting high or low impedance toward the output rail to be controlled. Magnetic amplifiers are used for medium and high power requirements.

The SSPR technique (Figure 3.49e) uses a semiconductor device as a switch, with a switch mode controller that is synchronized with the main PWM controller. Regulation of the voltage is achieved on the secondary side either by controlling the volt-second product across the output inductor for

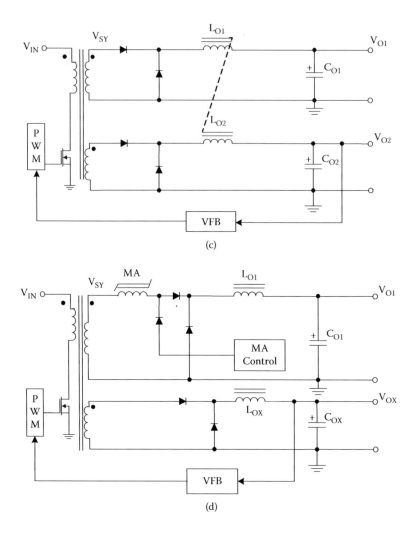

FIGURE 3.49 (continued)

buck-derived techniques or by controlling the amount of energy for boost- and flyback-derived topologies. For details, see Levin [148], Simopoulos [149], and Mammano [150].

3.5.13 Digital Control

Over the past 8 years, the concept of digital control in power supplies has become a frequent topic of discussion, while microcontroller and DSP suppliers were introducing low-cost programmable controllers. In switching power supplies, the analog control concepts are used to control the "on" and "off" states of a power switch, using PWM or soft switching techniques. If the function of the analog control circuits can be duplicated inside the

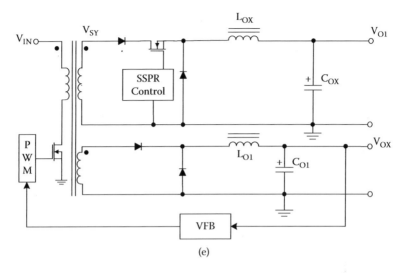

FIGURE 3.49 (continued)

software algorithms of a digital controller block, so that the power switch is driven by the intended PWM waveform, we can conceptually achieve fully digital control. Figure 3.50a indicates the concept in a simplified form for the case of a compound buck and boost configuration; the topology is shown in Figure 3.50b. In a typical example based on the above digital control concept [151], three basic steps are performed:

1. Analog output is sampled and converted to digital format using an ADC.
2. Processed digital information (analog input based) is subjected to the digital equivalent of the transfer function (which resides within the software algorithms).
3. Output of the transfer function drives the power switches to control the output voltage.

In Figure 3.50a, the signal I_L (the current in the inductors) is also fed into the digital controller, whereas input voltage is fed at another ADC input, which makes the system decide on the buck or boost requirement. Figure 3.50c shows the flow chart applicable to the system for the selection of the buck or boost mode and the overall power conversion–related flow chart (Figure 3.50d). In such a system, many other aspects of overall control, such as efficiency management, diagnostics, and interactive communications with the power supply, can be easily achieved [151].

Given the conceptual approach in the above example, the digital control concept in power supplies may have four different levels [152]:

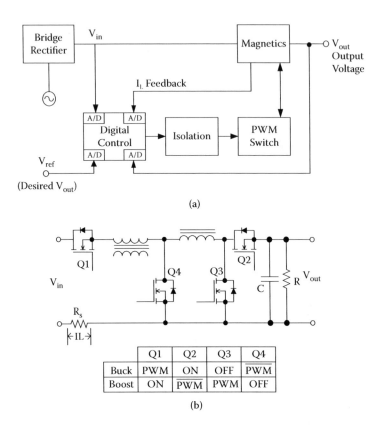

(a)

(b)

FIGURE 3.50
Digital control concept used in SMPS with buck and boost mode possibilities: (a) concept of digital control; (b) topology control; (c) compound buck boost flow chart; (d) basic flow chart. [Courtesy of PCIM; Vinsant et al. (151).]

- Level I—adds simple functions that are difficult to achieve with analog components (e.g., a ramping PWM waveform for the soft start function).
- Level II—secondary management function around a traditional analog circuit. The digital controller monitors the output parameters and uses existing external controls to enhance the functionality of the power supply, though the control loop is still analog.
- Level III—a higher level of integration where the switcher is integrated with the microcontroller, but the implementation of the feedback loop is still analog.
- Level IV—complete digital control where all parameters are digitized and analyzed by the controller to provide appropriate outputs. Usually requires a DSP with high-speed ADCs and PWM outputs.

Figure 3.51 shows the concept and implementation aspects of a digitally controlled buck converter based on a low-cost microcontroller such as the

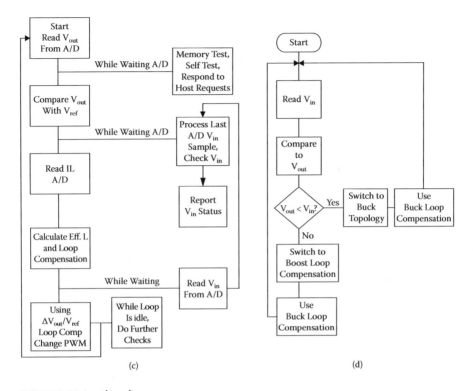

(c)   (d)

FIGURE 3.50 (continued)

PIC 16CXXX from Microchip Technology. Figure 3.51a shows the concept of absorbing the control function into the digital controller [153]. Figure 3.51b and Figure 3.51c show the final circuit and applicable flow charts for this pulse-skipping modulator-based design. The main and interrupt subroutines for timing generators are shown in the flow charts.

Using a simple low pin count microcontroller and a suitable driver IC, a more compact digitally controlled switcher can be developed. Figure 3.52a shows a digital power converter schematic based on a PIC microcontroller. Figure 3.52b shows the development environment for the PIC controller, including the in-circuit debugger (ICD) [154].

For more advanced control requirements, microcontroller manufacturers such as Microchip Technology have introduced 16-bit digital signal controllers (DSCs) with DSP functionality, low pin count, and low cost. A block diagram of a typical DSC from Microchip Corp. is shown in Figure 3.53a. Figure 3.53b shows the concept of implementing a synchronous buck converter, with the approximate timing involved. In such a system, sampling triggers and ADC capability are important so as not to miss critical parameters, as in Figure 3.53c, where asynchronous ADC sampling was used [155]. Analog comparators available within the DSC should be used for current limiting so that these comparators provide benefits that are not practicable or desirable to perform directly in the digital control loop (Figure 3.53d). With

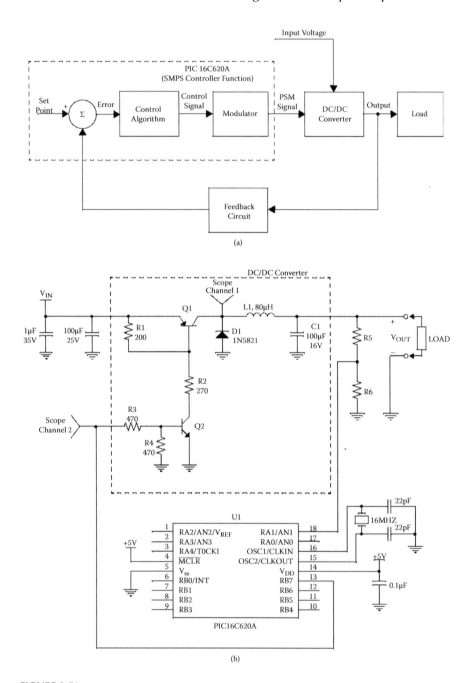

FIGURE 3.51
Digital control in a simple buck converter design: (a) absorbing the control function into the microcontroller; (b) circuit; (c) flow charts. (Courtesy of Microchip Technology, Inc.)

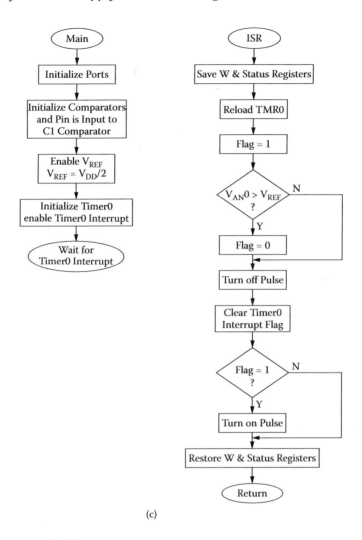

(c)

FIGURE 3.51 (continued)

the cost of DSPs dropping to $2 or less, use of a DSP in the control loop is becoming a viable option [156].

Caldwell [157], Kris [158], Etter and Fosler [159], Hagen and Freeman [160], and Pandola [161] provide some useful information for designing digital controllers for switching supplies, including the frequency response of the power stage [160]. Figure 3.54 shows a DSC-based approach for designing a full-bridge version with PFC where two DSC chips are used. Bramble and Holden [162] discuss details of a digital control system for a buck converter based on an 8051-compatible microcontroller.

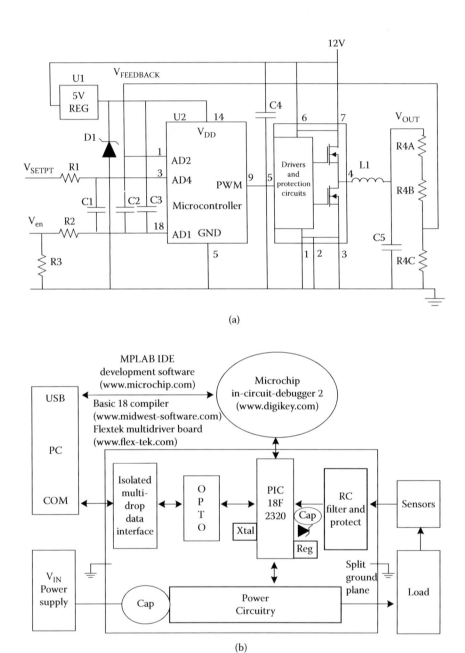

(a)

(b)

FIGURE 3.52
A PIC microcontroller–based power converter using a half-bridge driver IC: (a) circuit; (b) development environment. (Courtesy of *Power Electronics Technology*; Dharmawaskita [153].)

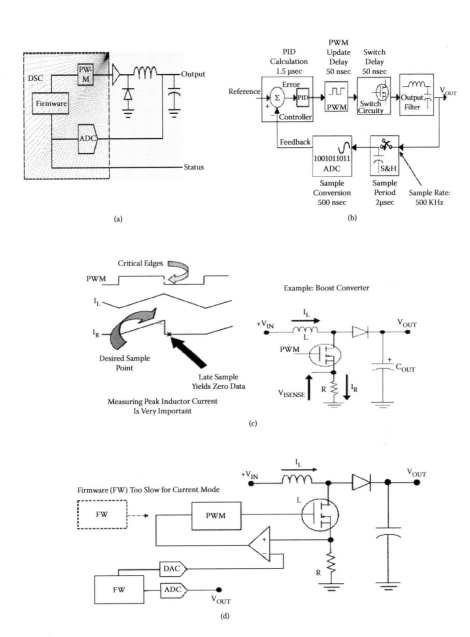

(a)

(b)

(c)

(d)

FIGURE 3.53

Use of a digital signal controller (DSC) in an SMPS design: (a) DSC block diagram; (b) example of a synchronous buck implementation; (c) need for asynchronous ADC sampling; (d) analog comparator used for current limiting outside the loop. (Courtesy of Power Electronics Technology; Hutchings [155].)

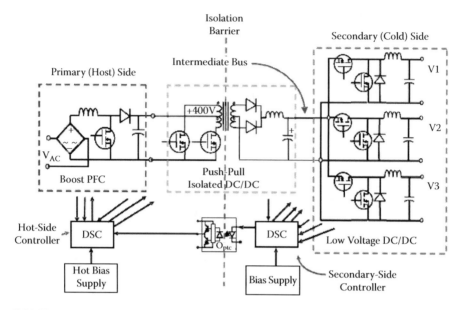

FIGURE 3.54
DSC-based approach for PFC-based full-bridge configuration with dual processors. (Courtesy of Microchip Corp.)

3.5.14   Power Supply Protection

Safety and reliability should be important aspects of power supply operation. Not only the load, but also the power supply and its input source, should be safe under all operating conditions. The following are a few important items to consider:

- Thermal design
- Overvoltage and overcurrent protection
- Protection against transients
- Long-term reliability aspects of output and input capacitors
- Age-related aspects

3.5.14.1   Thermal Design

In any electronic design, semiconductors, as well as the passive parts, have thermal limits for operation. For example, most manufacturers of silicon ICs specify the maximum junction temperature at about 150°C. Similarly, passive parts also have maximum temperature limits. Given such limits, the maximum power dissipation of an IC or a power semiconductor package is given by

$$P_{D(\max)} = \frac{T_{J(\max)} - T_{A(\max)}}{R_{\theta JA}}, \tag{3.9a}$$

TABLE 3.12

Thermal Characteristics of a Typical Power Package
(Example: CS8121 from ON Semiconductor)

Package	$R_{\theta JA}$ (°C/W)	$R_{\theta JC}$ (°C/W)
TO-220	50	3.5
14-lead surface outline (SO)	125	30
8-lead plastic dual inline package	10	52

where $T_{J(max)}$ is the maximum recommended junction temperature, $T_{A(max)}$ is the worst-case ambient temperature of the application, and $R_{\theta JA}$ is the junction-to-ambient thermal resistance of the package in degrees Celsius per watt (°C/W). A semiconductor's package determines its $R_{\theta JA}$ and quantifies how much the junction temperature will rise for each watt the device dissipates into still air. If a heat sink is used to mount the component, three components contribute to the thermal resistance, as given by

$$R_{\theta JA} = R_{\theta JC} + R_{\theta CS} + R_{\theta SA}, \qquad (3.9b)$$

where $R_{\theta JC}$, $R_{\theta CS}$, and $R_{\theta SA}$ are thermal resistances of the junction to case, case to heat sink, and heat sink to ambient, respectively. The thermal characteristics of a linear regulator such as CS8121 are shown in Table 3.12 for different possible packages, and it should be noted that the values can have a wide range of variations. Guidelines for designing in a linear regulator IC environment such as LDOs are given in Malley [163]. In compact electronic environments, copper foil of a PCB can be used to remove heat from a device; Figure 3.55 shows the thermal resistance versus PCB foil area.

The designer should be able to calculate the maximum possible power dissipation for a given environment and then translate that value to safe maximum output currents based on approximate relationships for the given circuit and the topology. Another aspect is the transient thermal response, and in this situation an electrical equivalent of an RC element combination can be used with a simulator such as SPICE to generate the thermal response of a power semiconductor [164]. Using a similar RC equivalent approach, when thermal simulation software is not available, or for quick calculations, designers can apply linear superposition to model the thermal performance of systems [165,166]. Within the last decade, many new cooling systems have been introduced, and designers should consider these systems, such as microchanneled heat sinks [167] and nonlinear fin pattern–based systems [168], in very high-power designs.

3.5.14.2 Overvoltage and Overcurrent Protection

3.5.14.2.1 Overvoltage Protection

A power supply should not generate any steady or transient overvoltages under all operating conditions, particularly in low-voltage, high-current

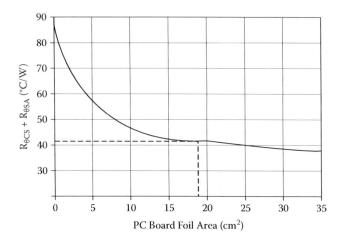

FIGURE 3.55
Thermal resistance from case to ambient with PC-board foil area. (Courtesy of *EDN Magazine*, © 1995, Reed Business Information, a division of Reed Elsevier. All rights reserved. [163].)

power supplies. Sometimes an unexpected component failure, such as a shorted power transistor in a buck converter, can generate a disastrous high-output voltage. Under a very peculiar and unwarranted load current transient, the power supply may generate a transient overvoltage condition.

In most situations, the easiest way to protect the load from overvoltage is to use crowbar circuits, which detect the overvoltage situation quickly and activate a short circuit across a fuse. To enable reliable protection, the overvoltage protection must be independent from the rest of the system's circuits; it must have its own voltage reference source and an independent power source. Figure 3.56a indicates a simple implementation of a crowbar circuit [169] in a nonisolated synchronous buck converter. Figure 3.56b indicates a case of diode-ORed redundant supplies with independent crowbar circuits.

In isolated power supplies, the transformer provides inherent protection against a switch failure; if any of the components in the feedback path opens accidentally, it can create a dangerous situation. Even in such situations, by using optoisolators, one can design crowbar protection circuits using parts similar to the case in Figure 3.56a [169].

3.5.14.2.2 Overcurrent Protection

One of the most important protection features in the power supply is current limiting. When designing a current limiter, one should think of two main aspects: measure the current and develop a limiting circuit using the current signal. Current limiter design necessitates tradeoffs among cost, complexity, reliability, and performance. There are several possible current-limiting schemes, as shown in Figure 3.57. In constant current limiting (Figure 3.57a), the output voltage drops sharply beyond the limit of the current. In an LDO or a common linear regulator, if such a scheme is applied at the limit, the

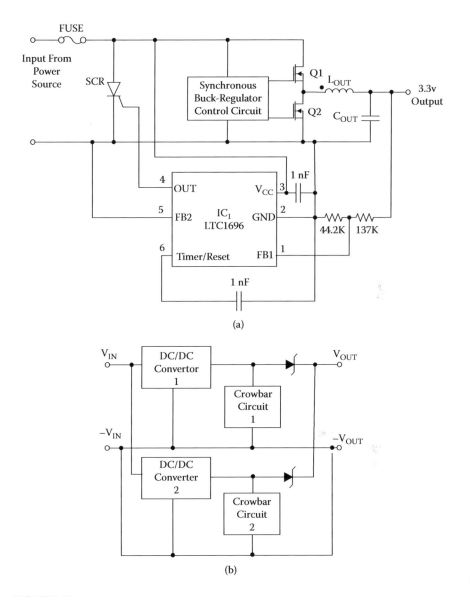

FIGURE 3.56
Crowbar protection for overvoltage conditions: (a) implementation of a simple crow bar circuit using limited components; (b) independent crow bar units in an ORed redundant power supply system. (Courtesy of *EDN Magazine*, © 2007, Reed Business Information, a division of Reed Elsevier. All rights reserved. [165].)

voltage across the pass element will exceed the normal operation value (from $[V_{in} - V_O]$ to $[V_{in}]$) and the dissipation limit of the transistor and the heat sink can be exceeded, and the designer should allow for such excess dissipation. Figure 3.57b shows foldback technique. An advantage in this scheme over the constant current limiting method is that dissipation within the regulator

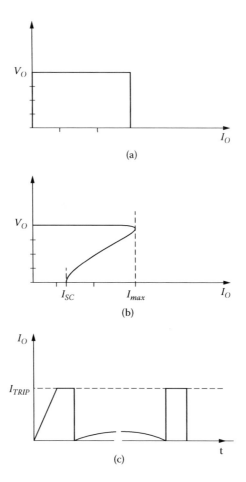

FIGURE 3.57
Current limiting methods: (a) constant current limiting; (b) foldback limiting; (c) hiccup current limiting.

circuits is minimized. The ratio of I_{SC}/I_{max} is an important parameter for this scheme, where a smaller value means better performance. However, unless the circuit is designed to reset automatically on removal of excess load, this scheme may need manual intervention at overcurrent. Figure 3.57c shows the concept of hiccup current limiting. It incorporates overcurrent shutdown but adds an automatic restart mechanism. The power supply shuts down for a limited period of time and automatically restarts after a timeout.

In all these schemes, the traditional approach is to use a low ohmic value current sense resistor in series with the load path. To limit dissipation, a lower-value resistor needs to be used, or an alternative such as a PCB track as a resistance is logical. In all these situations, resistance value variation over the operational temperature range creates an uncertainty in the limiting value. When a copper PCB track is used, the approximate trace length can be determined using the following relationship:

$$L_{trace} \approx 40I_oR_{sense},$$

where I_o is the current value and R_{sense} is the expected resistance value in milliohms for a width of 20 mils/A (for 1-oz. type where the thickness is about 1.3 mils [33 μm] at 25°C).

In current limiting schemes, the sensed current is used to activate the controller shutdown by an appropriate means of additional circuitry. In switching circuits with FETs, with the $R_{DS(on)}$ value for the FET available from the data sheets, one can develop a technique without a special-sense resistor, with knowledge of the converter topology [170]. For the techniques used to activate constant current limiting or foldback current limiting in linear regulators, see Malley [171]. Figure 3.58 shows different concepts to achieve current limiting in a buck regulator [172].

3.5.14.3   Protection against Input Transients

Due to acts of God, such as lightning or inductive load dumps on the AC input power supply, severe surge voltages may occur, and these transients are of very short duration, such as from 50 to 200 μs total. Almost all off-the-line power supplies need to be protected against such events where both common mode and differential (or transverse) mode transients can occur. In such situations, nonlinear devices such as metal oxide varistors and avalanche diodes can be combined with small inductors and capacitors, as per the representative schemes in Figure 3.59. It is important for the designers to subject these surge suppressor blocks, as well as the entire prototype of the power supply, to simulated surges, as specified under C62.41 or similar standards that specify surge-testing procedures and waveforms.

3.5.14.4   Reliability of Input/Output Capacitors

One major family of components that can affect the long-term reliability of a switching power supply is the smoothing capacitors, which carry large ripple currents. In these components, such as the output filter capacitors or input filter capacitors, high-frequency ripple currents can flow at the switching frequency, and this can generate heat due to the ESR of the capacitor. Designers tend to select filter capacitors based on the capacitance value rather than the ripple current. This approach can be catastrophic, because ripple current ratings can vary widely among capacitor technologies, manufacturers, and voltage ratings [173]. A typical example [173] is the Nichicon PL series capacitors, with voltage ratings from 6.3 V to 63 V, where the RMS ripple current rating can vary from about 950 mA to about 2.4 A. Power dissipation inside a capacitor is mainly determined by its ESR and is approximately given by

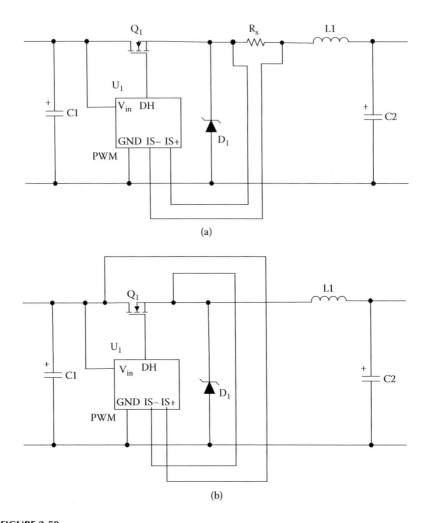

(a)

(b)

FIGURE 3.58
Current limiting in a synchronous buck regulator: (a) use of PCB trace for sensing; (b) use of
FET resistance for sensing. (Courtesy of Power Electronics Technology; [172].)

$$P_{Cap} = I_{rms}^2 ESR. \qquad (3.10)$$

The ESR varies widely with the temperature and operating frequency;
therefore, the operating frequency of a switching supply and the temperature inside the case can have a significant effect on the amount of heat generated within the capacitor [173]. The location of the capacitor within the
power supply can also have a significant impact. For example, if output
capacitors are placed closer to heat sinks or catch diodes, the heat generated in these external parts can also heat the capacitor. A capacitor's load life

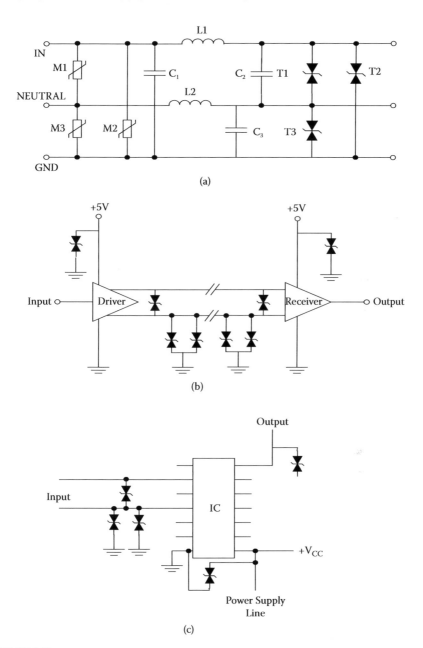

FIGURE 3.59
Transient protection for common and differential mode surges. (a) Multi-stage surge suppressor with MOVs (M_1 to M_3), avalanche diodes (T_1 to T_3) and passive parts; (b) line driver/receiver protection; (c) generic IC protection.

specification indicates how many hours the device is likely to operate under severe conditions, and the failure can be catastrophic or parametric. For aluminum electrolytic capacitors, the load life typically ranges between 1000 and 10,000 hr if operated around a maximum of 105°C. If such a capacitor is operated at a lower temperature, the load life is given by

$$L_{actual} = L_{105°} 2^{\frac{\Delta T_x}{10}} \tag{3.11}$$

where L_{actual} is the actual lifetime and ΔT_x is the temperature difference between the maximum allowable temperature and the actual temperature. The value of ΔT_x is given by

$$\Delta T_x = P_{Cap}/BA, \tag{3.12}$$

where B is the heat transfer coefficient (in W/cm²/°C) and A is the surface area of the capacitor. B may not be available in the data sheet but can be obtained from the manufacturer [173]. One problem arises in the calculation due to nonsinusoidal or DC-type current flowing inside the capacitor in the switching topology, and it varies with each topology. Huffman [173] provides these relationships based on idealized waveforms in each topology. To calculate the ESR at different temperatures, the ESR multiplier in Figure 3.42 can be used. In multiphase buck converters, ripple currents can be improved significantly compared to single-phase versions [174].

3.5.14.5   Age-Related Aspects

The reliability of a power supply cannot be solely designed in, tested in, or built in. It takes a team effort, starting with the definition and specification of the product, and does not end with the first shipment to the customer. The following points are key to achieving a high-reliability power supply:

- A rigorous specification review—attention to all aspects of reliability, including packaging, cooling, and connectors
- Proven topologies with proven component sets
- A comprehensive qualification—emphasis on areas of concern such as power line disturbances, parametric variations, lifetime considerations, and extensive testing to specifications
- A rigorous life test, including on/off cycling
- Extensive box testing for electromagnetic compatibility, etc.
- Qualification of the vendor's production
- Analysis of field problems on a continuous basis to understand real failures

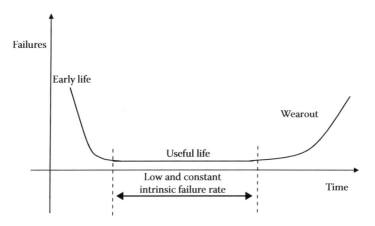

FIGURE 3.60
Bathtub curve as applied to reliability of products.

A power supply is a typical product or subsystem that exhibits the typical bathtub curve, as in Figure 3.60. The following tests are typical for achieving high levels of reliability:

- Thermal/shock cycling
- Temperature stress test
- Vibration
- EMI
- Thermal imaging (for an overall thermal profile)
- Stress analysis
- Power line disturbances
- Lifetime evaluation
- Test to specification

A discussion on these can be found in Forrester [37] and Pflueger [175].

3.5.15 Testing of Power Supplies

The power supply is the heart of an electronic system, and it is necessary for designers and test engineers to evaluate a supply's performance. This can be done at different stages, such as while refining the design or releasing the design. The most important basic tests are

- Load regulation (graph of load current versus load voltage under fixed input voltage)
- Line regulation (graph of load current versus input line voltage under a fixed load current)
- Load transient recovery

- Current limiting
- Startup time
- Output noise (or periodic and random deviations [PARD])

Other important tests are

- Inrush current
- Line current
- Efficiency
- VID control (for VRMs)

During the design stage, it may be necessary to measure the loop performance for tuning the power supply's loop behavior.

Essential tests can be easily done using common bench instruments, such as a 5½-digit or better digital multimeter, a digital storage scope, and a reasonable electronic load. Figure 3.61a indicates a simple setup for load and line regulation. Load transient recovery, which is very important in high-current power supplies for digital processors, indicates one of the most important behaviors of the system, and Figure 3.61b indicates the essential transient parameters to be measured using the scope. In this process, an electronic load, which allows us to program the current step (from high to low as well as low to high transition), is applied to the power supply being tested. Modern VRMs require very high slew rate capability, which can be in the range of 10 to 300 A/μs or even higher, and this kind of performance cannot be measured without a very high transient capability electronic load. Figure 3.61c shows the concept of an electronic load based on a MOSFET and a current sense resistor. (In practice, the electronic load itself may be an expensive piece of equipment.) A few parameters of the electronic load are critical to reliable measurements:

- Response speed of the control loop in the electronic load (T_r)
- Total resistance of the shunt resistor and the $R_{DS(on)}$ of the MOSFET (R)
- Total connector inductance and the inductance of the MOSFET (L)
- Turn-on and turn-off time of the MOSFET (T_f)
- The maximum transient loading level of the step loading (I)

Because the higher value of T_f or T_r determines the rise time of the electronic load (regardless of the load setting), this limits the transient capability of the electronic load. For example, if this parameter is 1 μs, the maximum possible slew rates can be 10-A/μs and 100-A/μs for 10 A and 100 A loads, respectively. Details on these aspects can be found in Romanchik [176] and Lee [177]. Such measurement may allow designers to tune the performance of a DC-DC converter [178]. In POL converters used for processor circuit blocks,

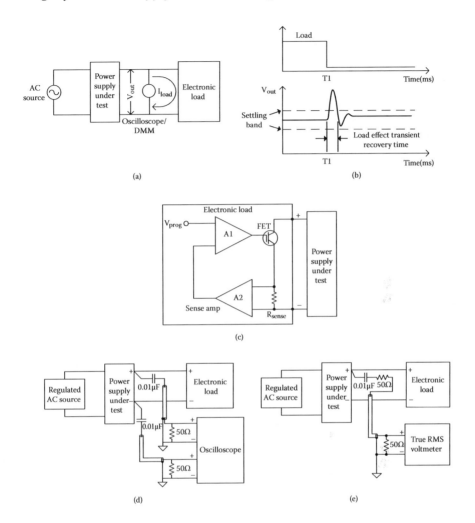

FIGURE 3.61
Power supply measurements: (a) basic test setup; (b) transient parameters for step load changes; (c) concept of an electronic load based on a MOSFET; (d) noise (or PARD) measurement using oscilloscope; (e) noise measurement using true rms voltmeter. (Courtesy of *Test & Measurement World*; [176].).

transient requirements can be in the range of 300 A/µs or more. Testing of these high-*di/dt* converters is discussed in Callanan [179].

Another important requirement is the measurement of noise or PARD, and this needs either a scope (for peak-to-peak measurement) or a true RF RMS voltmeter for RMS measurement. Figure 3.61d shows test setups for PARD measurement. In this process, if the scope inputs are ground referenced, the use of differential probes is necessary.

As discussed in Section 3.5.8, feedback loop measurements require breaking and injecting a small AC signal, and this may not be easy because monolithic ICs used as controllers do not allow opening up the loop. In such cases, alternative methods may be utilized [111,180]. Another recent development is that of automatic routines and software that allows the calibration of power supply design parameters. With component tolerances affecting a given design, one needs trimming of various voltage dividers and feedback loop components at different stages. Using PC-based calibration software, a calibration board containing ADCs, and switches, a power supply design can be automatically trimmed to refine the performance. Figure 3.62 shows such a setup for a power supply design based on a controller and a power management IC such as the ADM 1041 from Analog Devices, Inc. [137].

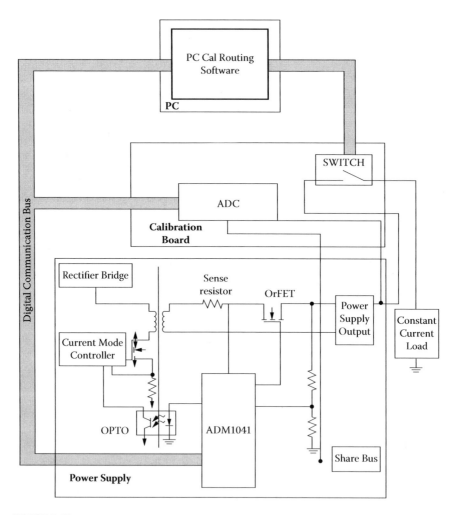

FIGURE 3.62
A power supply calibration set up for a power management and converter system designed around ADM 1041 controller from Analog Devices, Inc. (Courtesy of Power Electronics Technology. Source: Lee [177].)

Bibliography

Bose, B.K., *Modern Power Electronics and AC Drives*, Prentice Hall, Upper Saddle River, NJ, 2001.

Brown, M., *Power Supply Cookbook*, 2nd ed., Newnes, London, 2001.

Kassakian, S., Schlecht, M.F., and Verghese, G.C., *Principles of Power Electronics*, Addison-Wesley, Boston, 1991.

Mohan, N., Undeland, T., and Robbins, P., *Power Electronics: Converters, Applications, and Design*, John Wiley & Sons, NJ, 2003.

Rashid, M.H., *Power Electronics: Circuits, Devices and Applications*, 2nd ed., Prentice Hall, Upper Saddle River, NJ, 1993.

Sandler, S.M., *SMPS Simulation with SPICE 3*, McGraw-Hill, New York, 1997.

Shepherd, W., Hulley, L.N., and Liang, D.T.W., *Power Electronics and Motor Control*, 2nd ed., Cambridge University Press, New York, 1995.

Williams, B.W., *Power Electronics: Devices, Drives and Applications*, John Wiley & Sons, NJ, 1987.

References

1. Bierman, H., Power supplies get smaller and more reliable, *Electronics*, September 16, 42, 1985.
2. Lyman, J., Why makers are stepping up the pace in technology, *Electronics*, May 14, 93, 1987.
3. Kularatna, N., Powering systems based on complex ICs and the quality of utility AC source: an end to end approach to protection against transients, *Power Quality 2005 Proceedings*, October 25–27, 2005, Baltimore, MD (CD-ROM, Session PQS02).
4. Davis, S., High frequency power conversion in the next decade, *PCIM*, April, 18, 1989.
5. Curatolo, T., High density power components add flexibility to distributed power design, *Electronic Design*, June 12, 129, 2000.
6. Cassani, J.C., Hurd, J.J., Thomas, D.R., Hodgins, R.G. and Wittlinger, H.A., 80W, 1 MHz supply employs current controlled PWM power IC, *PCIM*, January, 7, 1993.
7. Small, C.H., Distributed power takes center stage, *EDN*, April 28, 54, 1994.
8. Smith, C., The top 10 keys to successful distributed power architecture design, *PCIM*, April, 12, 2000.
9. Hemena, W. and Malik, R., A distributed power architecture for PC industry, Proceedings of the PCIM Conference, October 1, 2000.
10. Alderman, A., Distribution vs. efficiency: power architect's dilemma, *Power Electronics Technology*, October, 42, 2003.
11. Bull, C. and Smith, C., Integrated building blocks for dual-output buck converter, *Power Electronics Technology*, October, 68, 2003.
12. Travis, B., Linear vs. switching supplies: weighing all the options, *EDN*, January 1, 40, 1998.

13. Khorshid, O., Selecting charge-pump DC/DC converters, *EDN*, August 17, 115, 2000.
14. Lam, C., Power-management techniques for multimedia mobile phones, *EDN*, April 13, 55, 2006.
15. Day, M., Integration saves time and board space, *Power Electronics Technology*, October, 64, 2003.
16. TPS65010 data sheet, Texas Instruments, Dallas, TX; available at http://www. datasheetcatalog.com/datasheets_pdf/T/P/S/6/TPS65010.shtml.
17. TPS61120 data sheet, Texas Instruments, Dallas, TX; available at http://www. datasheetcatalog.com/datasheets_pdf/T/P/S/6/TPS61120.shtml.
18. Armstrong, T., Wireless products signal new uses for VLDOs, *Power Electronics Technology*, July, 38, 2005.
19. Maxim Integrated Products, Linear regulators in portable applications, Application Note 751, May 24, 2001; available at http://www.maxim-ic.com/appnotes. cfm/appnote_number/751.
20. Brown, M., *Power Supply Cookbook*, Newnes, London, 2001.
21. Rubadue, J., Powering your core voltage, *EDN*, September 27, 69, 2003.
22. Williams, J., Astute designs improve efficiencies of linear regulators, *EDN*, August 17, 151, 1989.
23. Narveson, B., How many isolated DC-DC converters do you really need?, *Electronic Design*, January 6, 137, 1997.
24. Strassberg, D., Tiny titans: choose 'EM and use 'EM with care, *EDN*, May 2, 41, 2002.
25. Bindra, A., Eighth brick gains momentum, sixteenth gets proposed, *Power Electronics Technology*, March, 56, 2003.
26. Pietkiewicz, S., Speciallized circuits condition power for Palm top computers, *EDN Asia*, August, 55, 1993.
27. Goodenough, F., Advanced microprocessors demand amperes of current at <2V, *Electronic Design*, January 20, 44, 1997.
28. Mannion, P., VRMs: technological first aid, but for how long?, *Electronic Design*, January 20, 31, 1997.
29. Brown, S., Microprocessor controls its dedicated voltage regulator module, *PCIM*, April, 66, 1997.
30. Gentchev, A., Designing high-current VRM compliant CPU power supplies, *EDN*, October 26, 155, 2000.
31. Wong, P.-L., Lee, F.C., Xu, P., and Yao, K., Critical inductance in voltage regulator modules, *IEEE Transactions on Power Electronics*, 17, 485, 2002.
32. Cooper, D., Power management for high performance, *Power Electronics Technology*, July, 33, 2004.
33. Thornton, C., Auto-Track™ feature simplifies simultaneous power supply voltage sequencing, Analog Application Brief No. 6, Texas Instruments, Dallas, TX, 2003.
34. Kolinski, J., Chary, R., Henroid, A., and Press, B., *Building the Power-Efficient PC*, Intel Press, Santa Clara, CA, 2001.
35. Lakkas, G. and Duduman, B., The ACPI advantage for powering future generations of computers, *EDN*, September, 91, 2001.
36. Everett, C., External power supplies eliminate more than just heat from your design, *EDN*, April 15, 107, 1987.
37. Forrester, S., International power converter requirements, *PCIM*, December, 8, 1995.

38. Jovalusky, J., New energy standards banish linear supplies, *Power Electronics Technology*, March, 42, 2005.
39. Smith, K.L., DC supplies from AC sources—part 3, *Electronics and Wireless World*, February, 24, 1985.
40. Smith, K.L., DC supplies from AC sources—part 4, *Electronics and Wireless World*, May, 67, 1985.
41. Billings, K., *Switchmode Power Supply Handbook*, McGraw-Hill, New York, 1989.
42. Deshayes, R. and De Palma, J.F., High power semiconductor protection requires the appropriate fuses, *PCIM*, October, 58, 1996.
43. Bell, B., Active clamp resets transformers in converters, *Power Electronics Technology*, January, 26, 2004.
44. Zuk, P., Designing offline power supplies using power factor correction, *EDN*, September 1, 67, 2006.
45. Sandler, S., Hymowitz, C., and Eicher, H., Optimizing single stage power factor correction, *Power Electronics Technology*, March, 14, 2006.
46. Kularatna, N., *Power Electronics Design Handbook: Low-Power Components and Applications*, Newnes, London, 1998.
47. Mappus, S., Active clamp transformer reset: high or low side, *Power Electronics Technology*, July, 24, 2004.
48. King, B. and Gehrke, D., Active clamp control boosts forward converter efficiency, *Power Electronics Technology*, June, 52, 2003.
49. Hariharan, S. and Schie, D., Designing single switch forward converters, *Power Electronics Technology*, October, 50, 2005.
50. Khasiev, V., Moving forward converters to higher efficiency, *Power Electronics Technology*, May, 38, 2003.
51. Hancock, J.M., Improving the performance of flyback power supplies, *Power Electronics Technology*, September, 33, 2005.
52. Ruble, R. and Clarke, R., Designing flyback converters—part I, design basics, *PCIM*, January, 43, 1994.
53. Ruble, R. and Clarke, R., Designing flyback converters—part II, 48V dual output converter, *PCIM*, April, 23, 1994.
54. Leman, B.R., Finding the keys to flyback power supplies produces efficient design, *EDN*, April 13, 101, 1995.
55. TOPSwitch-GX® flyback design methodology, Application Note 32, Power Integrations, Inc., San Jose, CA, 2004; available at http://www.powerint.com/PDFFiles/an32.pdf.
56. Basso, C., Critical mode control stabilizes switch-mode power supplies, *EDN*, April 23, 171, 1998.
57. Huber, L. and Jovanovic, M.M., Optimizing flyback topologies for portable AC/DC adapter/charger applications: part I—adapter charger requirements, *PCIM*, August, 68, 1996.
58. Huber, L. and Jovanovic, M.M., Optimizing flyback topologies for portable AC/DC adapter/charger applications: part II—DC/DC converter design, *PCIM*, September, 34, 1996.
59. Gauen, K., Considerations for a two transistor, current mode forward converter—part I: overall design concepts, *PCIM*, April, 38, 1991.
60. Gauen, K., Considerations for a two transistor, current mode forward converter—part II: power circuit design, *PCIM*, May, 38, 1991.
61. Clemente, S. and Dubhashi, A., HV floating MOS gate driver IC, Application Note 978A, International Rectifier, El Segundo, CA, 1990.

62. McGinty, J., Gate drivers, *PCIM*, May, 28, 1998.

63. Motto, E. and Donlon, J., Hybrid ICs drive high-power IGBT modules, *Power Electronics Technology*, March, 24, 2005.

64. Danz, J.E., HIP4081, 80V high frequency H-bridge driver, Application Note 9325.3, Intersil Corporation, Milpitas, CA, 2003.

65. Product catalog, Payton America, Boca Raton, FL; available at http://www.paytongroup.com/catalogue/20-21.pdf.

66. TL494 data sheet, Texas Instruments, Dallas, TX, 2002; available at http://www.datasheetcatalog.com/datasheets_pdf/T/L/4/9/TL494.shtml.

67. Dixon, L., High power factor preregulator using the SEPIC converter, pp. 6-1 to 6-11, Unitrode Corporation, Merrimack, NH, 1993; available at http://focus.ti.com/lit/ml/slup103/slup103.pdf.

68. Dixon, L., High power factor preregulator using the SEPIC converter, pp. 7-1 to 7-7, Unitrode Corporation, Merrimack, NH, 1993.

69. Nuefeld, H., SEPIC design, *PCIM*, October, 36, 1998.

70. Rahban, T., Consider SEPIC topology for new designs, *Power Electronics Technology*, November, 18, 2002.

71. Dixon, L., High power factor preregulator using the SEPIC converter, pp. 8-1 to 8-4, Unitrode Corporation, Merrimack, NH, 1993.

72. Fichera, P., Choosing the power switch for off-line quasi resonant converters, *PCIM*, August, 20, 1990.

73. Basso, C., Determine free-running frequency of QR systems, *Power Electronics Technology*, October, 70, 2002.

74. Shennai, K. and Trivedi, M., Soft-switched, phase shifted topology cuts MOSFET switching stresses in FBC, *PCIM*, May, 42, 2001.

75. Andreycak, B., Designing a phase shifted zero voltage transition power converter, pp. 3-1 to 3-15, Unitrode Corporation, Merrimack, NH, 1993; available at http://focus.ti.com/lit/ml/slup101/slup101.pdf.

76. Andreycak, B., Design review: 500 Watt, 40W/in^3 phase shifted ZVT power converter, pp. 4-1 to 4-19, Unitrode Corporation, Merrimack, NH, 1993; available at http://focus.ti.com/lit/ml/slup102/slup102.pdf.

77. A 25W off-line flyback switching regulator, Application Note U-96A, pp. 9-47 to 9-51, Unitrode Corporation, Merrimack, NH, 1993.

78. Modeling, analysis and compensation of the current mode converter, Application Note U97, pp. 9-51 to 9-57, Unitrode Corporation, Merrimack, NH, 1993.

79. Dobrenko, D., Solving the problem of pulse skipping oscillations, *Power Electronics Technology*, October, 28, 2002.

80. Vosicher, E., Hysteretic controller fits processor needs, *PCIM*, January, 28, 2000.

81. Nowakowski, R. and Hodson, L., Hysteretic controller IC enables PC power supply to meet advanced CPU requirements, *PCIM*, July, 74, 2000.

82. Sherman, J. and Walters, M.M., Synchronous rectification: improving the efficiency of buck converters, *EDN*, March 14, 111, 1996.

83. Moore, A., Synchronous rectification improves the efficiency of a dual output supply, *PCIM*, July, 8, 1995.

84. Moore, B., Synchronous rectification aids low voltage power supplies, *EDN*, April 27, 127, 1995.

85. Christiansen, B., Synchronous rectification, *PCIM*, August, 14, 1998.

86. How, C.H., Synchronous rectifiers for DC-DC converters: better efficiency, but possible problems, *PCIM*, May, 74, 2000.

87. Bindra, A., Optimized synchronous rectification drives up DC-DC converter efficiency, *Electronic Design*, January 24, 58, 2000.

88. Yee, H.P., Synchronous rectifier controller IC simplifies and improves isolated DC-DC converter designs, *PCIM*, February, 44, 2000.

89. Bindra, A., Synchronous rectifier module eyes isolated power supplies, *Power Electronics Technology*, July, 56, 2003.

90. Elbanhawy, A., Buck converter losses under the microscope, *Power Electronics Technology*, February, 24, 2005.

91. Mappus, S., Predictive control maximizes synchronous converter efficiency, *Power Electronics Technology*, May, 44, 2003.

92. Dalal, D., Boosting power supply efficiency for desk top computers, *Power Electronics Technology*, February, 14, 2005.

93. Basso, C., Reducing wasted startup current in offline supplies, *Power Electronics Technology*, August, 54, 2004.

94. Fasching, M., Losses due to stray inductances in switch mode converters, *EPE Journal*, 6, 33, 1996.

95. Dadafshar, M., Inductor current sensing boosts regulator efficiency, *Power Electronics Technology*, April, 50, 2005.

96. Lin, F.C. and Chen, D.Y., Reduction of power supply EMI emission by switching frequency modulation, *IEEE Transactions on Power Electronics*, 9, 132, 1994.

97. Schindler, M., Proper layout and component selection control power supply EMI, *EDN*, October 26, 137, 2006.

98. Rogers, P., Board layout boosts power supply performance, *EDN*, November 5, 175, 1998.

99. Scolio, J., Basic switching regulator layout techniques, *EDN*, November 27, 79, 2003.

100. Davis, S., Spread spectrum ICs cut EMI, *Power Electronics Technology*, February, 70, 2003.

101. Armstrong, T., Alleviating noise concerns in handheld wireless products, *Power Electronics Technology*, October, 26, 2003.

102. Tse, K.K., Chung, S.H., Hui, S.Y., and So, H.O., Analysis and spectral characteristics of a spread spectrum technique for conducted EMI suppression, *IEEE Transactions on Power Electronics*, 15, 399, 2000.

103. Goodenough, F., Power supply designers trade off efficiency for noise with switcher ICs, *Electronic Design*, August 18, 40, 1997.

104. Wittenbreder, E.H., Power conversion synthesis—part I: buck converter design, *Power Electronics Technology*, March, 26, 2003.

105. Wittenbreder, E.H., Power conversion synthesis—part II: zero ripple converters, *Power Electronics Technology*, April, 54, 2003.

106. Wittenbreder, E.H., Power conversion synthesis—part III: near zero emissions, *Power Electronics Technology*, May, 35, 2003.

107. Wittenbreder, E.H., Power converter synthesis—part IV: near zero emissions, *Power Electronics Technology*, June, 46, 2003.

108. Hsiu, L., Goldman, M., Caristen, D., Witulski, A.F., and Kerwin, W., Characterization and comparison of noise generation for quasi-resonant and pulsewidth modulated converters, *IEEE Transactions on Power Electronics*, 9, 425, 1994.

109. Barrow, J., Reducing ground bounce in DC/DC converter applications, *EDN*, July 6, 73, 2006.

110. Chryssis, G., *High Frequency Switching Power Supplies*, 2nd ed., McGraw-Hill, New York, 1989.

111. Venable, D., Testing and stabilizing power supply feedback loops, *PCIM*, September, 8, 1997.

112. Mannas, S., Analysis of closed-loop DC-DC converters, *Power Electronics Technology*, October, 42, 2004.

113. Williams, J., Regulator IC speeds design of switching power supplies, *EDN*, November 12, 193, 1987.

114. Hesse, K., Battery powered applications require high performance buck converter ICs, *PCIM*, January, 20, 2001.

115. Sandler, S., Gain equalization improves flyback performance, *Power Electronics Technology*, July, 46, 2006.

116. King, B.M., Optimized LDO response to load transients requires the appropriate output capacitor and device performance, *PCIM*, September, 39, 2000.

117. Lee, M., Linear PNP regulator outperforms NPN types, *PCIM*, May, 36, 1989.

118. O'Malley, K., Compensation for linear regulators, Application Note SR003AN/D, Rev. 1, ON Semiconductor, Phoenix, AZ, 2001.

119. Simpson, C., Linear regulators: theory and operation and compensation, Application Note 1148, National Semiconductor, Santa Clara, CA, 2000.

120. Simpson, C., LDO regulators require proper compensation, *Electronic Design*, November 4, 99, 1996.

121. Goodenough, F., LDO controller handles 250A/μs load transients, *Electronic Design*, November 18, 162, 1996.

122. Ciscato, S., Low dropout voltage regulators survive in the automotive environment, *PCIM*, June, 10, 1997.

123. Schiff, T., High-power, high speed processors need stable, fast linear LDO regulators, *PCIM*, January, 57, 2001.

124. Rincon-Mora, G.A. and Allen, P.E., A low-voltage, low quiescent current, low dropout regulator, *IEEE Journal of Solid State Circuits*, 33, 36, 1998.

125. Kwok, K.C. and Mok, P.K.T., Pole-zero tracking frequency compensation for low dropout regulator, *IEEE International Symposium on Circuits and Systems*, 4, IV-735, 2002.

126. Chava, C.K. and Silva-Martinez, J., A robust frequency compensation scheme for LDO regulators, *IEEE International Symposium on Circuits and Systems*, 5, V-825, 2002.

127. Bontempo, G., Signorelli, T., and Pulvirenti, F., Low supply voltage, low quiescent current ULDO linear regulator, *8th IEEE International Conference on Electronics, Circuits and Systems*, 1, 409, 2001.

128. Ali, I., and Griffith, R., A fast response programmable PA regulator subsystem for dual mode CDMA/AMPS handsets, *IEEE MTT-S International Microwave Symposium Digest*, 1, 139, 2000.

129. Williams, J. and Owen, T., Exacting noise test ensures low-noise performance of low-dropout regulators, *EDN*, May 11, 149, 2000.

130. Paglia, P., Optimize output-voltage accuracy of adjustable low-dropout regulators, *EDN*, September 1, 105, 1998.

131. Wells, E., LDOs and hot swap power mangers, *PCIM*, January, 46, 1999.

132. Williams, J. and Owen, T., Performance verification of low noise low dropout regulators, Application Note AN83-1, Linear Technology Corporation, Milpitas, CA, 2000.

133. Christ, M., Extending the battery life using PMOS LDOs, *Electronic Engineering*, February, 61, 2001.

134. Williams, J. and Huffman, B., Switched capacitor networks simplify DC/DC converter design, *EDN*, November 24, 171, 1988.
135. Arimoto, K., Efficient power ICs help mobile products harness battery power, *AEU (Japan)*, July, 73, 2000.
136. Arimoto, K., Charge pumps shine in portable designs, Application Note 669, Maxim Integrated Products, Sunnyvale, CA, 2001.
137. Vitchev, V., Calculating essential charge pump parameters, *Power Electronics Technology*, July, 30, 2006.
138. Travis, B., Plugging efficiency leaks, *EDN*, June, 59, 2001.
139. Travis, B., The quest for high efficiency in low voltage supplies, *EDN*, September, 57, 2000.
140. Wei, C., High efficiency, high density, polyphase converters for high current applications, Application Note 77, Linear Technology Corporation, Milpitas, CA, 1999.
141. Benport, D., New design equations improve continuous mode interleaved flyback converter, *PCIM*, April, 36, 1995.
142. Peterson, A., Dual phase regulator improves performance of low voltage computer power supplies, *PCIM*, November, 22, 1996.
143. Harriman, P., New control method boosts multiphase bandwidth, *Power Electronics Technology*, January, 36, 2003.
144. O'Loughlin, M., Interleaved forward converters transform voltage regulation models, *EDN*, January 20, 69, 2005.
145. Bindra, A., Multiphase controllers heed call for more phases per chip, *Power Electronics Technology*, January, 56, 2004.
146. Betten, J. and Kollman, R., Interleaving DC/DC converters boost efficiency and voltage, *EDN*, October 13, 77, 2005.
147. Gallagher, J., Coupled inductors improve multiphase buck efficiency, *Power Electronics Technology*, January, 36, 2006.
148. Levin, G., Postregulation technique efficiently supplies multiple output voltages, *EDN*, January 15, 133, 1998.
149. Simopoulos, A., Linear postregulators for DC-DC converters, *Power Electronics Technology*, June, 28, 2004.
150. Mammano, B., Isolated power conversion: making the case for secondary side control, *EDN*, June 7, 123, 2001.
151. Vinsant, R., DiFore, J., and Clarke, R., Digitally controlled SMPS extends power system capabilities, *PCIM*, June, 30, 1994.
152. Duvenhage, F., The role of digital control in power supplies, *Power Electronics Technology*, August, 74, 2004.
153. Dharmawaskita, H., DC/DC converter controller using a PICmicro microcontroller, Application Note AN216, Microchip Technology, Chandler, AZ, 2002.
154. Caldwell, D., Microcontroller enables digital control in SMPS, *Power Electronics Technology*, February, 30, 2004.
155. Hutchings, B., Achieving high performance, reliable digital power supplies, *Proceedings of the PET Conference*, October 24–26, 2006 [PES01-CD ROM].
156. Choudhury, S. and Harrison, M., DSPs simplify digital control implementation of SMPS, *Power Electronics Technology*, July, 40, 2003.
157. Caldwell, D., Power goes digital, *EDN*, August 18, 75, 2005.
158. Kris, B., DSCs ease migration to digital control, *Power Electronics Technology*, November (supplement), 3, 2006.

159. Etter, B. and Fosler, R., Digital power control enables system identification, *Power Electronics Technology*, November (supplement), 6, 2006.

160. Hagen, M. and Freeman, D., Digital control measures in system response, *Power Electronics Technology*, November (supplement), 12, 2006.

161. Pandola, M., Explore the lesser known benefits of digital power, *Power Electronics Technology*, November (supplement), 16, 2006.

162. Bramble, S. and Holden, P., Digital feedback controls supply voltage accurately, *Power Electronics Technology*, January, 22, 2006.

163. Malley, K., Keep linear regulators in their safe zone, *EDN*, October 26, 137, 1995.

164. Walker, L. and Dashney, G., Spice generates thermal response models of a power semiconductor, *PCIM*, September, 57, 1996.

165. Stout, R., Linear superposition speeds thermal modeling: part I, *Power Electronics Technology*, January, 20, 2007.

166. Stout, R., Linear superposition speeds thermal modeling: part II, *Power Electronics Technology*, February, 28, 2007.

167. Solovitz, S.A., Stevanovic, L.D., and Beaupre, R.A., Microchannel heat sinks take heat sinks to the next level, *Power Electronics Technology*, November, 14, 2006.

168. Remsburg, R., Nonlinear fin patterns keep cold plates cooler, *Power Electronics Technology*, February, 22, 2007.

169. Percia, G., Overvoltage protection circuit saves the day, *EDN*, November 14, 93, 2002.

170. Pelletier, W. and Goder, D., Current limiting defuses the DC/DC time bomb, *EDN*, April, , 1998.

171. Malley, K., Linear regulator protection circuitry, Application Note SR005AN/D, Rev. 1, ON Semiconductor, Phoenix, AZ, 2001.

172. Rose, D., Low voltage, high current switching supplies for microprocessors require over current protection, *PCIM*, July, 95, 1997.

173. Huffman, B., Build reliable power supplies by limiting capacitor dissipation, *EDN*, March 13, 93, 1993.

174. Drew, J., Capacitor ripple current improvements, *Power Electronics Technology*, August, 33, 2004.

175. Pflueger, K.H., Power supply reliability: a practical improvement guide, *EDN*, March 3, 151, 1997.

176. Romanchik, D., Test verify power supply performance, *Test and Measurement World*, January, 43, 1995.

177. Lee, J., High slew rate electronic load checks new generation voltage regulator modules, *PCIM*, May, 52, 2001.

178. Venebale, D., New signal injection technique simplifies power supply feedback loop measurements, *PCIM*, September, 8, 1995.

179. Callanan, S., Testing high *di/dt* converters, *Power Electronics Technology*, March, 14, 2005.

180. Daly, B., Automatic routine speeds power supply calibration, *Power Electronics Technology*, March, 36, 2005.

181. Valentine, M., "PFC controller ICs continue to advance," *Power Electronics Technology*, September, 62–66, 2006.

4

Preprocessing of Signals

CONTENTS

4.1 Introduction 206
4.2 General Considerations 207
    4.2.1 Signal Range 207
    4.2.2 Noise and Error Considerations 208
    4.2.3 Bandwidth 211
4.3 Amplification 211
    4.3.1 Operational Amplifiers 211
        4.3.1.1 Input Imperfections 215
        4.3.1.2 Output Stage 216
        4.3.1.3 Differential to Single-Ended Conversion 216
        4.3.1.4 Noise in Op Amps 217
        4.3.1.5 Single-Rail Op Amps 217
        4.3.1.6 Current Feedback Op Amps 224
    4.3.2 Instrumentation Amplifiers 227
    4.3.3 Nonlinear Approaches 235
        4.3.3.1 Practical Log Amps 239
        4.3.3.2 Root Mean Square-to-DC Converter ICs 239
    4.3.4 Video and Communications Amplifiers 241
        4.3.4.1 Video Amplifiers 241
        4.3.4.2 Communications Amplifiers 243
4.4 Bridge Circuits 243
    4.4.1 Amplifying and Linearizing the Bridge Outputs 247
4.5 Filters 249
    4.5.1 Active and Passive Filters 251
    4.5.2 Switched Capacitor Filters 258
4.6 Switching and Multiplexing of Signals 265
    4.6.1 CMOS Analog Switches 265
    4.6.2 Error Sources of an Analog Switch 268
4.7 Signal Isolation 271
References 274

4.1 Introduction

In the analog world of signals, a raw signal available for further processing can contain many unwanted components, such as distortion, noise, nonlinearities, and offsets and crosstalk from other sources of signals. In a commercial design environment, the designer is under pressure to get the product or the system to the market in a short time. This pressure forces the design team to critically balance many different elements to achieve the optimum within the project and the lowest overall cost and time to market (TTM). This process demands a critical balance of mixing analog and digital design approaches. For this reason, experienced designers attempt to clean up the "raw signals" using analog circuits before they are processed digitally by the subsequent stages, because this significantly reduces the component costs as well as the development time. In this process, signal amplification, noise and crosstalk management, dynamic range considerations, electrical isolation, and signal filtering need to be effectively applied prior to transferring the signals to the data conversion stages and the processing stages. Designers of high-performance systems must define criteria for determining the preferred options from available alternatives. Figure 4.1 shows the possible signal-conditioning techniques at the system and device levels. This chapter considers some essential aspects of design in the analog domain of signal processing.

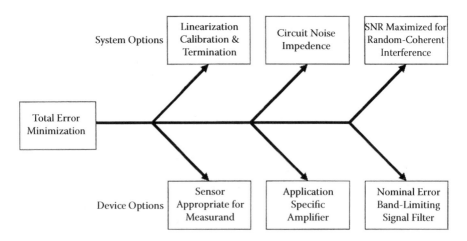

FIGURE 4.1
Options for error minimization at the system and device levels. (Reproduced by permission of CRC Press; Source: Garrett [1].)

4.2 General Considerations

In general, an output signal from a sensor or another stage of a circuit needs to be carefully analyzed and treated before the signal is passed to analog-to-digital conversion (ADC) stages. The primary aspects we need to carefully consider are

- Overall range of the signal in relation to the power rails and noise bed
- Bandwidth of the signal and the range of frequencies for faithful processing of the signals
- Noise aspects
- Electrical isolation aspects

A secondary aspect is the signal switching and multiplexing required to optimize the use of available resources such as data converters. In cases where many signals need to be processed by a digital processor subsystem, we need to carefully consider this option so that the total number of components can be reduced.

4.2.1 Signal Range

In many cases, a wide dynamic range is an essential aspect of a signal, something to be preserved at all costs. This is true, for example, in the high-quality reproduction of music and in communications systems. However, it is often necessary to compress the signal to a smaller range without any significant loss of information. Compression is often used in magnetic recording, where the upper end of the dynamic range is limited by tape saturation and the lower end is limited by the granularity of the medium. In professional noise-reduction systems, compression is "undone" by precisely matched nonlinear expansion during reproduction. Similar techniques are often used in conveying speech over noisy channels, where the performance is more likely to be measured in terms of word intelligibility than audio fidelity. The reciprocal processes of compressing and expanding are implemented using "compandors," and many schemes have been devised to achieve this function.

In terms of the signal voltage,

$$\text{Dynamic range (dB)} = 20 \log_{10} \frac{\text{Largest signal voltage}}{\text{Smallest signal voltage}} \tag{4.1}$$

In a linear impedance system, power is proportional to the signal voltage (or current) squared. Accordingly,

$$\text{Dynamic range (dB)} = 10 \log_{10} \frac{\text{Largest signal power}}{\text{Smallest signal power}} \qquad (4.2)$$

Also, it is useful to differentiate between the dynamic range of the signal and that of the processing system. The signal dynamic range is defined as

$$20 \log_{10} \frac{\text{Largest actual signal voltage}}{\text{Smallest actual signal voltage}}, \qquad (4.3)$$

whereas the system dynamic range is defined as

$$20 \log_{10} \frac{\text{Largest permissible signal voltage}}{\text{Smallest detected signal voltage}}. \qquad (4.4)$$

In a system design, one should be concerned with the system dynamic range, which should preferably match or exceed the signal dynamic range. In situations where a wide dynamic range of signal is encountered, use of nonlinear devices such as logarithmic amplifiers needs to be considered.

4.2.2 Noise and Error Considerations

The dynamic range of all signal-processing systems is limited by random noise, which sets a fundamental bound on the smallest signal that can be detected or otherwise utilized with an adequate signal-to-noise ratio (SNR). As discussed in Chapter 1, this noise may be generated by numerous mechanisms, including those associated with the source itself.

Noise cannot be discussed without reference to bandwidth, which will be unavoidably limited by the type(s) of amplifier(s) used. Deliberate filtering is often included in a signal-processing channel to reduce noise as well as to improve the separation of wanted from unwanted signals. This may take the form of band-pass, low-pass, or high-pass functions, or combinations of these, depending on the situation. Nonlinear filtering may also be used, for example, to minimize the disturbance of the signal path in the presence of impulsive noise. The noise powers of uncorrelated sources add, so noise voltages (or currents) must be added using a root sum of squares (RSS) calculation. This leads to some rather startling consequences. Suppose a system has a major voltage noise source of magnitude E_{an} and several minor noise sources that RSS sum to a magnitude of E_{bn}. Then, for the major source to contribute almost 90% of the total system noise, E_{an} needs to be only twice E_{bn}. When $E_{an}/E_{bn} = 5$, 98% of the noise is due to E_a.

It follows that the overall noise performance of a practical system can benefit greatly by minimizing the input-referred noise of the first stage and by using the highest possible gain in this stage. However, the second of these objectives is frequently unrealizable in systems that must handle signals of large dynamic range, because the high gain would then preclude distortion-free operation at maximum signal levels.

Noise spectral density (NSD) reflects the fact that the total noise power is directly proportional to the system's noise bandwidth, B_N. The NSD is therefore usually of interest in specifying a channel's input noise limitations. As the noise bandwidth is not, in general, equal to the -3-dB bandwidth of the amplifier, B_N can be viewed as the bandwidth of an equivalent system with a "brick wall" cessation of response at that frequency. A system with a single-pole low-pass corner at $f_0 = \frac{1}{2}\pi T$ has a B_N equal to $\pi f_0/2$, or $1.57 f_0$, whereas for two such real-pole low-pass sections in cascade, B_N is $\pi f_0/4$ (see Figure 4.2).

Noise signals are usually small, and therefore nonlinear effects are often negligible; in such circumstances, it is permissible to use superposition methods to evaluate each contributing source independently, followed by an RSS calculation to calculate the total noise. A notable exception is the logarithmic amplifier, where even very small noise voltages at the input can cause later stages in the amplifier to be in heavy limiting. Special approaches to both noise analysis and noise specification are required in such cases.

Whereas noise limits the low end of a system's dynamic range, performance at the upper end of the signal range is always degraded by the nonlinear aspects of the circuit behavior. In a typical circuit situation, such as a sensor output being amplified by an amplifier stage, noise sources can be represented by the circuit in Figure 4.3. The components of total input noise may be divided into external contributions associated with the sensor circuit

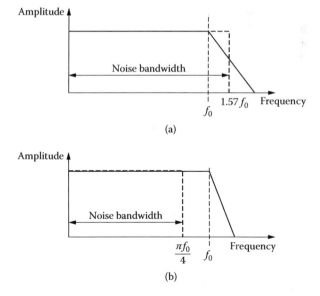

(a)

(b)

FIGURE 4.2
Filter noise bandwidths compared with -3 dB corner for frequency: (a) single-pole low-pass filter ($B_N = \pi f_0/2 = 1.57 f_0$); (b) two single-pole low-pass filter sections in cascade ($B_N = \pi f_0/4$).

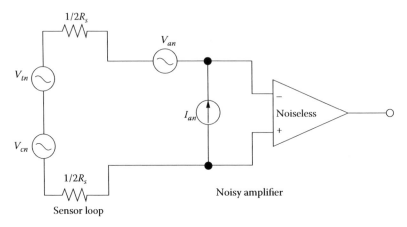

FIGURE 4.3
Sensor amplifier noise sources.

and internal amplifier noise sources referred to its input. Figure 4.3 indicates four noise sources: thermal noise (V_{tn}) of the source resistance of the sensor, noise associated with the sensor loop (V_{cn}) due to any DC current (I_{DC}) flowing in the loop via any contacts or junctions, and amplifier noise sources (V_{an} and I_{an} similar to case of amp noise sources as indicated in Figure 4.6b) referred to its input. From the discussions in Chapter 1, the first component is given by

$$V_{tn} = \sqrt{4kTR_s} \; V_{RMS}/\sqrt{Hz} , \tag{4.5}$$

where k is the Boltzmann constant (1.38×10^{-23}), T is the absolute temperature, and R_s is the source resistance (in ohms).

Noise associated with the loop and its DC is generally variable as the reciprocal of the signal frequency, $1/f$, and increases with the DC flowing in the loop based on the empirical relationship

$$V_{cn} = (0.57 \times 10^{-9})R_s\sqrt{\frac{I_{dc}}{f}} \; V_{RMS}/\sqrt{Hz} . \tag{4.6}$$

The effect of these uncorrelated noise sources can be combined by RSS, as discussed in Chapter 1. The above indicates that the source resistance values need to be kept small, and the loop DCs also need to be kept as small as possible.

The acquisition of a low-level analog signal that represents some measurand in the presence of appreciable interference is a frequent requirement in design. In this process we use many different kinds of signal-conditioning circuits together with closed-form expressions available for determining the error of a signal corrupted by random Gaussian noise or coherent sinusoidal interference. These are expressed in the SNR with probability functions. For a useful discussion on these, see Chapter 4 of Garrett [1].

4.2.3 Bandwidth

All signals we come across in real-world systems are band limited, and many sensor output signals are baseband type, where a low range of frequencies from DC to a finite maximum value are output by the sensor. Occasionally, some signals are output within a selected frequency range. In many sensors that are capable of responding to a wider range of input signal frequencies, the actual output level can be dependent on the signal frequency. Given these real-world conditions, two basic factors come into the design specifications: the total bandwidth of the signal to be processed and the signal dependency on the frequency. Because of these two situations, one has to make sure that the amplifiers and signal-conditioning circuits used with the input signals are capable of faithfully processing the input signal and that some equalization is employed to compensate for any frequency dependencies of the signal.

4.3 Amplification

There are four general classes of amplifier, characterized by the transfer characteristics expressed in volts per volt, amperes per ampere, volts per ampere, or amperes per volt. These four cases are summarized in Table 4.1. In circuit analysis, these four cases are used to ideally represent amplifier circuits.

An important building block used in analog circuits is the differential amplifier, shown in Figure 4.4. This basic circuit can be built by a matched pair of bipolar junction transistors (BJTs) or field-effect transistors (FETs), as in Figure 4.4a and Figure 4.4b. This circuit is useful in many applications, such as instrumentation and linear power supply control circuits, and also becomes an integral part of common operational amplifier (op amp) internal circuits. Transfer characteristics for the pairs are shown in Figure 4.4c and Figure 4.4d. Detailed analysis of these circuits is available in Sedra and Smith [2].

4.3.1 Operational Amplifiers

The most common circuit block used in building various types of amplifiers and filters—the operational amplifier—is a combination of a differential pair and a gain stage, as shown in Figure 4.5, which simplifies a common op amp such as the 741. In different types of op amps, either BJT differential pairs or FET differential pairs are combined with additional stages to give the overall conditions of ideal performance: (1) infinite open loop gain, (2) infinite input

TABLE 4.1

Generic Types of Amplifiers and Simple Examples

Type of Amplifier	Figure	Input/ Output Relationship	Input/Output Resistance		A Simple Example
			R_{in}	R_{out}	
Voltage amplifier		$V_0 = A \cdot V_{in}$	∞	0	Triode vacuum tube
Current amplifier		$I_0 = A \cdot I_{in}$	0	∞	Bipolar transistor

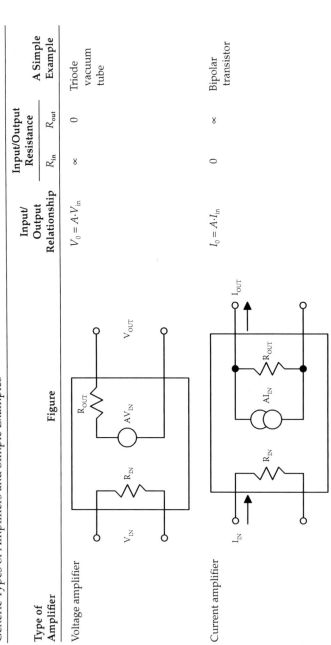

Transimpedance
amplifier

V_{IN} R_{IN} $G_M V_{IN}$ R_{OUT} I_{OUT}

$I_0 = G_M \cdot V_{in}$ ∞   ∞   Field effect
transistor

Transconductance
amplifier

I_{IN} R_{IN} $R_M I_{IN}$ R_{OUT} V_{OUT}

$V_0 = R_M \cdot I_{in}$ 0   0   No simple
device

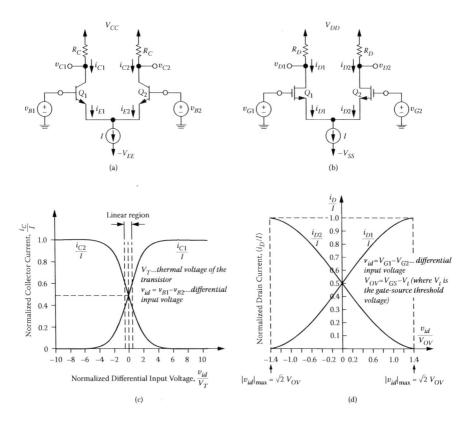

FIGURE 4.4
Differential amplifier: (a) BJT implementation; (b) FET implementation; (c) transfer characteristics for the BJT case; (d) transfer characteristics for the FET pair.

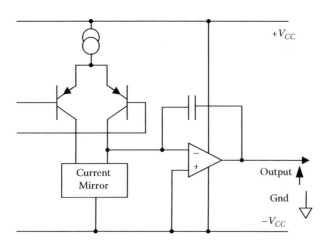

FIGURE 4.5
Simplified real op amp showing single-ended output and the differential input.

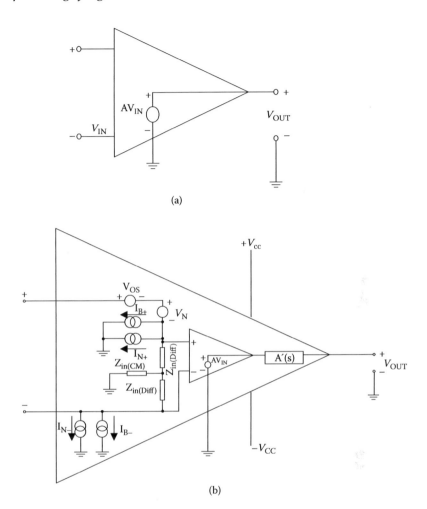

FIGURE 4.6
Ideal and practical op amps: (a) ideal case; (b) practical op amp.

resistance at inverting and noninverting inputs, (3) infinite bandwidth, and, most importantly, (4) single-ended output.

Given the above details related to op amps, now we can appreciate the real-world conditions of the op amp [3,4]. Figure 4.6a depicts a real op amp based on a voltage amplifier. Let's briefly discuss the input imperfections and output obstacles as depicted in Figure 4.6b.

4.3.1.1 Input Imperfections

The actual characteristics of real op amps are considerably more complicated. Each input contains a DC source (I_B, the bias current) and a DC voltage source (V_{OS}, the offset voltage) in series with the inputs. The amplifier has differential and common mode input impedances ($Z_{in(DIFF)}$ and $Z_{in(CM)}$, respectively) which usually consist of a resistance and a capacitance in parallel.

There are also three uncorrelated noise sources: two current sources (I_N) and a voltage source (V_N) that appear differentially. Finally, the amplifier has gain with regard to common mode signals (the average value of the two inputs), which the ideal amplifier does not, so its common mode rejection ratio (CMRR) needs to be specified.

4.3.1.2 Output Stage

The output side of the model is also not ideal. There is a finite value output impedance (R_o) in series with the voltage source. This output impedance is shown in Figure 4.6b as $A'(s)$. The gain is both finite and a function of frequency in a real amplifier, which also has a finite slew rate (the slew rate is the rate of rise of the output voltage per microsecond) and limited output voltage and current capabilities.

4.3.1.3 Differential to Single-Ended Conversion

One fundamental requirement of a simple op amp is that an applied signal that is fully differential at the input must be converted to a single-ended output, that is, with respect to the often-neglected fourth terminal, which is the ground. To see how this can lead to difficulties, take a look at Figure 4.5. The approach illustrated in Figure 4.5 is used in several early versions of op amps 101, 741, 748, and others.

The circuit first transforms a differential input voltage into a differential current. This input stage function is represented by PNP transistors in Figure 4.5. The current is then converted from a differential to single-ended form by a current mirror that is connected to the negative supply rail. The output from the current mirror drives a voltage amplifier and power output stage that is connected as an integrator. Most descriptions of this simplified model do not emphasize that the integrator has, of course, a differential input. It is biased positive by a couple of base emitter voltages, but the noninverting integrator input is referred to the negative supply. It should be apparent that most of the voltage difference between the amplifier output and the negative supply appears across the compensation capacitor. If the negative supply voltage is changed abruptly, the integrator amplifier will force the output to follow the change. When the entire amplifier is in a closed-loop configuration, the resulting error signal at its input will tend to restore the output, but the recovery will be limited by the slew rate of the amplifier. As a result, an amplifier of this type may have outstanding low-frequency power supply rejection, but the negative supply rejection is fundamentally limited at high frequencies. Because it is the feedback signal to the input that causes the output to be restored, the negative supply rejection will approach zero for signals at frequencies above the closed-loop bandwidth. This means that high-speed, high-level circuits can "talk to" low-level circuits through the common impedance of the negative supply line. The above phenomenon demands some special cases in decoupling and grounding, and details are discussed in Brokaw [5].

Table 4.2 summarizes the important parameters in op amps. Figure 4.7 shows the settling time and the slew rate. Given the internal structure of an op amp similar to the case in Figure 4.5, there is a maximum limit applicable to common mode voltages (at the input and the output). Another set of important design considerations includes the input and output common mode dynamic ranges. As shown in Figure 4.8, a dual-rail op amp could have its input or output swing within the positive and negative voltage rails. In a practical case, this common mode value will have a range that is usually less than the difference between the two rail voltages. The maximum possible output voltage will generally be within the limits of $+(V_S - V_{Sat(Hi)})$ and $-(V_S - V_{Sat(Lo)})$, where $V_{Sat(Hi)}$ and $V_{Sat(Lo)}$ are the transistor saturation voltages of the output stages. For example, in the case of rail voltages of ±15 V, the output range or the swing may be ±13 V if the saturation voltages are 2 V each. Similarly, the input common mode dynamic range may be less than the rail-to-rail value, and typically it varies, as in the example in Figure 4.8b. In general, the lower the rail values, the smaller the range. In newer versions of single-rail op amps, the situation can be much different, as discussed later [3].

4.3.1.4 Noise in Op Amps

In common op amps (voltage feedback types), the noise generally consists of white noise and the $1/f$ component of noise. (Sometimes the $1/f$ component is called flicker noise.) Typical noise spectral density characteristics are as per Figure 4.9. The general shape is shown in Figure 4.9, and the corner frequency of the voltage and current noise do not necessarily have the same values. For an op amp, the $1/f$ corner frequency is a figure of merit (the lower, the better).

4.3.1.5 Single-Rail Op Amps

In the previous paragraphs we considered only the dual- or split-rail op amps, where signals could take either negative or positive values with respect to ground. With battery-powered, low-voltage systems proliferating, this dual-rail convenience has gradually diminished and many, semiconductor companies have developed and introduced single-rail op amps. These single-rail op amps do not have the convenient ground reference that dual supply systems have; thus, biasing needs to be employed to ensure that the output voltage swings between the correct values. Input sources connected to ground are actually connected to a supply rail in single supply systems. Also, in single supply systems, the output and input signal swings are quite limited compared to dual-rail versions, and the SNR performance may also degrade due to lower signal swings. More details and comparisons can be found in Jung [3].

TABLE 4.2

Important Op Amp Parameters

Parameter	Description	Remedy
Offset voltage (V_{OS})	Defined as the voltage that must be applied to the input to cause the output to be zero. It is the result of mismatch in the base emitter voltages of the differential input transistors (gate-source voltage mismatch in FET input amplifiers). Dependent on the temperature.	This offset can be trimmed to zero with a potentiometer, but is affected by temperature variations.
Input bias current (I_B)	Result of the base current requirement of the input stage BJTs (or the leakage currents of the gate-source of input FETS). Dependent on the temperature.	The input source of the op amp needs to supply this current.
Open-loop voltage gain (A_{OL})	Gain of the ideal op amp. Real op amps have very large A_{OL} values, but not infinite.	Design calculations should account for this nonideal situation after conceptually developing the circuit configuration. Nonideal case calculations should allow the comparison of discrepancies from the ideal case.
Frequency response	Many op amps have a single-pole frequency response with a limited bandwidth compared to the ideal (internally compensated op amps). Some op amps can have a second pole and this could create instability (wide bandwidth versions).	For single-pole op amps, gain bandwidth is a constant.
Slew rate	Rate at which the op amp can change its output from zero to maximum based on an input step (see Figure 4.7). An important consideration in large signal circuits such as buffers, where input can be a large step.	Selection of the op amp needs to be based on the slew rates of the signals to be processed by the op amp. For high-frequency circuits, fast slew rate devices need to be selected.
Settling time	Defined as the time that elapses between the application of a step input to the time at which the amplifier output enters and remains within a specified error band symmetrical about the final output value (see Figure 4.7). Settling time is determined by both the linear and the nonlinear characteristics of the amplifier. It varies with the input signal level and is greatly affected by impedances external to the amplifier. In applications such as multiplexers and S/H circuits in ADCs, this parameter deter-mines the operational limits of the circuit.	Careful interpretation of the data sheet parameters is needed, particularly in interfacing high-resolution ADCs. The best way to know how fast an amplifier will settle in a particular application is to measure it.

TABLE 4.2 (continued)

Important Op Amp Parameters

Parameter	Description	Remedy
Common mode rejection range (CMRR)	Ratio of the differential voltage gain to the common-mode voltage gain. Ideal value should be infinity. CMRR falls off as frequency increases. Common source of CM noise is 50 Hz/60 Hz ac noise.	Care must be used to ensure that CMRR of op-amp is not degraded by other circuit components.

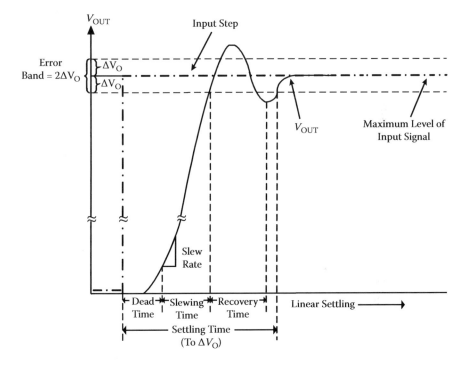

FIGURE 4.7
Op amp slew rate and settling time.

4.3.1.5.1 Design Calculations with Single Supply Op Amps

This section briefly discusses the special design approaches required in single-rail op amps. Compared to the case of dual or split supply inverting amplifiers, shown in Figure 4.10a, if the signal source to be amplified is not referenced to ground (Figure 4.10b), the output will contain the signal to be amplified as well as the reference voltage. Figure 4.10c shows a way to eliminate the effect of the V_{ref} value, considering it as a common mode voltage. In dual or split supply–based op amp circuits (or in the case where a single rail is used but signal inputs are with reference to $\frac{1}{2}V_S$, where this approach can be used to simplify the dual-rail requirement), the signal can swing positive

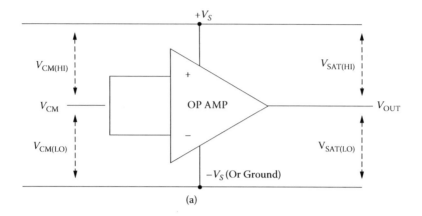

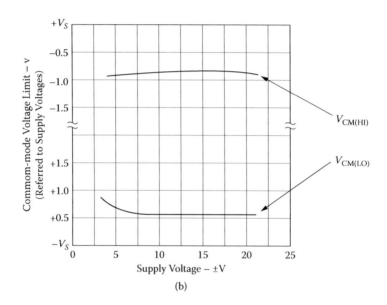

FIGURE 4.8
Common mode voltage considerations in op amps: (a) input and output common mode ranges; (b) graphical display of input common mode ranges. [Courtesy of Analog Devices, Norwood, MA; Source: Jung [3].)

or negative within limits without any inaccurate impact on the output, but in a single supply situation as in Figure 4.10d, this is not the case. Use of a single supply limits the polarity of the output voltage, and the output will generally be within $0 < V_{out} < V_S$, which precludes negative output voltages. However, the circuit may allow a negative input voltage, as long as it does not create a negative voltage value at the op amp input. (Op amp inputs are highly susceptible to reverse voltage breakdown.)

In order to use a single-rail op amp for cost and power supply simplicity, the designer needs to consider special circuit design approaches based

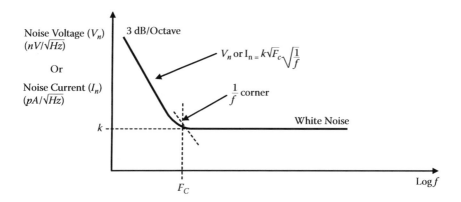

FIGURE 4.9
Noise spectral density of op amps.

on the basic approach in Figure 4.10c. A good approach proposed by Texas Instruments [4] treats the output requirements based on the four possible cases of the straight line at the output, which is based on the different possibilities of input given by

$$V_{out} = +mV_{in} + b, \tag{4.7a}$$

$$V_{out} = +mV_{in} - b, \tag{4.7b}$$

$$V_{out} = -mV_{in} + b, \tag{4.7c}$$

$$V_{out} = -mV_{in} - b. \tag{4.7d}$$

In a given design requirement where two values of input and corresponding output voltages are known, we can solve two simultaneous equations and get the corresponding values of m and b. By the use of a circuit such as in Figure 4.11a, the case in Equation 4.7a can be developed. Decoupling capacitors shown in the circuit reduce noise and provide increased noise immunity. In some situations, more extensive filtering may be required. As the power supply rail is used to generate the reference source required, power supply noise will be amplified by the circuit gain, and precautions may be necessary with add-on filtering. For the case in Figure 4.11a,

$$V_{out} = V_{in}\left(\frac{R_2}{R_1 + R_2}\right)\left(\frac{R_F + R_G}{R_G}\right) + V_{Ref}\left(\frac{R_1}{R_1 + R_2}\right)\left(\frac{R_F + R_G}{R_G}\right). \tag{4.8}$$

Using Equation 4.7a and Equation 4.8,

$$m = \left(\frac{R_2}{R_1 + R_2}\right)\left(\frac{R_F + R_G}{R_G}\right) \tag{4.9a}$$

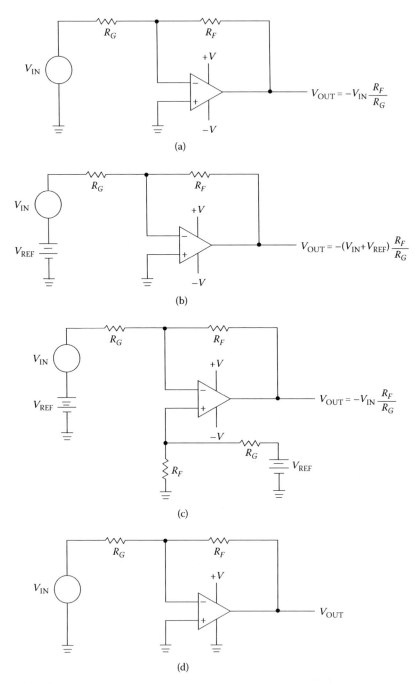

FIGURE 4.10
Transition to single-rail op amp from the dual-rail cases by inserting a reference source:
(a) dual-rail inverting op amp circuit; (b) dual-rail inverting op amp circuit with reference voltage input; (c) dual-rail inverting op amp circuit with common mode voltage inserted; (d) single supply situation. (Adapted from [4].)

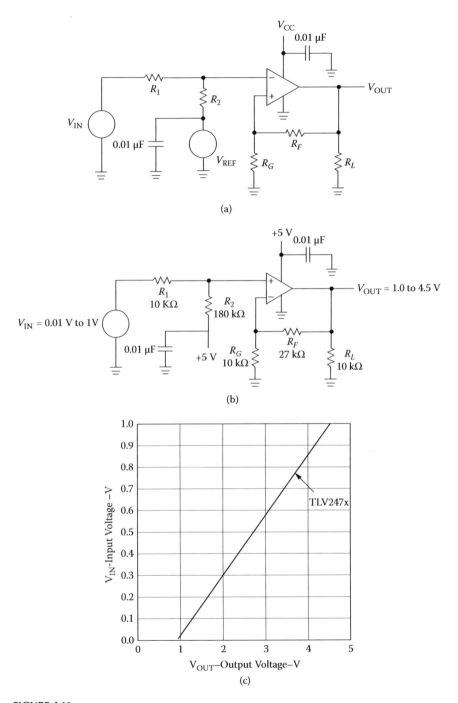

(a)

(b)

(c)

FIGURE 4.11
Single-rail amplifier based on reference source approach: (a) circuit configuration; (b) actual circuit; (c) measured performance. (Reprinted with permission from Elsevier Science & Technology Books; Source: Mancini [4].)

$$b = V_{Ref} \left(\frac{R_1}{R_1 + R_2} \right) \left(\frac{R_F + R_G}{R_G} \right). \qquad (4.9b)$$

Based on the above approach, for a case such as $V_{out} = 1$ for $V_{in} = 0.01$ V, and $V_{out} = 4.5$ for $V_{in} = 1$ V with $R_L = 10k$ and resistors with 5% tolerance and $V_{CC} = 5$ V, the circuit in Figure 4.11b can be achieved. Use of an op amp such as the TLV247X series from Texas Instruments will give a measured transfer curve, as in Figure 4.11c. Similar cases can be developed for the other three possibilities of Equation 4.7. Details are available in Mancini [4].

4.3.1.6 Current Feedback Op Amps

The discussion in the previous sections was based on the common op amp architecture called the voltage feedback amplifier (VFA) and traditionally based on a differential input stage, a high gain stage with a single-pole response, and an output stage designed for wide voltage swing. Figure 4.12a shows the internal concept (related to Figure 4.5); a simplified conceptual version of the op amp without any feedback circuit is shown in Figure 4.12b. When the circuit is connected in the noninverting amplifier configuration, as in Figure 4.12c, the closed loop gain is given by

$$\frac{V_{out}}{V_{in}} = \frac{1 + \dfrac{R_2}{R_1}}{1 + \dfrac{1}{A(s)} \left[1 + \dfrac{R_2}{R_1} \right]} = \frac{1 + \dfrac{R_2}{R_1}}{1 + \dfrac{1}{A(s)\beta}}. \qquad (4.10)$$

It is important to note that the error signal developed because of the feedback network and the finite open-loop gain is, in fact, a small voltage, v, in Figure 4.12c.

During the 1990s, a new version of op amps—current feedback op amps (CFAs)—was proliferating due to the demand of high-bandwidth applications such as video and digital subscriber loops as shown in Figure 4.12d. CFAs do not have the traditional differential input structure and thus sacrifice the parameter matching inherent to the VFA structure. The CFA configuration prevents them from obtaining the precision of VFAs, but the circuit allows wider bandwidth and a higher slew rate. Constant gain bandwidth restriction applied to the VFA is removed from the CFA, and its bandwidth is relatively independent of the closed-loop gain. The slew rate of CFAs is much improved from their counterpart VFAs because their structure (shown in Figure 4.12d) enables the output stage to supply the slewing current until the output reaches the final value.

In general, VFAs are used for precision and general-purpose applications, whereas CFAs are used for high-frequency applications, usually above 100 MHz. The fundamental concept is based on the fact that in bipolar

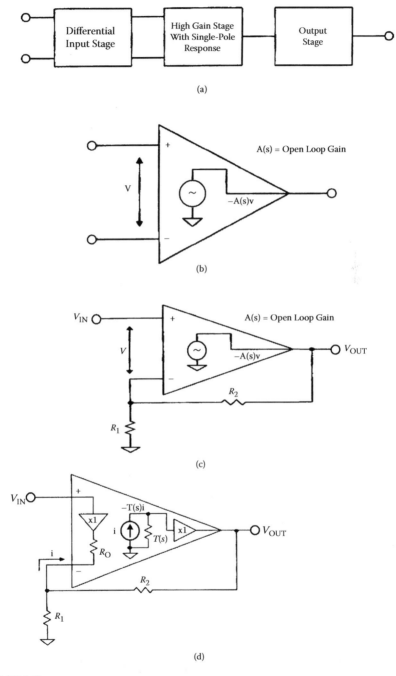

FIGURE 4.12
Voltage feedback op amp and the noninverting amplifier: (a) simplified block approach to internal of the VFA; (b) simplified VFA indicating open loop gain function $A(s)$; (c) case of the noninverting amplifier indicating the effect of the feedback loop and the open loop gain; (d) architecture of the current feedback op amp (CFA).

transistor circuits, currents can be switched more quickly than voltages, all other things being equal. For the CFA structure in Figure 4.12d, a unity gain buffer connects the noninverting input to the inverting input. In the ideal case, the output impedance of this buffer is zero (i.e., $R_o = 0$) and the error signal is a small current, i, which flows into the inverting input. This error current is mirrored into a high impedance, $T(s)$, and the voltage developed across $T(s)$ is equal to $T(s)i$. The quantity $T(s)$ is generally referred to as the open-loop transimpedance gain. This voltage is then buffered and is connected to the op amp output. It is easy to derive the expression of the closed-loop gain as

$$\frac{V_{out}}{V_{in}} = \frac{1 + \dfrac{R_2}{R_1}}{1 + \dfrac{R_2}{T(s)}\left[1 + \dfrac{R_o}{R_1} + \dfrac{R_o}{R_2}\right]} . \tag{4.11}$$

If R_o is assumed to be zero, Equation 4.11 reduces to a case similar to the VFA:

$$\frac{V_{out}}{V_{in}} = \frac{1 + \dfrac{R_2}{R_1}}{1 + \dfrac{R_2}{T(s)}} . \tag{4.12}$$

Current feedback op amps are often called transimpedance op amps because the open-loop transfer function is an impedance, as described above. From the above simple model, several important CFA characteristics can be deduced:

- CFAs do not have balanced inputs. The inverting input is low impedance, and the noninverting input is high impedance (compared to the common ideal op amp cases [VFAs], where we assume infinite input impedances at both inputs).
- The open-loop gain of a CFA is measured in units of ohms (rather than V/V for the VFAs).
- For a fixed-value feedback resistor R_2, the closed-loop gain of a CFA can be varied by changing R_1 without significantly affecting the closed-loop bandwidth. This can be easily seen by Equation 4.12, because the denominator determines the overall frequency response.

Figure 4.13a shows the internal equivalent circuit for a CFA-based circuit connected as an inverting amplifier. Components R_T and C_T provide the combination for the transimpedance of the circuit. Compared to a VFA, the frequency response of a CFA is shown in Figure 4.13b and Figure 4.13c. For

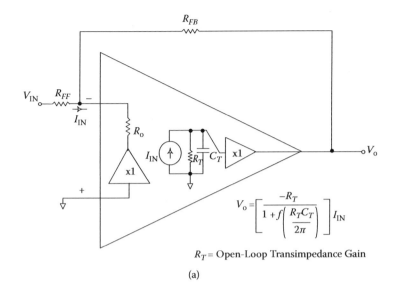

$$V_o = \left[\frac{-R_T}{1 + f\left(\frac{R_T C_T}{2\pi} \right)} \right] I_{IN}$$

R_T = Open-Loop Transimpedance Gain

(a)

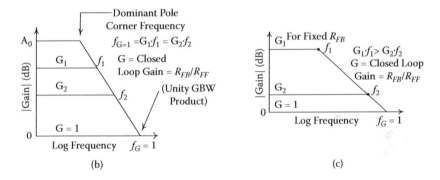

(b)   (c)

FIGURE 4.13
Inverting CFA configuration and comparison of performance with VFA: (a) inverting circuit based on a CFA; (b) VFA gain; (c) CFA gain.

more information, see Jung [3] and Mancini [4]. Table 4.3 summarizes some common op amp configurations.

4.3.2 Instrumentation Amplifiers

The most popular among the specialty amplifiers is the instrumentation amplifier (or in-amp). It is widely used in many industrial and measurement applications where DC precision and gain accuracy must be maintained within a noisy environment and where large common mode signals (such as the AC line frequency) are present. The acquisition of accurate measurement signals, especially low-level signals in the presence of interference, requires

TABLE 4.3

Common Op Amp Configurations

Function	Configuration	Remarks
Inverting amplifier $V_{out} = -\dfrac{R_F}{R_G} V_{in}$		
Noninverting amplifier $V_{out} = \left[1 + \dfrac{R_F}{R_G}\right] V_{in}$		In-phase signal amplification
Low-pass filter/integrator		Limits bandwidth of signal

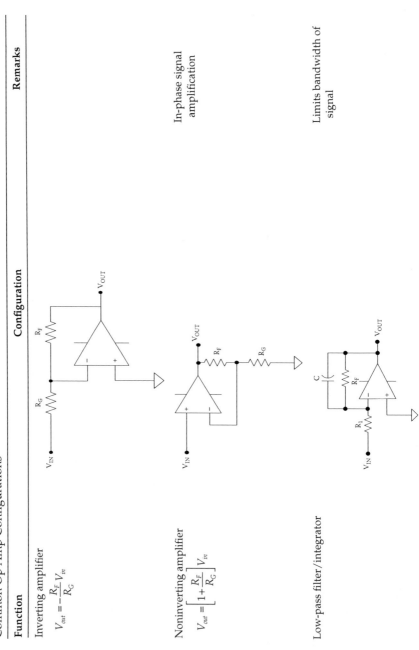

High-pass filter/differentiator

$$V_{out} \approx -\frac{dV_{in}}{dt}$$

Blocks DC
Amplifies AC

Voltage follower

$$V_{out} = V_{in}$$

Provides very high
input impedance to
the input
Low resistance to load
(output) side

Voltage adder

$$V_{out} = -R_F \left[\frac{V_1}{R_1} + \frac{V_2}{R_2} + \frac{V_3}{R_3} + \cdots \frac{V_N}{R_N} \right]$$

Sum multiple input
voltage sources

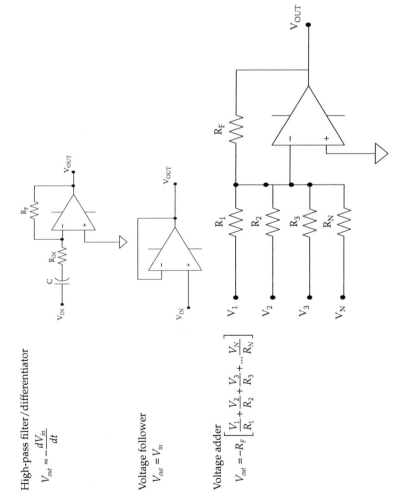

(continued on next page)

TABLE 4.3 (continued)

Common Op Amp Configurations

Function	Configuration	Remarks
Differential amplifier $$V_{Out(Diff)} = \left[\frac{R_F}{R_G}\right] V_{in}$$	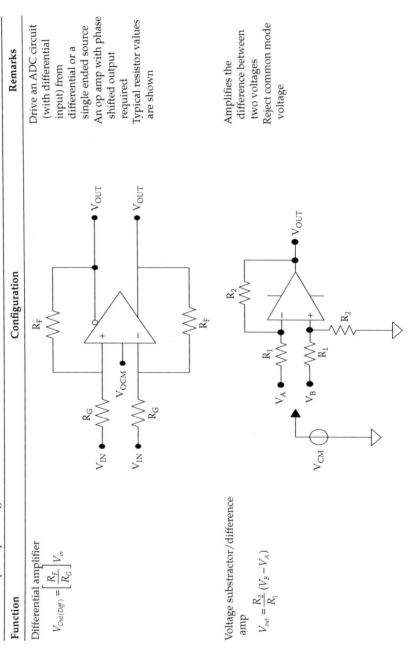	Drive an ADC circuit (with differential input) from differential or a single ended source An op amp with phase shifted output required Typical resistor values are shown
Voltage substractor/difference amp $$V_{out} = \frac{R_2}{R_1}(V_B - V_A)$$		Amplifies the difference between two voltages Reject common mode voltage

Amplify low level
differential signal
Reject common mode
signals

Instrumentation amplifier

$$V_{out} = \left[1 + \frac{2R_F}{R_G}\right]V_{in} + V_{REF}$$

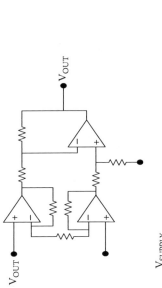

Current sensing amplifier

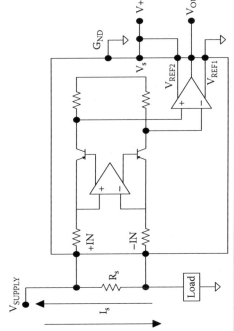

amplifier performance beyond the typical signal-conditioning capabilities of op amps. An in-amp is usually the first circuit block encountered by a sensor in a signal-conditioning channel, and in large part it is responsible for data accuracy. Instrumentation amplifiers should provide

- Adequate linearity
- Stability
- Low noise and low offset
- Sensitivity for microvolt-order signals
- Very high CMRR for interference rejection
- Very high gain values

A generic in-amp is shown in Figure 4.14a. An in-amp is a precision closed-loop gain block with a pair of differential terminals and a single-ended output that works with respect to a reference or common terminal, as in Figure 4.14a. The input impedances should be typically greater than 1 GΩ. Unlike an op amp, an in-amp uses an internal feedback resistor network plus one gain set resistor, R_G. Also, the internal resistance network and R_G should be isolated from the input terminals. The output voltage is usually referenced to a pin, usually designated the "reference" or V_{ref}. This is usually connected to ground or to any other voltages dependent on the requirement. In single supply cases, this could be the mid-value of the supply voltage. To be effective, an in-amp needs to have a high gain simultaneously rejecting common mode signals, so that the common mode error source (common mode error referred to input (RTI) = V_{CM}/CMRR) in Figure 4.14a is relatively negligible. Typical values for the CMRR are from 70 to 100 dB. The CMRR should have high values from DC to higher values, particularly covering the AC line frequencies such as 50 to 60 Hz.

A simple subtractor or a difference amplifier circuit that can be used for limited applications is shown in Figure 4.14b. It is not a true in-amp satisfying the above discussion, but it is suitable for simple differential to single-ended conversion. For this circuit, the following features and relationships are valid:

$$V_{out} = (V_2 - V_1)\frac{R_2}{R_1}.$$ (4.13a)

For a high CMRR,

$$\frac{R_2}{R_1} = \frac{R'_2}{R'_1}$$ (4.13b)

$$CMRR = 20\log_{10}\left[\frac{1+\dfrac{R_2}{R_1}}{K_r}\right],$$ (4.13c)

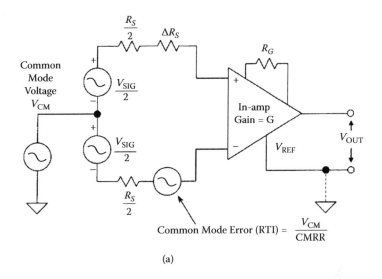

(a)

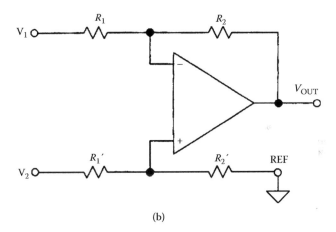

(b)

FIGURE 4.14
In-amp requirements and implementations: (a) generic in-amp; (b) op amp subtractor (difference amp); (c) two op amp solution; (d) three op amp solution.

where K_r is the total fractional mismatch between the values of R_1/R_2 to R_1'/R_2'. (See Equation 4.14c.)

For the above circuit, the 0.1% value of K_r CMRR is about 66 dB. A few limitations of the circuit are that

- Input impedance seen by inputs is not balanced (input 1 sees a value of R_1 and the second input sees $R_1' + R_2'$).
- A small source impedance mismatch can degrade the CMRR.
- The CMRR is primarily dependent on the resistor ratio matching (but absolute resistor values are not important to the CMRR).

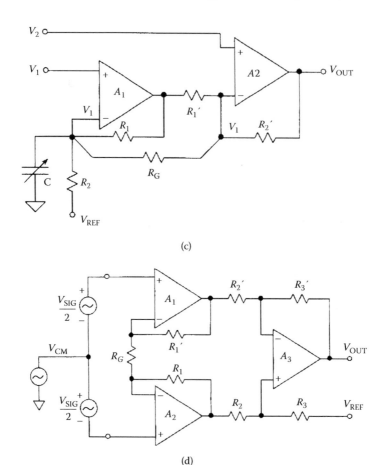

(c)

(d)

FIGURE 4.14 (continued)

For these reasons, discrete implementation of the circuit may not be suitable for many precision circuits, and complete monolithic solutions such as AMP03 and AD629 from Analog Devices provide better solutions [3].

Given the limitations of the above solution, true in-amps based on op amps are better solutions for high-performance requirements. A two op amp solution is shown in Figure 4.14c, and a three op amp solution is shown in Figure 4.14d. For the case of the two op amp solution in Figure 4.13,

$$V_{out} = (V_2 - V_1)\left[1 + \frac{R_2}{R_1} + \frac{2R_2}{R_G}\right] + V_{Ref} \tag{4.14a}$$

$$CMRR \leq 20\log_{10}\left[\frac{G100}{K_r}\right] \tag{4.13b}$$

$$K_r = \frac{\left(\dfrac{R_1'}{R_2'} - \dfrac{R_1}{R_2}\right)}{\dfrac{R_1}{R_2}} 100 \qquad (4.14c)$$

$$G = 1 + \frac{R_2}{R_1} + \frac{2R_2}{R_G}. \qquad (4.14d)$$

For this configuration, dual supply op amps are a better solution because V_{ref} can be connected to ground, compared to the case of half the power supply value of a single supply. In a single supply situation, there will be restrictions of the input voltage range that depend on the value of G [3]. The use of a trim capacitor C could improve the AC CMRR value. AD627 is an example of an improved version of a single supply two op amp version of the in-amp [3].

For the three op amp in-amp architecture in Figure 4.14d, the key relationships are

$$V_{Out} = V_{sig} \frac{R_3}{R_2}\left[1 + \frac{2R_1}{R_G}\right] + V_{Ref}, \qquad (4.15a)$$

if $R_2 = R_3$,

$$G = 1 + \frac{2R_1}{R_G} \qquad (4.15b)$$

$$CMRR \le 20\log\left[\frac{G.100}{\%K_r}\right], \qquad (4.15c)$$

where K_r is the percentage mismatch ratio between R_3/R_2 and R_3'/R_2'.

AD620 from Analog Devices is an example of such a monolithic device [3]. Figure 4.15 shows details of a simplified AD620 schematic and its CMRR performance for different gains. By combining another suitable op amp stage, AD620 can be used for single supply situations. Figure 4.15c shows such a circuit. In-amp noise performance and offset issues are discussed in Jung [3]. Avoiding common application problems in in-amp circuits is discussed in Kitchin and Counts [6]. A useful designer's reference is Kitchin and Counts [7].

4.3.3 Nonlinear Approaches

In an electronic design world, where a digital design approach is preferred in many instances, there is still much room for analog computation techniques, particularly in situations where there is wide dynamic range of the signal to be processed or where only a basic calculation with a limited accuracy

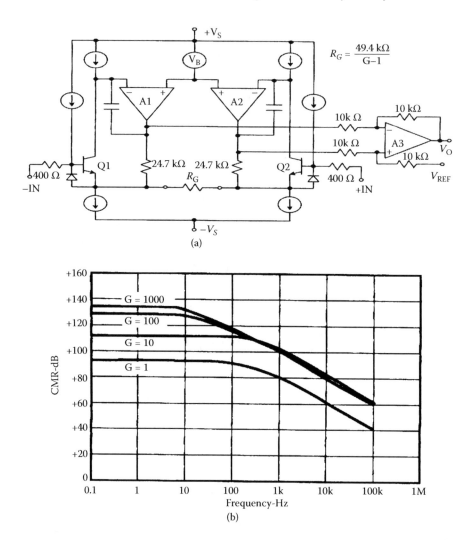

FIGURE 4.15
An example of a three op map IC, AD620: (a) simplified schematic; (b) CMRR at different gain values; (c) single supply conversion using a junction gate FET (JFET) op amp AD822. (Courtesy of Analog Devices, Inc.)

is required without complicated circuits. For example, a simple AC power meter may be constructed very easily with a single analog multiplier, where the moving coil meter can act as the integrator. A digital power meter would require conversion of both voltage and current to digital form, with considerable attention to the timing of the conversions because the relative phase of the two signals is of critical importance. Another simple example is that of the complexity of the digital approach in calculating the first derivative of a varying analog signal using digital techniques compared to a simple C-R

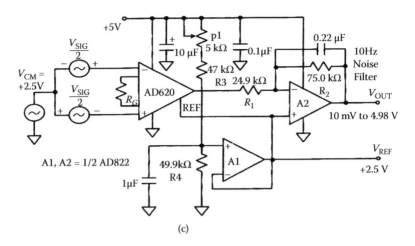

(c)

FIGURE 4.15 (continued)

network for analog differentiation. Digitizing a signal with wide dynamic range is also expensive. If such a signal is digitized with a 16-bit ADC (which is a very expensive device), the ratio of a least significant bit (LSB) to full scale is 96 dB, whereas if the signal is first applied to a logarithmic converter (frequently misnamed a logarithmic amplifier), then a dynamic range approaching 120 dB is practical with an 8-bit ADC. See Figure 4.16.

The operation of many analog computational circuits depends on the logarithmic properties of silicon junctions such as diodes and transistors. A simple grounded-base transistor-based circuit, as in Figure 4.17, provides a dynamic range of about 1,000,000:1 (6 decades) or more; the only disadvantage of such a circuit is that the signals can have only a single polarity.

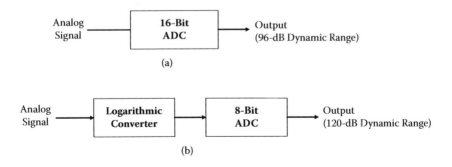

FIGURE 4.16
Analog-processing advantage of signals with high dynamic range: (a) 16-bit ADC—the expensive way of giving 96-dB dynamic range; (b) inexpensive 8-bit ADC with logarithmic converter, yielding a 120-dB dynamic range.

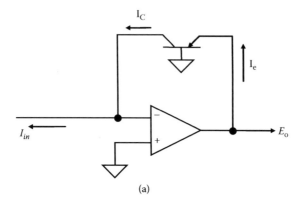

(a)

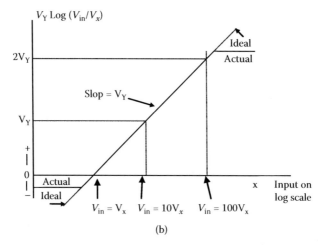

(b)

FIGURE 4.17
A simple log converter and transfer function: (a) simple implementation; (b) transfer function.

For the simple case of Figure 4.17, the simplified relationship for E_o is

$$E_o = \frac{kT}{q} \ln\left(\frac{I_{IN}}{I_{ES}}\right), \tag{4.16}$$

where k is the Boltzmann constant, T is the absolute temperature, and q is the electron charge. I_{ES} is the emitter saturation current of the transistor.

Such logarithmic converters are temperature sensitive. kT/q has a temperature coefficient (TC) of 0.34%/°C at about 25°C, and I_{ES} doubles for every 10°C temperature rise and varies with device size and geometry. Many of these basic concepts are used in refined forms or in combination with other compensation circuits in many nonlinear devices such as logarithmic converters [8,9]. A practical log amp has the graph of transfer characteristics shown in Figure 4.17b and has the transfer function

$$V_{out} = V_y \log_{10}(V_{in}/V_x). \tag{4.17}$$

This is valid over some range of input values that may vary from 100:1 (40 dB) to more than 1,000,000:1 (120 dB). The scale of the horizontal axis (the input) is logarithmic, and the ideal transfer characteristic is a straight line. V_x is known as the intercept voltage of the log amp because the graph crosses the horizontal axis at this value of V_{in}. With inputs very close to zero, log amps cease to behave logarithmically, and most then have a linear V_{in}/V_{out} law. This behavior is often lost in device noise, and the noise often limits the dynamic range of a log amp.

4.3.3.1 Practical Log Amps

The logarithm function is indeterminate for negative values of x. Log amps can respond to negative inputs in three ways:

1. A full-scale negative output, as shown in Figure 4.18a. This basic log amp saturates with the negative inputs.
2. An output that is proportional to the log of the absolute value of the input and disregards its sign, as shown in Figure 4.18b. This type of log amp can be considered to be a full-wave detector with a logarithmic characteristic and is often referred to as a detecting log amp.
3. An output that is proportional to the log of the absolute value of the input and has the same sign as the input, as shown in Figure 4.18c. This type of log amp can be considered to be a video amp with a logarithmic characteristic and may be known as a log video amplifier or sometimes a true log amp.

There are practical logarithmic amplifiers and their variations in IC form available from analog IC manufacturers such as Analog Devices [10]. Based on the same design concepts, some advanced versions, such as analog computation units (ACUs), allow multiplication and division functions with reasonable accuracies. A summary is provided in Kularatna [11]. The key parameters of log amps are noise (RTI), dynamic range, frequency response, slope, intercept point, and log linearity [10,11]. Variable gain amplifiers (VGAs) and voltage control amplifiers (VCAs) are discussed in Jung [3] and Israelsohn [12].

4.3.3.2 Root Mean Square-to-DC Converter ICs

The root mean square (RMS) is a fundamental measurement of the magnitude of an AC signal. Defined practically, the RMS value assigned to the AC signal is the amount of DC required to produce an equivalent amount of heat in the same load. Defined mathematically, the RMS value of a voltage is defined as the value obtained by squaring the signal, taking the average, and

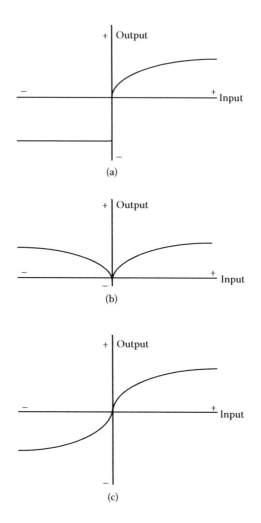

FIGURE 4.18

Log amps with negative values of x input: (a) basic log amp; (b) detecting log amp; (c) true log amp or log video amp.

then taking the square root. The averaging time must be sufficiently long to allow filtering at the lowest frequencies of operation desired. There are two basic techniques used in IC forms of RMS-to-DC converters, namely explicit and implicit methods.

4.3.3.2.1 Explicit Method

The explicit method is shown in Figure 4.19a. The input signal is first squared by a multiplier. The average value is then taken by using an appropriate filter, and the square root is taken using an op amp with a second squarer in the feedback loop. This circuit has limited dynamic range because the stages following the squarer has to deal with a signal that varies enormously in

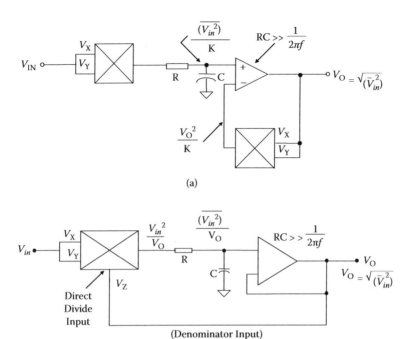

(a)

(b)

FIGURE 4.19
RMS-to-DC conversion techniques: (a) explicit method; (b) implicit method. (Reproduced by permission of Analog Devices, Norwood, MA.)

amplitude. This limits the method to inputs that have a maximum dynamic range of approximately 10:1 (20 dB). However, excellent bandwidth (greater than 100 MHz) can be achieved with high accuracy.

4.3.3.2.2 Implicit Method

Figure 4.19b shows the circuit for computing the RMS value of a signal using the implicit method. Here, the output is fed back to the direct-divide input of a multiplier. In this circuit, the output of the multiplier varies linearly (instead of as the square) with the RMS value of the input. This considerably increases the dynamic range compared to the explicit method. The disadvantage of this approach is that it generally has less bandwidth than the explicit computation. For details of these devices and applications, see Kitchin and Counts [13].

4.3.4 Video and Communications Amplifiers

4.3.4.1 Video Amplifiers

In video applications, amplifier circuits are expected to process signals within about 100 MHz or even higher. The bandwidth of the op amps and other circuits used in video applications must be sufficient so that the video signal is not attenuated or shifted in phase significantly. In many applications, gain flatness is expected to be within about 0.1 dB beyond about 50 MHz. Circuit parasitics and load impedance can significantly affect this 0.1-dB flatness. For these reasons, specially designed video signal–capable VFA types or CFAs are required. Figure 4.20 shows the gain and gain flatness of a typical video amplifier, such as the AD8075 optimized for driving source- and load-terminated 75-Ω cables. Another useful property may be the case of similar gains for large signals (2 Vpp) and small signals (200 mVpp) within the useful bandwidth of the amplifier. Figure 4.20a shows the gain and gain flatness

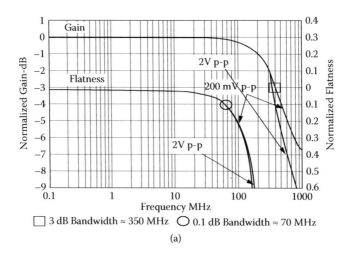

(a)

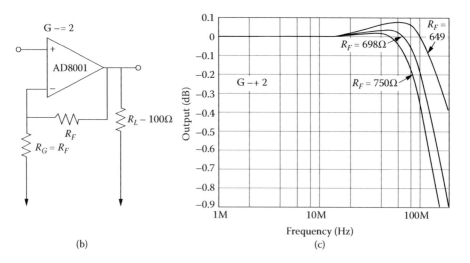

(b)   (c)

FIGURE 4.20
Typical video amplifier characteristics: (a) performance of AD8075 VFA-based video buffer; (b) AD8001 current feedback amplifier; (c) bandwidth flatness versus feedback resistor value for circuit in (b). (Courtesy of Analog Devices, Inc.)

of the AD8075 VFA case, and Figure 4.20b shows the case of a video amp based on an AD8001 CFA. In the case of the current feedback op amp case of Figure 4.20b, the bandwidth flatness is significantly dependent on the feedback resistor value (Figure 4.20c). For this reason it is important to consider the resistor parasitics, PCB layout, etc. More details are available in Jung [3] and DiSanto [14].

4.3.4.2 Communications Amplifiers

Components used in the signal path in communications systems must have wide dynamic range at high frequencies. Dynamic range is primarily limited by distortion and noise introduced by active elements in amplifiers, mixers, etc. Modern op amps have sufficient bandwidths of hundreds of megahertz and have become popular building blocks in communications systems.

The op amps usable in communications systems need to be specified not only for AC specifications (such as bandwidth, slew rate, and settling time) but also in terms of communications-specific specifications. These additional specifications include performance for harmonic distortion, spurious free dynamic range (SFDR), intermodulation distortion, intercept points, and noise and noise figures. A discussion of these is beyond the scope of this chapter, and more information can be found in Jung [3].

4.4 Bridge Circuits

In the real world of signal processing, we have to treat many physical variables from a wide range of sensing elements. Most sensing elements are based on resistance, capacitance, voltage, or current outputs. Resistive elements are the most common transducers, and they can be made sensitive to temperature, strain (by pressure or flex), or light signals. Table 4.4 provides some examples.

A balanced bridge is one of the most useful circuits for sensing low-voltage signals from a transducer. When applied properly, bridge amplifiers

TABLE 4.4

Typical Sensing Elements and Their Output Variables

Transducer Type	Variable
Resistance temperature device (RTD)	Resistance change
Strain gauges	Resistance change
Piezoelectric devices	Charge output/capacitance
Thermocouples	Voltage output
Photodiode detectors	Current output
Thermistors	Resistance

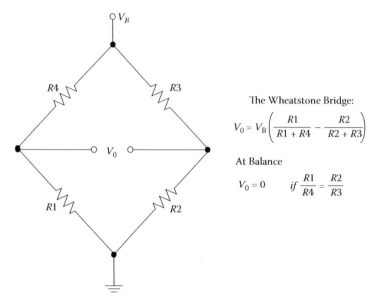

The Wheatstone Bridge:

$$V_0 = V_B\left(\frac{R1}{R1 + R4} - \frac{R2}{R2 + R3}\right)$$

At Balance

$$V_0 = 0 \quad if \quad \frac{R1}{R4} = \frac{R2}{R3}$$

Figure 4.21
The basic Wheatstone bridge.

reproduce transducer signals accurately and reliably. Most bridges employ resistive elements for measuring temperature, pressure, or strain. Figure 4.21 shows the common Wheatstone bridge circuit, which is attractive for measuring small changes of resistance in circuits. There are two ways of operating this kind of bridge circuit. One is to operate it as a null detector, where the bridge measures resistance indirectly by comparison with a similar standard resistance. The other is to use it as a device that reads a resistance difference directly, as a proportional voltage output. For the case of Figure 4.21, the output is given by

$$V_o = V_B\left[\frac{R_1}{R_1 + R_4} - \frac{R_2}{R_2 + R_3}\right]. \tag{4.18a}$$

At balance,

$$V_o = 0, \text{ if } R_1/R_4 = R_2/R_3. \tag{4.18b}$$

Under the conditions of Equation 4.18b, the circuit is said to be at null, irrespective of the mode of excitation (voltage or current, which could be AC or DC), the magnitude of the excitation, the mode of readout (current or voltage), or the impedance of the detector. Therefore, if the ratio of R_2/R_3 is fixed at K, a null is achieved when $R_1 = KR_4$.

Table 4.5 shows commonly used bridge circuits and their output relationships with linearity errors. (Note that the linearity here is only for the bridge

TABLE 4.5

Different Bridge Circuits and Properties

Drive Type	Type of Bridge	Output Voltage	Linearity Error
Constant voltage drive	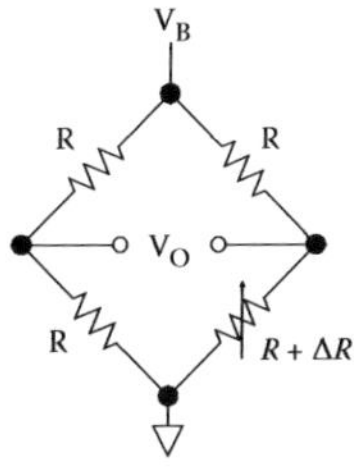 (a) Single Element Varying	$V_o = \dfrac{V_B}{4}\left[\dfrac{\Delta R}{R + \dfrac{\Delta R}{2}}\right]$	0.5%/%
	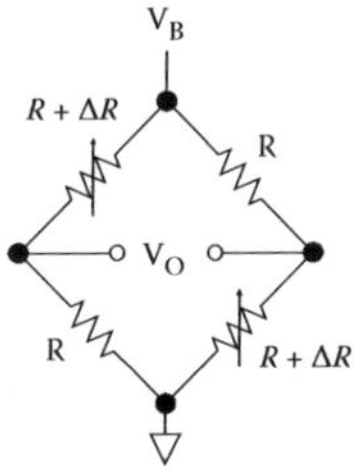 (b) Two Element Varying (1)	$V_o = \dfrac{V_B}{2}\left[\dfrac{\Delta R}{R + \dfrac{\Delta R}{2}}\right]$	0.5%/%
	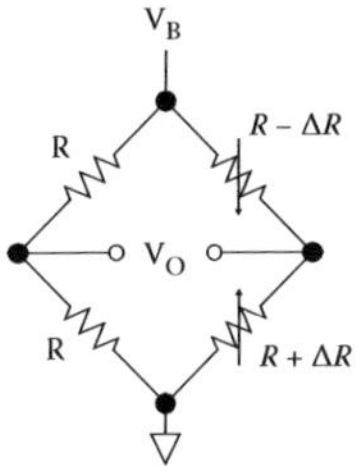 (c) Two Element Varying (2)	$V_o = \dfrac{V_B}{2}\left[\dfrac{\Delta R}{R}\right]$	0
	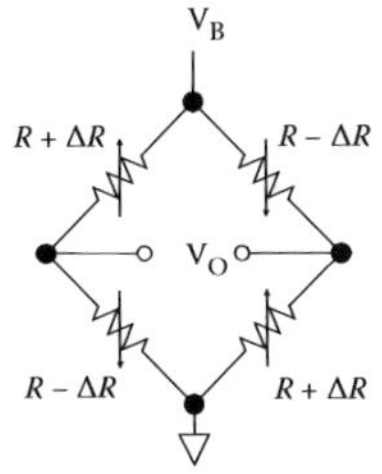 (d) All Element Varying	$V_o = V_B\left[\dfrac{\Delta R}{R}\right]$	0

(a) Single Element Varying

(b) Two Element Varying (1)

(c) Two Element Varying (2)

(d) All Element Varying

(continued on next page)

TABLE 4.5 (continued)

Different Bridge Circuits and Properties

Drive Type	Type of Bridge	Output Voltage	Linearity Error
Constant current drive	(e) Single Element Varying	$V_o = \dfrac{I_B}{4}\left[\dfrac{\Delta R}{R + \dfrac{\Delta R}{4}}\right]$	0.25%/%
	(f) Two Element Varying (1)	$V_o = \dfrac{I_B}{2}\left[\Delta R\right]$	0
	(g) Two Element Varying (2)	$V_o = \dfrac{I_B}{2}\left[\Delta R\right]$	0
	(h) All Element Varying	$V_o = I_B\left[\Delta R\right]$	0

performance, assuming a perfectly linear sensor is used.) Current drive, although not as popular as voltage drive, does have advantages when the bridge is located remotely from the source of excitation. One advantage is that the wiring resistance does not introduce errors in the measurement; another is the simple and inexpensive cabling. Regardless of the absolute level, the stability of the excitation voltage or current directly affects the overall accuracy of the bridge output, as is evident from the relationships in Table 4.5. Therefore, either stable references or ratiometric drive techniques are required. Ratiometric here means using the same excitation voltage for the reference input of the ADC circuits. Details can be found in Jung [3].

4.4.1 Amplifying and Linearizing the Bridge Outputs

Op amp–based amplifier circuits are commonly used with bridge configurations to amplify the bridge outputs. Figure 4.22a shows a few alternative amplifier arrangements using op amps for a single-element varying bridge. Although attractive because of its simplicity, it has relatively poor overall performance. Its gain predictability and accuracy are poor, and it can unbalance the bridge due to loading from R_F and the op amp bias current. The R_F resistors must be carefully chosen and matched to maximize the common mode rejection (CMR). However, a useful feature of the circuit is that it is capable of single supply operation with a solitary op amp. When the one end of the R_F resistor connected to the noninverting input is returned to $V_S/2$, both positive and negative ΔR values can be accommodated.

As indicated in Figure 4.22b, a much better method is to use an in-amp for the required gain. This technique provides better gain accuracy, with the in-amp gain usually set with a single resistor, R_G. Because the amplifier provides dual, high-impedance loading to the bridge nodes, it does not unbalance or load the bridge. Using modern in-amps (such as the AD62X series) with gains from 10 to 1000, excellent CMR and gain accuracy can be achieved. However, due to the intrinsic characteristics of the bridge (see output relationship in Table 4.5), the output is still nonlinear. The circuit can be used in single-rail mode as well [3].

Various techniques are available to linearize bridge outputs, but it is important to distinguish between the linearity of the bridge equation and the sensor response linearity. For example, if the active sensor is a resistance temperature detector (RTD), even if the bridge used to implement the measurement is perfectly linear, the output can still be nonlinear due to the RTD's intrinsic nonlinearity. Figure 4.22c shows the case of using an op amp circuit to linearize the bridge output. For this circuit,

$$V_o = -V_B \left[\frac{\Delta R}{2R} \right]. \tag{4.19}$$

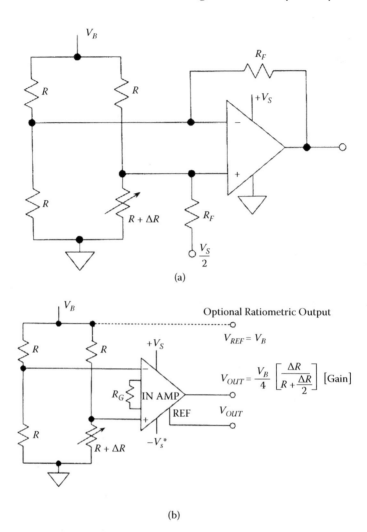

(a)

(b)

FIGURE 4.22
Bridge amplifiers: (a) simple precision op amp based; (b) instrumentation amplifier based; (c) amplifier circuit for linearizing the single-element varying bridge; (d) two op amp version for linearizing the single-element varying bridge. (Courtesy of Analog Devices, Inc.)

For this circuit, dual supplies are required. The key point is that the bridge's incremental resistance/voltage output becomes linear, even for large values of ΔR. A much more flexible circuit is shown in Figure 4.22d, with better CMR performance [3]. Two element varying bridge circuits, remote bridge circuits, and system offset minimization are detailed in Jung [3]. Another useful reference is [30].

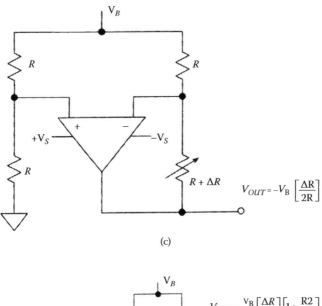

$$V_{OUT} = -V_B \left[\frac{\Delta R}{2R} \right]$$

(c)

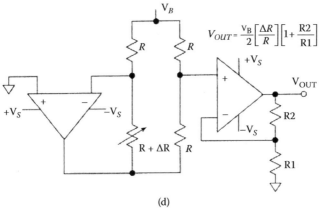

$$V_{OUT} = \frac{V_B}{2} \left[\frac{\Delta R}{R} \right] \left[1 + \frac{R2}{R1} \right]$$

(d)

FIGURE 4.22 (continued)

4.5 Filters

Filters are frequency-dependent networks that can pass or block a range of frequencies. They can come in the form of active or passive analog filters, or in completely digital form. Four basic types of filters exist. They are low pass, high pass, band pass, and band reject, as shown in Figure 4.23.

Five key parameters of filters are shown in Figure 4.24. Compared to the cases in Figure 4.23, real-world filters have gradual transitions from one band to the other. The cutoff frequency (F_c) is the frequency at which the fil-

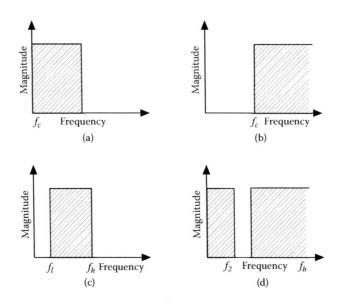

FIGURE 4.23
Ideal cases of filter responses: (a) low pass; (b) high pass; (c) band pass; (d) band reject (notch).

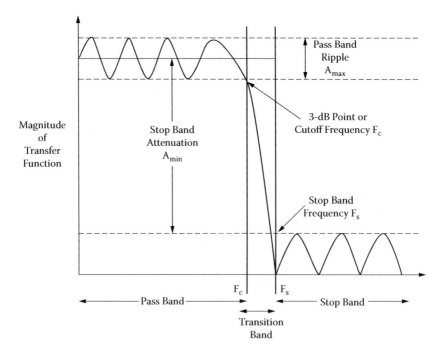

FIGURE 4.24
Key filter parameters for a low-pass filter.

ter response leaves the error band (or the −3-dB point for Butterworth filters). The stop-band frequency (F_s) is the frequency at which the minimum attenuation in the stop band is reached. The pass-band ripple (A_{max}) is the variation (error band) in the pass-band response. The minimum pass-band attenuation (A_{min}) defines the minimum signal attenuation within the stop band. The steepness of a filter is defined as the order (M) of the filter, and this parameter is also the number of poles of the filter transfer function. All filters may not have all these features. For example, an all-pole configuration will have no ripple in the stop band. A filter affects the phase of a signal. For example, a single-pole filter will have a 90° phase shift at the crossover frequency, whereas a pole pair will create a phase shift of 180° at the crossover frequency. The quality factor (Q) of the filter determines the rate of change of the phase. Table 4.6 shows standard second-order filter responses and pole locations. The phase response of filters is shown in Figure 4.25.

Even though digital filters are very predictable, precise, simulatable, flexible, and accurate, they have one unavoidable shortcoming. They are unable to answer the Nyquist criteria's need for strict bandwidth limiting before signal processing any sampled data. It is always necessary to precede digital signal processing with an analog low-pass filter, as a mandatory line of defense against aliasing due to an input signal bandwidth that is greater than one-half the sampling rate. Another disadvantage of a digital filter is its relative cost due to system resources. For these reasons, analog filters, particularly the active filters, play a significant preprocessing role in most designs.

4.5.1 Active and Passive Filters

Filters can be implemented using passive components. They use no gain elements and require no power supply. They usually generate only the thermal noise from their resistive components. However, for instrumentation type applications where lower frequencies are involved, passive filters require large C and L values and inductor losses become appreciable. For this reason, active filters using op amps are quite popular in many instrumentation and similar systems where frequencies involved are within a few hundred kHz or less.

There are many active filter configurations; some of the common ones are Butterworth (or maximally flat), Chebyshev (or equal ripple), Bessel, and Cauer (or elliptical). In the Butterworth type, response is nearly flat in the pass band and rolls off smoothly and monotonically; it has no ripple in either the pass band or stop band. This type is considered the best compromise between attenuation and phase response. In contrast, the Chebyshev filter has ripple in the pass band, although one can design this ripple to be as small as possible, but at a cost. It has faster roll-off near the cutoff value for the same number of poles than a Butterworth filter. The transient performance of the Chebyshev filter is also inferior. The Bessel or Thompson filter has good linear phase response in the pass band and thus appears as a low-pass delay line. Although the pass band and roll-off region characteristics are smooth

TABLE 4.6

Standard Second-Order Filter Responses, Pole Locations, and Transfer Functions

Filter Type	Magnitude Response	Transfer Function	Pole Location
Low pass		$$\dfrac{\omega_0^2}{s^2+\dfrac{\omega_0}{Q}s+\omega_0^2}$$	
Band pass		$$\dfrac{\dfrac{\omega_0}{Q}s}{s^2+\dfrac{\omega_0}{Q}s+\omega_0^2}$$	
Notch (band reject)		$$\dfrac{s^2+\omega_z^2}{s^2+\dfrac{\omega_0}{Q}s+\omega_0^2}$$	
High pass		$$\dfrac{s^2}{s^2+\dfrac{\omega_0}{Q}s+\omega_0^2}$$	
All pass		$$\dfrac{s^2-\dfrac{\omega_0}{Q}s+\omega_0^2}{s^2+\dfrac{\omega_0}{Q}s+\omega_0^2}$$	

(similar to the Butterworth filter), the rate of roll-off is slower. The elliptical filter has ripple in both the pass and stop bands, nonlinear phase response, and the fastest roll-off from pass band to stop band.

Figure 4.26 shows the Butterworth low-pass amplitude and phase response, and Table 4.7 shows its polynomial coefficients. The characteristics of the filter are defined by Equations 4.20a and 4.20b, where f_c is the cutoff frequency.

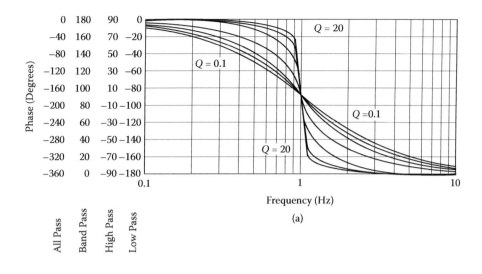

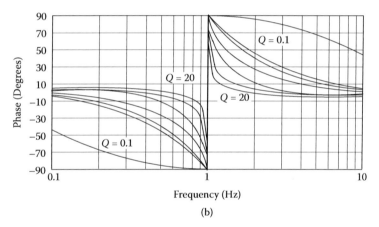

FIGURE 4.25

Phase response of filters: (a) low pass, high pass, band pass, and all pass; (b) notch filter. (Courtesy of Analog Devices, Norwood, MA; Source: Jung [3].)

Similar cases for Butterworth high-pass and other types of filters in general can be found in Garrett [1].

$$A(f) = \frac{b_0}{\sqrt{B(s)B(-s)}} = \frac{1}{\sqrt{1 + \left(\dfrac{f}{f_c}\right)^{2n}}} \tag{4.20a}$$

$$B(s) = \left[j\frac{f}{f_c} \right]^n + b_{n-1} \left[j\frac{f}{f_c} \right]^{n-1} + \dots b_0 \tag{4.20b}$$

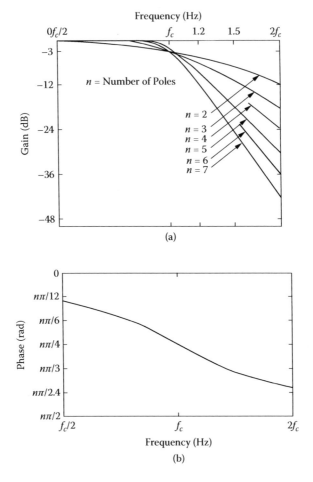

FIGURE 4.26
Butterworth low-pass response: (a) amplitude; (b) phase. (Garret [1].)

TABLE 4.7

Butterworth Polynomial Coefficients

Poles, n	b_0	b_1	b_2	b_3	b_4	b_5
1	1.0					
2	1.0	1.414				
3	1.0	2.0	2.0			
4	1.0	2.613	3.414	2.613		
5	1.0	3.236	5.236	5.236	3.236	
6	1.0	3.864	7.464	9.141	7.464	3.864

Source: Garrett, P., *High Performance Instrumentation and Automation*, CRC Press, Boca Raton, FL, 2005.

There are many active filter topologies based on op amp stages. The theory is well established, and there are many good textbooks and application notes on the subject [10,15–17]. One can achieve different transfer functions with different topologies. An important practical design criterion is the filter sensitivity to changes in op amp or other passive component parameters. The sensitivity function is mathematically defined as

$$S_x^y = \lim_{\Delta x \to 0} \frac{\Delta y / y}{\Delta x / x}. \tag{4.20c}$$

This allows us to use the approximate interpretation as

$$S_x^y \approx \frac{\Delta y / y}{\Delta x / x}. \tag{4.20d}$$

Only a limited number of topologies are least sensitive to component drift. Of these, unity gain and multiple feedback networks are of particular value in implementing low-pass and band-pass filters, respectively, to Q values up to 10. Another one is the low-sensitivity bi-quad resonator, which can provide stable Q values up to 200. Figure 4.27 indicates these examples. More detailed discussion can be found in Garrett [1], and design guidelines can be found in Jung [3].

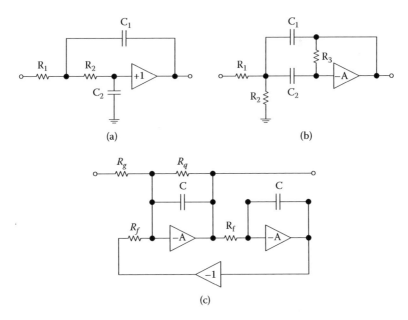

(a) (b) (c)

FIGURE 4.27
A few useful active filter networks: (a) unity gain (Butterworth); (b) multiple feedback; (c) bi-quad.

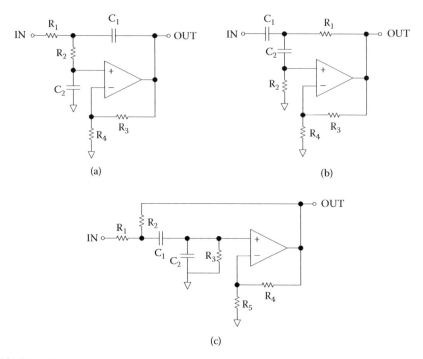

FIGURE 4.28
Sallen-Key implementations using a single op amp: (a) low pass; (b) high pass; (c) band pass.

The Sallen-Key or voltage-controlled voltage source provides good phase response and relative independence from op amp specifications, with the weakness that its frequency response and Q are sensitive to the gain setting. Figure 4.28 shows the Sallen-Key implementations using single op amps.

The multiple feedback design is more sensitive to op amp specifications but less sensitive to passive component values with the difficulty of obtaining a high Q performance. The state variable filter provides independent control over gain, cutoff frequency, Q, and other parameters at the cost of more passive parts. As another advantage, the topology yields low-pass, high-pass, or band-pass response. Another useful topology is the frequency-dependent negative resistance, or impedance converter, which removes the op amps from the signal path. Although it requires more parts, it is relatively insensitive to variations in their values [17]. Table 4.8 provides a summary of op amp parameter sensitivities for the above filter topologies.

The latest advances in analog ICs simplify the filter implementation and provide required performance. For example, the MAX 29X series provides Butterworth response with cutoff frequencies adjustable up to 25 kHz. Figure 4.29 shows the basic topology of the state variable filter and a circuit implementation suggestion from Analog Devices based on digitally variable resistors that allows the designer to change the filter parameter under processor control. There is much useful active filter design software available from various sources [18,19].

TABLE 4.8

Filter Sensitivities to Real Op Amp Characteristics

Topology	Characteristics
Sallen-Key	Uses op amp as a gain block.CFAs can be used.
	Least dependent on op amp frequency response.
	Needs an op amp with flat response slightly beyond the filter stop-band frequency.
Multiple feedback	Integrators need 20-dB minimum loop gain (open loop gain 10 times the closed loop gain) to avoid Q enhancement.
State variable	Uses op amp as amplifiers and integrators.
	Needs an op amp with flat response slightly beyond the filter stop-band frequency.
	Integrators need 20-dB minimum loop gain (open loop gain 10 times the closed loop gain) to avoid Q enhancement.
Frequency-dependent negative resistance	Loop gain should be at least 20 dB at resonant frequency.
	Use dual devices for matched performance in each leg.
	Use FET op amps for low-bias currents.

A useful approach to designing filters is the use of a general filter transfer function and a spreadsheet for calculations [20]. In this kind of approach, one can begin with the general second-order universal filter transfer function,

$$H(s) = \frac{h_{HP}\left(\dfrac{s}{2\pi f_0}\right)^2 + h_{BP}\left(\dfrac{s}{2\pi f_0}\right) + h_{LP}}{\left(\dfrac{s}{2\pi f_0}\right)^2 + d\left(\dfrac{s}{2\pi f_0}\right) + 1}. \tag{4.21}$$

The transfer function's parameters—f_0, d, h_{HP}, h_{BP}, h_{LP}—allow for construction of all filter types. The roll-off frequency, f_0, is the frequency at which the s term begins to dominate. Designers consider frequencies below this value as low and above as high. At about this frequency, the level is considered in band. Damping, d, is a measure of how a filter changes from lower frequencies to higher frequencies. It is an index of the filter's tendency to oscillate. Practical values range from 0 to 2 (where $d < 0$ is unstable; $d = 0$, oscillator; $d = 1.414$, critically damped; $d = 2$, fully damped; and $d > 2$, excessively damped). The high-pass coefficient, h_{HP}, is the coefficient in the numerator that dominates for frequencies greater than the roll-off frequency. Similarly, the band-pass coefficient, h_{BP}, dominates near the roll-off frequency, and the low-pass coefficient, h_{LP}, of the numerator dominates at frequencies lower than f_0. Using this to generate a low-pass filter, one can set the h_{HP} and h_{BP} to zero and obtain the transfer function as

$$H(s)_{LP} = \frac{h_{LP}}{\left(\dfrac{s}{2\pi f_0}\right)^2 + d\left(\dfrac{s}{2\pi f_0}\right) + 1}. \tag{4.22}$$

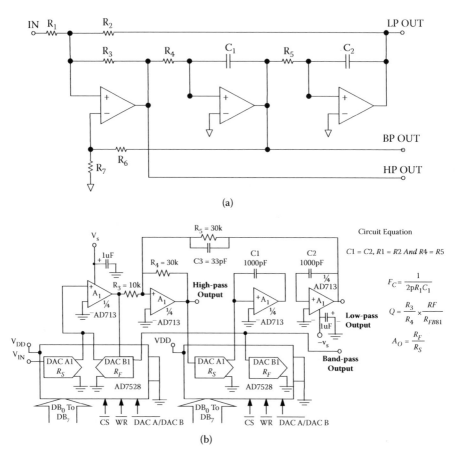

(a)

(b)

FIGURE 4.29

State variable filter and a digital implementation for variable parameters: (a) basic topology; (b) programmable configuration using digitally variable resistors. (Reprinted by permission of *EDN Magazine* © 1997, Reed Business Information, a division of Reed Elsevier, Inc. All rights reserved.)

A useful spreadsheet (FilterPLot.xls) that demonstrates the use of this approach is provided by Ess [20]. Figure 4.30 shows the cases of a low-pass and a band-reject version derived from a spreadsheet. For more details, see Ess [20]. For a practically useful discussion on filter theory, see Sedra and Smith [2].

4.5.2 Switched Capacitor Filters

Active RC filter circuits need large-value capacitors with accurate RC time constants. This makes their production in monolithic form almost impossible. Switched capacitor filters are an alternative approach to RC filters and resistor-less systems, allowing monolithic implementation. The switched capacitor filter technique is based on the realization that a capacitor switched between two circuit nodes at a sufficiently high rate is equivalent

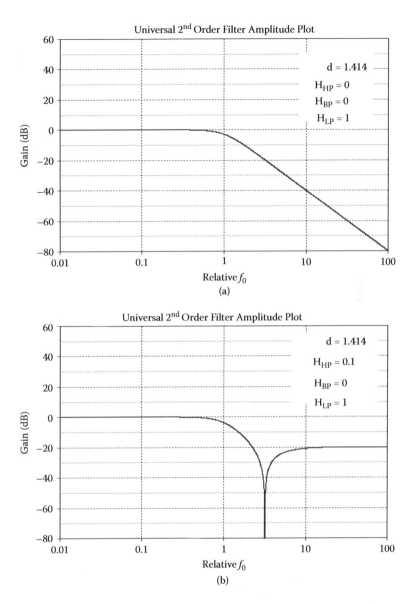

FIGURE 4.30
Low-pass and band-reject filters using a filter plot Excel spreadsheet: (a) low-pass filter; (b) band-reject filter. (Source: ESS [21].) (Courtesy of EDN; [21].)

to a resistor connecting between these two nodes. Figure 4.31 illustrates the principle where a simple integrator and the equivalent switched capacitor integrator are compared. Two MOS switches are driven by a nonoverlapping two-phase clock, as in Figure 4.31c. This situation creates the charge–discharge configurations of Figure 4.31d, and the overall circuit performs the

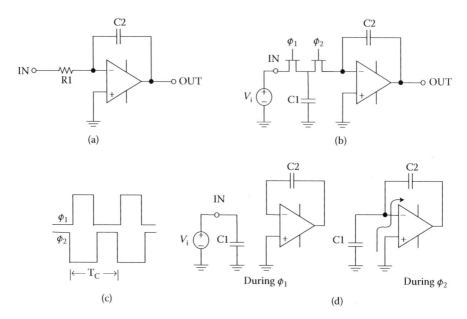

FIGURE 4.31
Switched capacitor implementation: (a) simple integrator; (b) its switched capacitor filter implementation to eliminate the resistor; (c) two-phase nonoverlapping clock used for φ_1 and φ_2; (d) charge and discharge process during clock phases.

same function as the RC integrator, provided the clock frequency, f_c, is much higher than the frequency of the signal being filtered.

For the process in Figure 4.31, the average current flowing between the input node and the virtual ground node is given by

$$i_{av} = \frac{C_1 v_i}{T_c}.$$ (4.23a)

Thus,

$$R_{eq} \equiv \frac{v_i}{i_{av}} = \frac{T_c}{C_1}.$$ (4.23b)

This provides an equivalent time constant (τ) for the switched capacitor circuit as

$$\tau = C_2 R_{eq} = T_C \frac{C_2}{C_1}.$$ (4.23c)

This simple concept of high-frequency switching of the capacitors can be easily used in either the discrete form or the IC process for building all the types of filters discussed in the previous section. Although switched capacitor filters are discrete time filters, they are not digital filters. For this reason, passive and active filters are sometimes called continuous time filters.

Interest in switched capacitor filters is due to their useful attributes:

- The filter's performance is decided by capacitor ratios rather than absolute values (this fits well with IC processes).
- Higher-frequency implementations are easy (100- to 250-kHz implementations are quite easy).
- These can be combined with other CMOS process–based IC implementations for audio and video requirements.
- The filters are less sensitive to temperature changes.
- The cutoff frequency can be adjusted by changing the clock frequency.
- There is excellent low-frequency response without size and cost penalties of large reactive parts.

However, switched capacitor filters have the following disadvantages:

- Poorer noise performance than active and passive filters (limits the dynamic range of the signals)
- Clock feed-through due to the sampled data technique
- Limits high-frequency operation

Figure 4.32a illustrates the possibility of using a commercial switched capacitor IC such as the LMF100 from National Semiconductor [15] to implement a universal state variable filter with very few external components. In this situation, the LMF100 IC implements the two integrators required in a state variable filter (compare Figure 4.32a with Figure 4.29a), where the capacitor required in Figure 4.29a is now absorbed by the internal circuits of LMF100 (as per Figure 3.42b) by second-order switched capacitor filter circuits, with the possibility of the filter cutoff frequencies' being adjusted by the clock signal.

Following is a guide to the procedure for designing a low-pass filter to achieve the example in Figure 4.32c. The essential characteristics (related to Figure 4.24) required are given below:

- Cutoff frequency (f_c): 2.5 kHz
- Pass-band ripple (A_{max}): 3 dB
- Stop frequency (f_s): 6.25 kHz
- Stop-band attenuation (A_{min}): 30 dB
- Pass-band characteristics: maximally flat
- Stop-band characteristics: monotonic

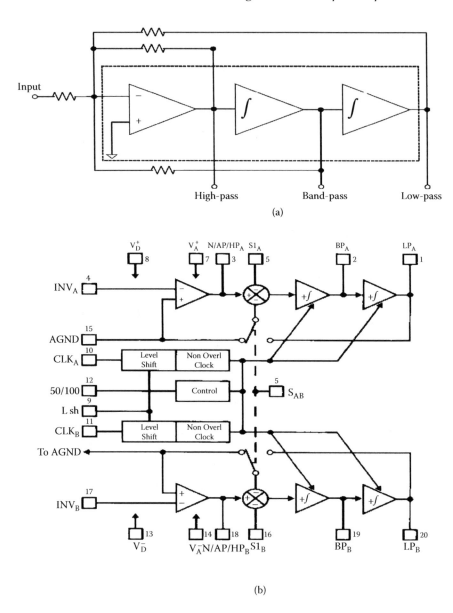

(a)

(b)

FIGURE 4.32
Practical implementation of a state variable filter using a commercial switched capacitor filter
IC: (a) simplified version of a universal state variable filter; (b) block diagram of National Semiconductor's LMF100; (c) Butterworth filter order nomograph; (d) implementation of a fourth-
order Butterworth filter. (Adapted from [21].)

The above requirements suggest that the designer should consider a Butter-
worth-type filter. Use of a Butterworth filter order nomograph (Figure 4.32c)
indicates that the filter order (n) should be 4. As a filter whose order is greater
than 2 can be built by cascading second-order filters, the option is to use two

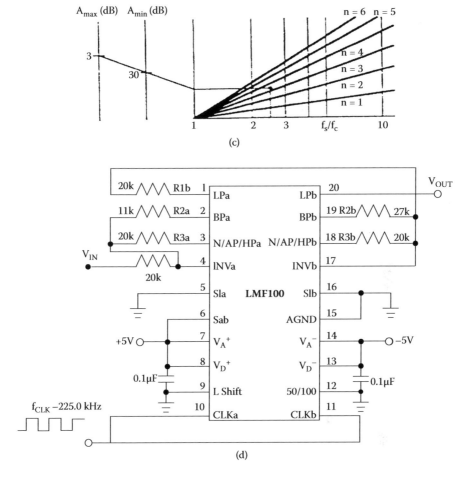

FIGURE 4.32 (continued)

cascaded stages. Then, using common filter design procedures, we end up with the circuit in Figure 4.32d. Details of this procedure are discussed in Hoskins [21].

A designer should be aware of the special performance issues in switched capacitor filter implementation, considering them as sampled data systems. (A more detailed discussion of sampled data systems is provided in Chapter 5.) Figure 4.33a illustrates the basic frequency domain response for both a theoretical eighth-order elliptic filter and a sampled data system based on a clock frequency of f_{CLK}. The sampling process in a switched capacitor filter effectively multiplies the elliptic filter response by the sampled data response. The final result is the situation in Figure 4.33b, where some image frequencies are added to the basic frequency domain response of the elliptic filter. The two image components in Figure 4.33b are due to the sampling process (mixing of f_{CLK} and f_{in}, which generates components $f_{CLK} - f_{in}$ and

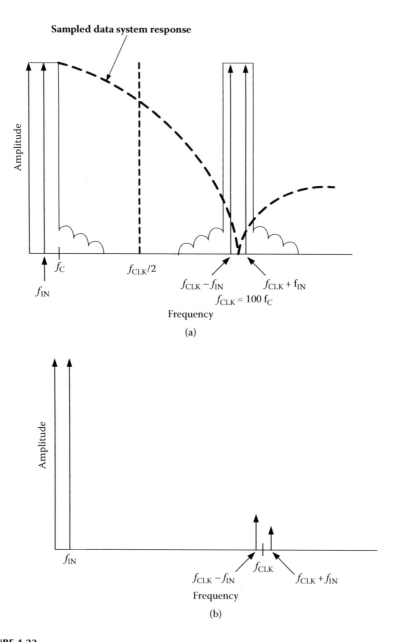

FIGURE 4.33

Frequency spectrum of a switched capacitor filter: (a) theoretical sampled data system response and elliptic filter response; (b) overall response with image components due to sampling process occurring at multiples of the clock frequency.

$f_{CLK} + f_{in}$). This discussion leads us to be careful with the effects of aliasing [22,23], which makes the spectrum fold around one-half the sampling frequency.

4.6 Switching and Multiplexing of Signals

In most designs, to save hardware resources we attempt to switch and multiplex signals toward an ADC or a processor. In this process, analog switches and multiplexers are used as common building blocks. Most analog switches are based on CMOS technology, and nonidealities of these devices can degrade the overall specifications of a design. An ideal analog switch has the following characteristics:

- Zero on resistance
- Off impedance is infinite at all frequencies
- Zero switching time
- Zero leakage switching
- Zero power dissipation
- Infinite mean time between failures

Figure 4.34a illustrates a typical n-channel multiplexer with switch S_1 turned on and switches S_2 to S_n turned off. The on resistance (R_{on}) of S_1 forms a voltage divider with R_S and R_{in}, thereby producing an input-to-output error. The sum of the off-state leakage currents, I_L, also induces an error voltage in the output load resistance. Here the load resistance (R_L) is the parallel combination of R_{in} and $R_S + R_{on}$. There is a tradeoff in the error budget for the circuit in Figure 4.34a. If you make R_{in} too high to reduce the error accruing from R_{on}, then the error from the sum of the leakage currents could be unacceptably high. On the other hand, if R_{in} is made too low to minimize the leakage-induced error, then the voltage divider due to the tree resistors could produce large errors. If the source impedance R_S is very low, the leakage term in the error budget is negligible, and we can choose an arbitrarily high value for R_{in}. If the source is a high-impedance transducer, the leakage effects can be objectionable. In that case, one could make a tradeoff by equalizing the on- and off-state errors. With reasonable approximations, for equal on- and off-state errors, it can be shown that [24]

$$R_L \cong \sqrt{V_{IN} \frac{R_{ON}}{I_L}}. \tag{4.24}$$

Figure 4.34b gives the curves of the load resistances that equalize the on- and off-state errors as a function of R_{on} and I_L. The foregoing discussion is equally valid for analog switches as well. For more details, see Tavis [24] and Connor [25].

4.6.1 CMOS Analog Switches

Figure 4.35 shows a basic CMOS switch and the R_{on} versus signal voltage for the p-channel MOSFET (PMOS), n-channel MOSFET (NMOS), and CMOS

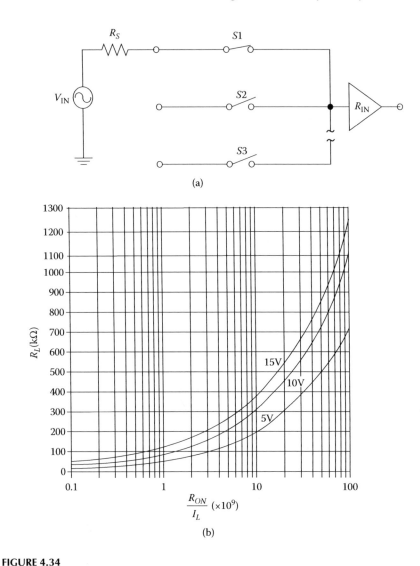

FIGURE 4.34
Typical n-channel multiplexer and its conditions required for equal on- and off-state errors: (a) equivalent circuit; (b) relationship of load resistance to R_{on} value. (Courtesy of *EDN*; Source: Connor [25].) (Reprinted by permission of *EDN Magazine* © 1990, Reed Business Information, a division of Reed Elsevier, Inc. All rights reserved.)

elements. Figure 4.35b shows the nonlinear nature of the R_{on} value for the PMOS and NMOS elements. This nonlinear property can cause errors in DC accuracy as well as AC distortion. The bilateral CMOS switch in Figure 4.35a minimizes the variation. Parasitic latch-up is another secondary characteristic of CMOS analog switches, and this can cause destructive situations if the voltages applied at the switch terminals are above the positive rail value or below the negative rail [26]. One way to protect against this possibility

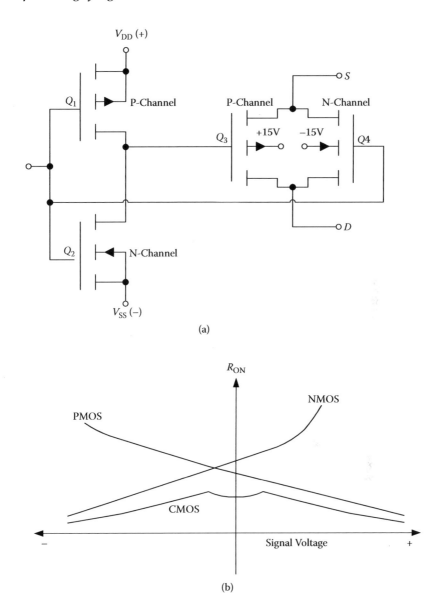

FIGURE 4.35
CMOS switch: (a) basic CMOS analog switch; (b) resistance of NMOS, PMOS, and CMOS elements. (Courtesy of Analog Devices, Inc., [26].)

is to use diodes in series with the power rails, as in Figure 4.36. Latch-up protection is different from overvoltage protection, and this situation must be tackled by other means.

Another important design criterion may be to protect the switch stages against possible overcurrent. If the switch drives a high-impedance load, a series resistor can protect the situation.

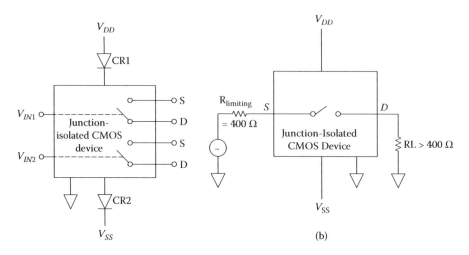

FIGURE 4.36
Protection against possible CMOS latch-up problem and overcurrents: (a) for latch-up; (b) for overcurrent. (Courtesy of Analog Devices, Inc. [26].)

4.6.2   Error Sources of an Analog Switch

It is important to understand the total error sources of analog switches. Figure 4.37a shows the equivalent circuit of two adjacent analog switches, where junction capacitances and leakage currents are shown. DC performance is affected mainly by the R_{on} value and leakage. Figure 4.37b illustrates the case of the on condition; the off condition is shown in Fig 4.37c.

When the switch is on,

$$V_{OUT} = V_{IN}\left[\frac{R_L}{R_S + R_{ON} + R_L}\right] + I_{LKG}\left[\frac{R_L(R_{ON} + R_S)}{R_S + R_{ON} + R_L}\right]. \qquad (4.25a)$$

If R_S is very small,

$$V_{OUT} = V_{IN}\left[\frac{R_L}{R_{ON} + R_L}\right] + I_{LKG}\left[\frac{R_L(R_{ON})}{R_{ON} + R_L}\right]. \qquad (4.25b)$$

When the switch is off,

$$V_{out} = I_{LKG}R_L. \qquad (4.25c)$$

If an AC signal is transferred via the CMOS switch, parasitic components in Figure 4.37a affect the AC performance, further degraded by the external

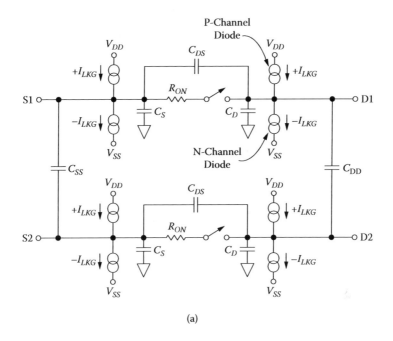

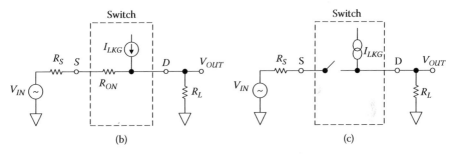

FIGURE 4.37
Equivalent circuits of the CMOS analog switch: (a) general; (b) DC performance under on condition; (c) DC performance under off condition. (Courtesy of Analog Devices, Norwood, MA.)

capacitances. Figure 4.38a and Figure 4.38b illustrate the equivalent circuit and the corresponding Bode plot for the on condition. Figure 4.38c and Figure 4.38d illustrate the same for the off condition.

Transfer functions and DC gain under the on condition are given by

$$A(s)_{ON} = \left[\frac{R_L}{R_L + R_{ON}}\right]\left[\frac{(1 + sR_{ON}C_{DS})}{1 + s(C_L + C_D + C_{DS})\left(\frac{R_L R_{ON}}{R_L + R_{ON}}\right)}\right] \qquad (4.26a)$$

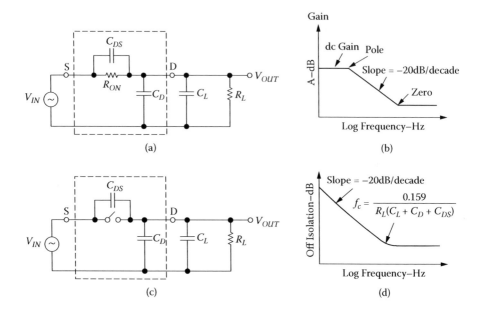

FIGURE 4.38
The AC performance of an analog switch: (a) on-equivalent circuit; (b) Bode plot under on condition; (c) off-equivalent circuit; (d) Bode plot for off condition. (Courtesy of Analog Devices.)

$$\text{DC gain} = \left[\frac{R_L}{R_L + R_{ON}} \right]; \quad f_{zero} = \frac{0.159}{R_{on}C_{DS}};$$

$$f_{pole} = \frac{0.159}{(C_L + C_D + C_{DS}) \left(\dfrac{R_L R_{ON}}{R_L + R_{ON}} \right)} \tag{4.26b}$$

Under off conditions,

$$A(s)_{OFF} = \frac{s(R_L C_{DS})}{1 + sR_L (C_L + C_D + C_{DS})}. \tag{4.27}$$

Typical off isolation versus frequency for a practical CMOS switch (e.g., ADG51X from Analog Devices) is shown in Figure 4.39. Crosstalk versus frequency has a similar behavior. More discussion on the applications aspects and dynamic performance of analog devices is provided by Kester [26].

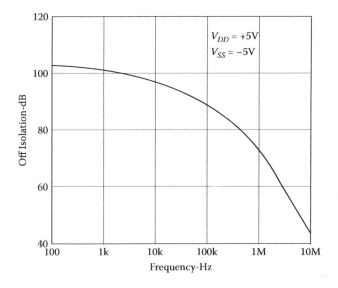

FIGURE 4.39
Off isolation and crosstalk in a typical analog switch. (Courtesy of Analog Devices, Norwood, MA.)

4.7 Signal Isolation

Isolation is an important requirement in many data acquisition and signal-processing applications. In a practical sense, isolation allows two sides of a signal-processing system to be electrically separated so that there are no currents possible across the two ground terminals or ground planes. For example, in a simple case such as a transformer with two windings, if the ends of each winding are shorted and then we apply a voltage source across the two nodes (shorted coils), there will be no current flow. However, if the voltage of the applied source is increased to a very high value, there may be a small current possible due to parasitic resistances across the windings. If the source is an AC one, at a higher frequency or higher voltage than the normal source voltagev, there will be some AC leakage current possible due to stray capacitances across the windings. This simple example explains the "isolation" across two sides of a circuit or a system for electrical signals.

A key industrial function that requires isolation is intrinsic safety. In intrinsically safe applications, the devices and wiring in the hazardous area must be incapable of leaking (or incapable of releasing enough electrical energy) under normal or fault conditions. One classic example is an electrocardiograph (ECG) system in medical applications, where both the patient and the ECG machine need protection against electrical shock and high voltages.

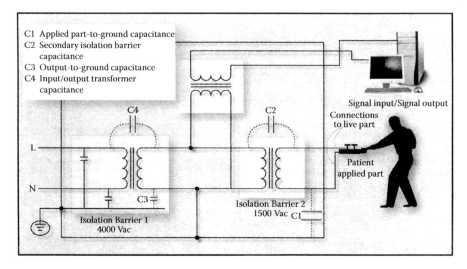

C1 Applied part-to-ground capacitance
C2 Secondary isolation barrier
    capacitance
C3 Output-to-ground capacitance
C4 Input/output transformer
    capacitance

FIGURE 4.40
Isolation barriers required in medical environments. (Courtesy of *Power Electronics Technology*; Blyth [29].)

Figure 4.40 illustrates the scenario and the required isolation barriers in a medical environment. In general, isolation is an important feature for any applications where common mode potentials pose a threat to the integrity of data.

A few examples where isolation amplifiers need to be used are

- Avoiding a case where the sensor is at high potential relative to other processing circuitry, or developing a faulty condition on the sensor side that creates a high voltage toward other circuitry
- Where the sensor is not expected to carry dangerous voltages irrespective of faults of other circuitry (examples are patient environments or explosive environments)
- Where ground loops need to be broken

In a practical product environment, to achieve isolation it is necessary to isolate signal paths as well as power supply paths simultaneously.

Isolation can be achieved using three basic means: (1) across a magnetic coupling, (2) across an optical coupling, and (3) across an electric field. The first example is a transformer where analog accuracy of 12 to 16 bits and bandwidths up to several 100 kHz are possible. However, the maximum isolation rating may be less than 10 kV. Optical isolators are fast and inexpensive, with high isolation ratings (4 to 7 kV), but they have poor linearity, particularly for large signals. They are not useful for direct coupling of precision signals. Capacitively coupled isolation amplifiers have low accuracy, less than about 12 bits, and they have low bandwidths and lower voltage ratings in general. In selecting an isolation stage, isolation voltage and linearity are not the only important considerations. Power is also a prime consideration, as both input and output must be powered without disturbing the strategy of galvanic isolation.

For applications that require both power and signal isolation, transformer coupled isolation amplifiers, such as the AD204 from Analog Devices, can be used. Figure 4.41a illustrates a typical example where a temperature sensor (AD22100A) output is sent via a transformer secondary. AD204 is the element that provides power and signal transfer via an isolation barrier. ADM663 is a voltage regulator that supplies power to the temperature sensor.

Figure 4.41b illustrates a case where a transducer signal is sent to a voltage-to-frequency converter (VFC) and then transferred for further processing via an optoisolator. The VFC is used here because the linearity issue of a photo

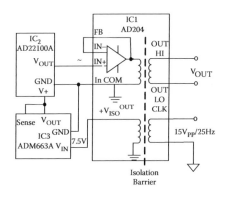

(a)

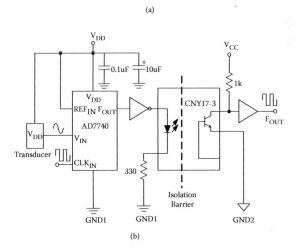

(b)

FIGURE 4.41
Examples of different isolation techniques: (a) transformer based; (b) optoisolator voltage to frequency converter based; (c) optoisolator with positive feedback for better frequency response; (d) digital isolator. (Reprinted by permission of *EDN Magazine*, © 2001, Reed Business Information, a division of Reed Elsevier, Inc. All rights reserved.)

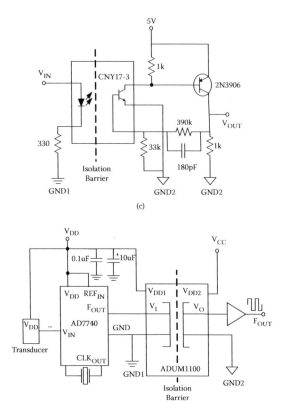

(c)

(d)

FIGURE 4.41 (continued)

diode can be indirectly overcome by measuring the frequency at the output side. The frequency range of optocouplers is generally limited to a few tens of kilohertz. A few techniques, such as differential line drivers and positive feedback to the base of the output optotransistor, can be used to improve the frequency response. Figure 4.41c illustrates the case of differential line driving after a VFC to increase the frequency capability of the circuit from about 5 to 32 kHz. For more details on these techniques, see Stapleton and O'Grady [27] and Kester [28].

References

1. Garrett, P., *High Performance Instrumentation and Automation*, CRC Press, Boca Raton, FL, 2005.
2. Sedra, A.S. and Smith, K.C., *Microelectronic Circuits*, 5th ed., Oxford University Press, New York, 2004.

3. Jung, W.G., ed., *Op Amp Applications*, Analog Devices, Norwood, MA, 2002.
4. Mancini, R., Op amps for everyone, Texas Instruments, Dallas, TX, 2002.
5. Brokaw, P., An IC amplifier user's guide to decoupling, grounding and making things go right for a change, Application Note AN202, Analog Devices, Norwood, MA, 1990.
6. Kitchin, C. and Counts, L., The right way to use instrumentation amplifiers, *EDN*, September 15, 69, 2005.
7. Kitchin, C. and Counts, L., *A Designer's Guide to Instrumentation Amplifiers*, 3rd ed., Analog Devices, Norwood, MA, 2006.
8. *Linear Design Seminar*, Analog Devices, Norwood, MA, 1987.
9. Sheingold, D.H., *Nonlinear Circuits Handbook*, Analog Devices, Norwood, MA, 1976.
10. *Linear Design Seminar*, Analog Devices, Norwood, MA, 1995.
11. Kularatna, N., *Modern Component Families and Circuit Block Design*, Newnes, London, 2000.
12. Israelsohn, J., Gain control, *EDN*, August 8, 38, 2002.
13. Kitchin, C. and Counts, L., *RMS to DC Conversion Application Guide*, 2nd ed., Analog Devices, Norwood, MA, 1986.
14. DiSanto, G., Proper PC board layout improves dynamic range, *EDN*, September 11, 93, 2004.
15. Lacanette, K., A basic introduction to filters—active, passive and switched capacitor, Application Note 779, National Semiconductor, Santa Clara, CA, 1991.
16. Ghausi, M.S. and Laker, K.R., *Modern Filter Design: Active RC and Switched Capacitors*, Prentice Hall, Upper Saddle River, NJ, 1981.
17. Lancaster, D., *Active Filter Cookbook*, Newnes, London, 1996.
18. Schweber, B., Analog filters: even more essential in the digitized world, *EDN*, April 24, 43, 1997.
19. Swager, A.W., Design software links active-filter performance with real devices, *EDN*, April 9, 45, 1992.
20. Ess, D.V., Filters in a nutshell: spreadsheet promotes intuitive feel, *EDN*, June 9, 79, 2005; available at http://www.edn.com/article/ca605510.html.
21. Hoskins, K., Switched-capacitor filters cut component count and size, *PCIM*, November, 58, 1989.
22. Markell, R., Knowledge of subtleties aids switched capacitor filter design, *EDN*, August 2, 121, 1990.
23. Yager, C., High frequency complex filter design, Application Note 4, Microlinear Corporation, San Jose, CA, 1988.
24. Tavis, B., Take account of errors in designs using analog switches and multiplexers, *EDN*, January 4, 61, 1996.
25. Conner, D., Analog switches and multiplexers, *EDN*, March 15, 131, 1990.
26. Kester, W., ed., *System Application Guide*, Analog Devices, Norwood, MA, 1993.
27. Stapleton, H. and O'Grady, A., Isolation techniques for high-resolution data acquisition systems, *EDN*, February 1, 113, 2001.
28. Kester, W., ed., *Practical Design Techniques for Sensor Signal Conditioning*, Analog Devices, Norwood, MA, 1999.
29. Blyth, P., Converters address medical equipment compliance, *Power Electronics Technology*, March, 38, 2006.
30. Williams, J., Good bridge-circuits design satisfies gain and balance criteria, *EDN*, October 25, 161, 1990.

5

Data Converters

CONTENTS

5.1 Introduction 278
5.2 Sampled Data Systems 279
    5.2.1 Discrete Time Sampling of Analog Signals 280
    5.2.2 Implications of Aliasing 281
    5.2.3 High-Speed Sampling 282
    5.2.4 Baseband Antialiasing Filters 283
    5.2.5 Undersampling 286
5.3 ADC Resolution and Dynamic Range Requirement 287
    5.3.1 Effective Number of Bits of an ADC 288
    5.3.2 Analog Bandwidth, Spurious Components, and
           Harmonics 289
           5.3.2.1 Analog Bandwidth 289
           5.3.2.2 Spurious-Free Dynamic Range 290
5.4 ADC Errors 292
    5.4.1 Differential Nonlinearity 295
    5.4.2 Integral Nonlinearity 297
    5.4.3 Gain Error and Offset Error 297
    5.4.4 Testing of ADCs 298
5.5 Effects of Sample-and-Hold Circuits 298
5.6 Basic SHA Operation 299
    5.6.1 Track Mode Specifications 299
    5.6.2 Track-to-Hold Mode Specifications 299
    5.6.3 Aperture and Aperture Time 301
    5.6.4 Hold Mode Droop 304
    5.6.5 Dielectric Absorption 304
    5.6.6 Hold-to-Track Transition Specification 305
    5.6.7 SHA Architectures 305
5.7 ADC Architectures 306
    5.7.1 Successive Approximation ADCs 306
    5.7.2 Flash Converters 308
    5.7.3 Integrating ADCs 309
    5.7.4 Subranging ADCs (Half-Flash ADCs) 311
    5.7.5 Two-Step Architectures 313

    5.7.6   Sigma-Delta Converters 315
            5.7.6.1   Key Concepts behind Σ-Δ ADCs................................. 315
            5.7.6.2   Block Diagram of a Σ-Δ ADC................................... 318
    5.7.7   Self-Calibration Techniques 322
    5.7.8   Figure of Merit for ADCs... 322
5.8   DACs 322
    5.8.1   General Considerations.. 323
    5.8.2   Performance Parameters and Data Sheet Terminology 324
5.9   Principles Used in DACs.. 329
    5.9.1   Voltage Division 329
    5.9.2   Current Division 329
    5.9.3   Charge Division 329
5.10 DAC Architectures.. 333
    5.10.1 Resistor Ladder DAC Architectures.................................... 333
    5.10.2 Ladder Architecture with Switched Subdividers 333
    5.10.3 Intermeshed Ladder Architecture...................................... 335
    5.10.4 Current-Steering Architecture 335
    5.10.5 R-2R Network-Based Architectures 336
    5.10.6 Other Architectures.. 337
5.11 Data Acquisition System Interfaces.. 337
    5.11.1 Signal Source and Acquisition Time................................... 337
    5.11.2 The Amplifier–ADC Interface... 339
References 340

5.1   Introduction

The workhorses in the mixed-signal world, or the link between the digital-processing world and the analog real world, are the analog-to-digital converter (ADC) and digital-to-analog converter (DAC) ICs, which generally are grouped as the data converters. Until about 1988, engineers had to stockpile their most innovative ADC designs because available manufacturing processes simply could not economically implement those designs onto monolithic chips. Prior to 1988, except for the introduction of successive approximation and integrating and flash ADCs, the electronics industry saw no major changes in monolithic ADCs. Since then, manufacturing processes have caught up with the technology and several successful techniques, such as successive approximation, flash, sigma-delta, and integrating types, have been implemented on monolithic chips. These basic architectures and their variations currently come in monolithic form at reasonable prices.

Different types of equipment need different types of ADCs. For example, a digital storage scope needs high digitizing speeds but can sacrifice resolution.

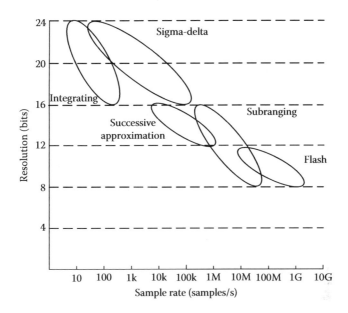

FIGURE 5.1
ADC architectures covering different ranges of resolution and sample rates.

Some PC plug-in digitizers and radio frequency (RF) test equipment use subranging ADCs, which provide better resolution than flash types, but at the expense of speed. General-purpose data acquisition systems generally use successive approximation types or sigma-delta types, whereas digital multimeters use integrating types. In general, a designer has to select the type of converter based on many specifications, as summarized in the rest of the chapter, and design constraints such as cost, packaging, and power consumption. Figure 5.1 provides approximate ranges of resolution and sample speed applicable to different techniques.

5.2 Sampled Data Systems

To specify intelligently the ADC portion of the system, one must first understand the fundamental concepts of sampling and quantization and their effects on the signal. Let's consider the traditional problem of sampling and quantizing a baseband signal whose bandwidth lies between DC and an upper frequency of interest, f_a. This is often referred to as Nyquist or sub-Nyquist sampling. The topic of super-Nyquist sampling (sometimes called undersampling), where the signal of interest falls outside the Nyquist bandwidth (DC to $f_s/2$), is treated later. Figure 5.2a shows the key elements of a sampled data system.

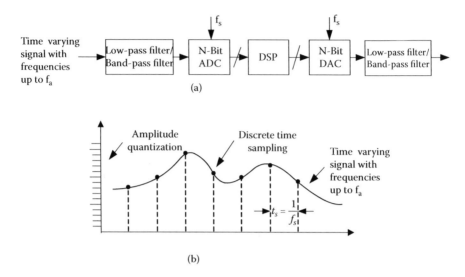

(a)

(b)

FIGURE 5.2
Key elements of a sampled data system and sampling process: (a) block diagram; (b) sampling and quantization.

5.2.1 Discrete Time Sampling of Analog Signals

Figure 5.2b shows the concept of discrete time and amplitude sampling of an analog signal. The continuous analog data must be sampled at discrete intervals, t_s, which must be carefully chosen to ensure an accurate representation of the original analog signal. It is clear that the more samples that are taken (faster sampling rates), the more accurate the digital representation; if too few samples are taken, a point is reached where critical information about the signal is actually lost.

To discuss the problem of losing information in the sampling process, it is necessary to recall Shannon's information theorem and Nyquist's criteria. Shannon's information theorem states:

- An analog signal with a bandwidth of f_a must be sampled at a rate of $f_s > 2f_a$ to avoid the loss of information.
- The signal bandwidth may extend from DC to f_a (baseband sampling) or from f_1 to f_2, where $f_a = f_2 - f_1$ (undersampling, or super-Nyquist sampling).

Nyquist's criteria are

- A signal with a maximum frequency f_a must be sampled at a rate of $f_s > 2f_a$, or information about the signal will be lost because of aliasing.
- Aliasing occurs whenever $f_s < 2f_a$.

- A signal that has frequency components between f_a and f_b must be sampled at a rate $f_s > 2\ (f_b - f_a)$ in order to prevent alias components from overlapping the signal frequencies.

Aliasing is simply the event of unwanted and originally nonexistent signals falling into the bandwidth of interest. Aliasing is sometimes used to advantage in undersampling applications.

5.2.2 Implications of Aliasing

To understand the implications of aliasing in both the time and frequency domains, first consider the case of a time domain representation of the sampled sine wave signal shown in Figure 5.3. In Figure 5.3a and Figure 5.3b,

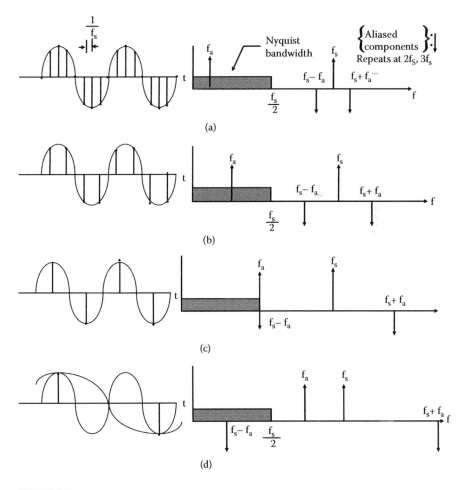

FIGURE 5.3

Time and frequency effects of aliasing : (a) $f_s = 8f_a$; (b) $f_s = 4f_a$; (c) $f_s = 2f_a$; (d) $f_s = 1.3f_a$.

it is clear that an adequate number of samples have been taken to preserve the information about the sine wave. Figure 5.3c represents the ambiguous limiting condition where $f_s = 2f_a$. If the relationship between the sampling points and the sine wave is such that the sine wave is being sampled at precisely the zero crossings (rather than at the peaks, as shown in the illustration), then all information regarding the sine wave would be lost. Figure 5.3d represents the situation where $f_s < 2f_a$, and the information obtained from the samples indicates a sine wave having a frequency lower than $f_s/2$. This is a case where the out-of-band signal is aliased into the Nyquist bandwidth between DC and $f_s/2$. As the sampling rate is further decreased and the analog input frequency f_a approaches the sampling frequency f_s, the aliased signal approaches DC in the frequency spectrum.

Let us look at the corresponding frequency domain representation of each case. From each case of frequency domain representation, we make the important observation that, regardless of where the analog signal being sampled happens to lie in the frequency spectrum, the effects of sampling will cause either the actual signal or an aliased component to fall within the Nyquist bandwidth between DC and $f_s/2$. Therefore, any signals that fall outside the bandwidth of interest, whether they be spurious tones or random noise, must be adequately filtered before sampling. If unfiltered, the sampling process will alias them back within the Nyquist bandwidth, where they will corrupt the wanted signals.

5.2.3 High-Speed Sampling

Now let us discuss the case of high-speed sampling, analyzing it in the frequency domain. First, consider the use of a single-frequency sine wave of frequency f_a sampled at a frequency f_s by an ideal impulse sampler (see Figure 5.4a). Also assume that $f_s > 2f_a$, as shown. The frequency domain output of the sampler shows aliases or images of the original signal around every multiple of f_s, that is, at frequencies equal to

$$\pm K f_s \pm f_a, \text{ where } K = 1, 2, 3, 4, \dots, n. \tag{5.1}$$

The Nyquist bandwidth, by definition, is the frequency spectrum from DC to $f_s/2$. The frequency spectrum is divided into an infinite number of Nyquist zones, each having a width equal to $0.5f_s$, as shown.

Now consider a signal outside the first Nyquist zone, as shown in Figure 5.4b. Notice that even though the signal is outside the first Nyquist zone, its image (or alias), $f_s - f_a$, falls inside. Returning to Figure 5.4a, it is clear that if an unwanted signal appears at any of the image frequencies of f_a, it also will occur at f_a, thereby producing a spurious frequency component in the Nyquist zone. This is similar to the analog mixing process and implies that some filtering ahead of the sampler (or ADC) is required to remove frequency components that are outside the Nyquist bandwidth but whose aliased components fall inside it. The filter performance will depend

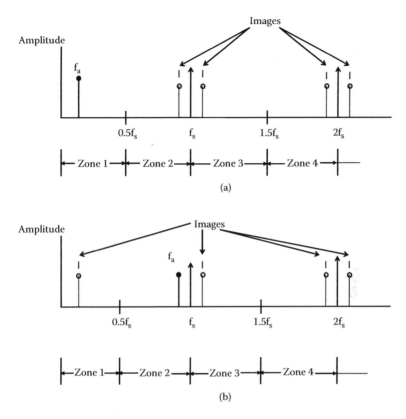

FIGURE 5.4
Analog signal at frequency f_a sampled at f_s: (a) signal lying within $f_s/2$; (b) signal lying between $f_s/2$ and f_s.

on how close the out-of-band signal is to $f_s/2$ and the amount of attenuation required.

5.2.4 Baseband Antialiasing Filters

Baseband sampling implies that the signal to be sampled lies in the first Nyquist zone. It is important to note that with no input filtering at the input of the ideal sampler, any frequency component (either signal or noise) that falls outside the Nyquist bandwidth in any Nyquist zone will be aliased back into the first Nyquist zone. For this reason, an antialiasing filter is used in almost all sampling ADC applications to remove these unwanted signals.

Properly specifying the antialiasing filter is important. The first step is to know the characteristics of the signal being sampled. Assume that the highest frequency of interest is f_a. The antialiasing filter passes signals from DC to f_a while attenuating signals above f_a. Assume that the corner frequency of the filter is chosen to be equal to f_a. The effect of the finite transition from

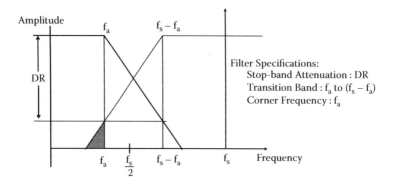

FIGURE 5.5
Effects of antialiasing filters on system dynamic range.

minimum to maximum attenuation on system dynamic range (DR) is illustrated in Figure 5.5.

Assume that the input signal has full-scale components well above the maximum frequency of interest, f_a. The diagram shows how full-scale frequency components above $f_s - f_a$ are aliased back into the bandwidth DC to f_a. These aliased components are indistinguishable from actual signals and therefore limit the DR to the value on the diagram, which is shown as DR.

The antialiasing filter transition band there is determined by the corner frequency f_a, the stop-band frequency $(f_s - f_a)$, and the stop-band attenuation DR. The required system DR is chosen based on the requirement for signal fidelity.

Filters have to become more complex as the transition band becomes sharper, all other things being equal. For instance, a Butterworth filter gives 6-dB attenuation per octave for each filter pole. Achieving 60-dB attenuation in a transition region between 1 and 2 MHz (1 octave) requires a minimum of 10 poles. This is not a trivial filter and definitely a design challenge. Therefore, other filter types generally are better suited to high-speed applications where the requirement is for a sharp transition band and in-band flatness coupled with linear phase response. Elliptic filters meet these criteria and are a popular choice.

From this discussion we can see how the sharpness of the antialiasing transition band can be traded off against the ADC sampling frequency. Choosing a higher sampling rate (oversampling) reduces the requirement on transition band sharpness (hence the filter complexity) at the expense of using a faster ADC and processing data at a faster rate. This is illustrated in Figure 5.6a and Figure 5.6b, which show the effects of increasing the sampling frequency while maintaining the same analog corner frequency, f_a, and the same DR requirement.

Based on this discussion, one could start the design process by selecting a sampling rate of two to four times f_a. Filter specifications could be determined from the required DR based on cost and performance. If such a filter is not realizable, a high sampling rate with a faster ADC will be required.

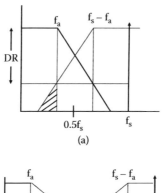

(a)

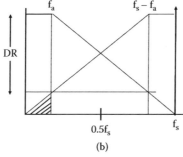

(b)

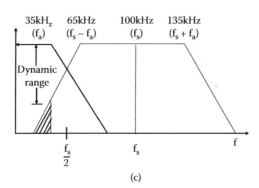

(c)

FIGURE 5.6
The relationship between sampling frequency and the antialiasing filter requirement: (a) low sampling rate with a sharper filter; (b) high sampling rate with a relaxed filter specification; (c) a numerical example.

The antialiasing filter requirements can be relaxed somewhat if it is certain that there never will be a full-scale signal at the stop-band frequency, $f_s - f_a$. In many applications, it is improbable that full-scale signals will occur at this frequency. If the maximum signal at the frequency $f_s - f_a$ will never exceed x dB below full scale, the filter stop-band attenuation requirement is reduced by that amount. The new requirement for stop-band attenuation at $f_s - f_a$ based on this knowledge of the signal now is only DR $- x$ dB. When making this type of assumption, be careful to treat any noise signals that may occur above the maximum signal frequency, f_a, as unwanted signals that also alias back into the signal bandwidth.

Properly specifying the antialiasing filter requires a knowledge of the signal's spectral characteristics as well as the system's DR requirements. Consider the signal in Figure 5.6c, which has a maximum full-scale frequency content of f_a = 35 kHz sampled at a rate of f_s = 100 kS/s. Assume that the signal has the spectrum shown in Figure 5.6c and is attenuated by 30 dB at 65 kHz ($f_s - f_a$). Observe that the system DR is limited to 30 dB at 35 kHz because of the aliased components. If additional DR is required, an antialiasing filter must be provided to provide more attenuation at 65 kHz. If a DR of 74 dB (12 bits) at 35 kHz is desired, then the antialiasing filter attenuation must go from 0 dB at 35 kHz to 44 dB at 65 kHz. This is an attenuation of 44 dB in approximately one octave; therefore, a seven-pole filter is required.

One must consider that broadband noise may be present with the signal, which also can alias within the bandwidth of interest. This is especially true with wideband op amps that provide low distortion levels. For some practical details about antialiasing filter designs, see Garrett [1] and Holdaway [36].

5.2.5　Undersampling

Thus far we have considered the case of baseband sampling; that is, all the signals of interest lie within the first Nyquist zone. Figure 5.7a shows such a

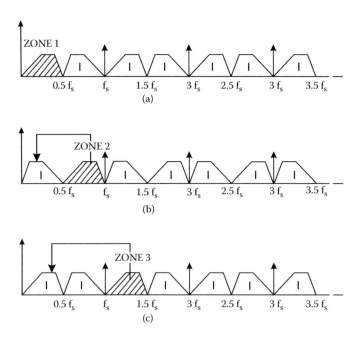

FIGURE 5.7
Undersampling and frequency translation: (a) signal in Nyquist zone 1; (b) signal in Nyquist zone 2; (c) signal in Nyquist zone 3.

case, where the band of sampled signals is limited to the first Nyquist zone and images of the original band of frequencies appear in each of the other Nyquist zones.

Consider the case shown in Figure 5.7b, where the sampled signal band lies entirely within the second Nyquist zone. The process of sampling a signal outside the first Nyquist zone is often referred to as undersampling, harmonic sampling, or band-pass sampling. Note that the image that falls in the first Nyquist zone contains all the information in the original signal, with the exception of its original location. (The order of the frequency components within the spectrum is reversed, but this is easily corrected by reordering the output of the fast Fourier transform [FFT].)

Figure 5.7c shows the sampled signal restricted in the third Nyquist zone. Note that the image that falls in the first Nyquist zone has no frequency reversal. In fact, the sampled signal frequencies may lie in any unique Nyquist zone, and the image falling into the first Nyquist zone is still an accurate representation (with the exception of the frequency reversal that occurs when the signals are located in even Nyquist zones). More details are available in Kester [2], Hill [3], Fonte [4], Analog Devices [5], and Steer [6].

5.3 ADC Resolution and Dynamic Range Requirement

Having discussed the sampling rate and filtering, we next discuss the effects of dividing the signals amplitude into a finite number of discrete quantization levels. Table 5.1 shows relative bit sizes for various-resolution ADCs for a full-scale input range chosen as approximately 2V, which is popular for higher-speed ADCs. The bit size is determined by dividing the full-scale range (2.048 V) by 2^n.

The selection process for determining the ADC resolution should begin by determining the ratio between the largest signal (full-scale) and smallest sig-

TABLE 5.1

Bit Sizes, Quantization Noise, and Signal-to-Noise Ratio (SNR) for 2.048-V Full-Scale Converters

Resolution (n bits)	1 LSB = q	Percent FS	RMS Quantization Noise $q/\sqrt{12}$	Theoretical Full-Scale SNR (dB)
6	32 mV	1.56	9.2 mV	37.9
8	8 mV	0.39	2.3 mV	50.0
10	2 mV	0.098	580 mV	62.0
12	500 µV	0.024	144 µV	74.0
14	125 µV	0.0061	36 µV	86.0
16	31 µV	0.0015	13 µV	98.1

nals one wishes the ADC to detect. Convert this ratio to decibels and divide by 6. This is the minimum ADC resolution requirement for DC signals. More resolution will actually be needed to account for extra signal headroom, because ADCs act as hard limiters at both ends of their range. Remember that this computation is for DC or low-frequency signals and that the ADC performance will degrade as the input signal slew rate increases. The final ADC resolution will actually be dictated by dynamic performance at high frequencies. This may lead to the selection of an ADC with more resolution at DC than is required.

Table 5.1 also indicates the theoretical RMS quantization noise produced by a perfect *n*-bit ADC. In this calculation, the assumption is that quantization error is uncorrelated with the ADC input. With this assumption, the quantization noise appears as random noise spread uniformly over the Nyquist bandwidth, DC to $f_s/2$, and it has an RMS value equal to $q/\sqrt{12}$. Other cases may be different, and some practical explanation is given in the *Linear Design Seminar* [5].

5.3.1   Effective Number of Bits of an ADC

Table 5.1 shows the theoretical full-scale signal-to-noise ratio (SNR) calculated for the perfect *n*-bit ADC, based on the relationship

$$\text{SNR} = 6.02N + 1.76 \text{ (in decibels)}. \tag{5.2}$$

Various error sources in the ADCs cause the measured SNR to be less than the theoretical value shown in Equation 5.2. These errors are due to various secondary effects and noise sources. In addition, the errors are a function of the input slew rate and therefore increase as the input frequency gets higher. In calculating the RMS value of the noise, it is customary to include the harmonics of the fundamental signal. This sometimes is referred to as signal-to-noise-and-distortion, $S/(N + D)$, or SINAD, but usually simply as SNR.

This leads to the definition of another important ADC dynamic specification, the effective number of bits (ENOB). The effective bits are calculated by first measuring the SNR of an ADC with a full-scale sine wave input signal. The measured SNR (SNR_{actual} or SINAD) is substituted into the equation for SNR, and the equation is solved for N as

$$\text{ENOB} = \frac{SINAD - 1.76 \text{ dB}}{6.02}. \tag{5.3}$$

A typical ADC, the AD676 from Analog Devices (a 16-bit ADC), is shown in Figure 5.8. For this device, the SNR value of 88 dB corresponds to approximately 14.3 effective bits (for 0-dB input), but it drops to 6.4 ENOB at 1 MHz. The methods for calculating ENOB, SNR, and other parameters are described in various publications [7,8].

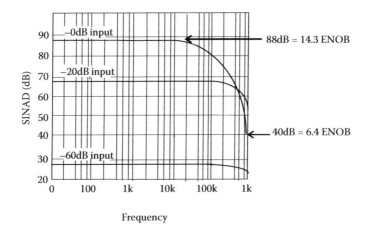

FIGURE 5.8
SINAD and ENOB for AD676. (Reproduced by permission of Analog Devices, Norwood, MA.)

In testing ADCs, the SNR is usually calculated using digital signal processing (DSP) techniques while applying a pure sine wave signal to the input of the ADC. A typical test system is shown in Figure 5.9a. The FFT processes a finite number of time samples and converts them into a frequency spectrum such as the one shown in Figure 5.9b for AD676. The frequency spectrum is then used to calculate the SNR as well as the harmonics of the fundamental input signal. The RMS value of the signal is computed first. The RMS value of all other frequency components over the Nyquist bandwidth (this includes not only noise but also distortion products) is also computed. Various error sources in the ADC cause the measured SNR to be less than the theoretical value (6.02N + 1.76 dB).

It is important to realize that the FFT output of testing an ADC has serious correlation to the ratio of input frequency/sampling frequency, and this is illustrated in Figure 5.10a, where f_a/f_s is exactly 32 versus 4096/127 (which is 32.25196850394) for the case of a 4096-point FFT for a 12-bit ADC. The noise floor of the case is shown in Figure 5.10b, where the FFT noise floor and quantization noise are shown. Details are available in Chapter 2 of the *Linear Design Seminar* [5].

5.3.2 Analog Bandwidth, Spurious Components, and Harmonics

5.3.2.1 Analog Bandwidth

The analog bandwidth of an ADC is that frequency at which the spectral output of the fundamental swept frequency (as determined by the FFT analysis) is reduced by 3 dB. It may be specified in either a small-signal bandwidth (SSBW) or a full-scale-signal or full-power bandwidth (FPBW), so there can be a wide variation in specifications between manufacturers.

Like an amplifier, the analog bandwidth specifications of a converter do not imply that the ADC maintains good distortion performance up to its

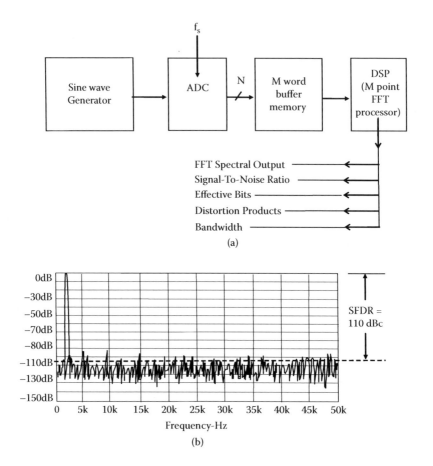

FIGURE 5.9
Testing an ADC for its performance parameters: (a) test system; (b) typical FFT output AD676. (Reproduced by permission of Analog Devices, Norwood, MA.)

bandwidth frequency. In fact, the SINAD (or ENOB) of most ADCs will begin to degrade considerably before the input frequency approaches the actual 3-dB bandwidth frequency. Figure 5.11 shows the ENOB and full-scale frequency response of an ADC with an FPBW of 1 MHz; however, the ENOB begins to decrease rapidly above 100 kHz.

5.3.2.2 Spurious-Free Dynamic Range

Probably the most significant specification for an ADC used in communications applications is its spurious-free dynamic range (SFDR). The SFDR of an ADC is defined as the ratio of the RMS signal amplitude to the RMS value of the peak spurious spectral context measured over the bandwidth of interest. Unless otherwise stated, the bandwidth is assumed to be the Nyquist bandwidth, DC to $f/2$.

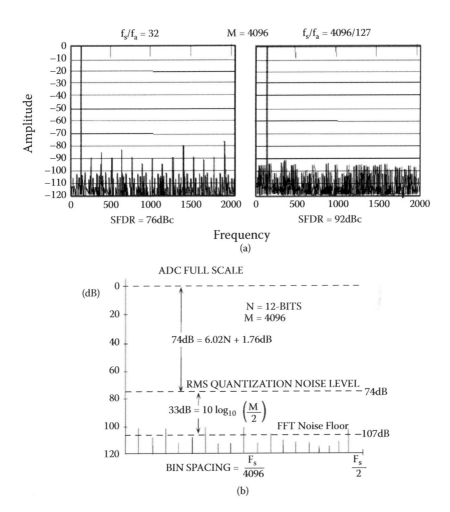

FIGURE 5.10
Testing of 12-bit ADC using a 4096-point FFT: (a) effect of ratio f_a/f_s on SFDR; (b) noise floor and relationships. (Courtesy of Analog Devices, Norwood, MA.)

Occasionally, the frequency spectrum is divided into an in-band region (containing the signals of interest) and an out-of-band region (signals here are filtered out digitally). In this case there may be an in-band SFDR specification and an out-of-band SFDR specification, respectively.

The SFDR is generally plotted as a function of signal amplitude and may be expressed relative to the signal amplitude (dBc) or the ADC full scale (dBFS), as shown in Figure 5.12. For a signal near full scale, the peak special spur is generally determined by one of the first few harmonics of the fundamental. However, as the signal falls several decibels below full scale, other spurs generally occur that are not direct harmonics of the input signal. This is because of the differential nonlinearity of the ADC transfer function,

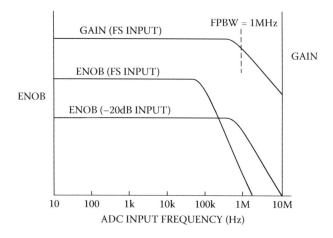

FIGURE 5.11
ADC gain (bandwidth) and ENOB versus frequency. (Courtesy of Analog Devices, Norwood, MA.)

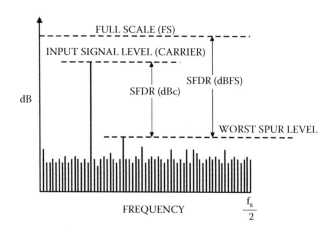

FIGURE 5.12
Spurious-free dynamic range.

as discussed earlier. Therefore, the SFDR considers all sources of distortion, regardless of their origin. More details on testing are available in the *Linear Design Seminar* [5] and Demler [9].

5.4   ADC Errors

First, let us look at how bits are assigned to the corresponding analog values in a typical ADC. The method of assigning bits to the corresponding analog value of the sampled point is often referred to as quantization (see Figure 5.13a). As the analog voltage increases, it crosses transitions of "decision levels," which causes the ADC to change state. In an ideal ADC, transitions are at half-unit levels, with Δ representing the distance between the decision levels. The Δ is often referred to as the bit size or quantization size. The fact that Δ always has a finite size leads to uncertainty, because any analog value within the finite range can be represented. This quantization uncertainty is expressed as plus or minus half the least significant bit (LSB), as shown in Figure 5.13b. As this plot shows, the output of an ADC may be thought of as the analog signal plus some quantizing noise. The more bits the ADC has, the less significant this noise becomes.

Certain parameters limit the rate at which an ADC can acquire a sample of the input waveform. These include the acquisition turn-on delay, acquisition time, sample or track time, and hold time. Figure 5.13c shows a graphic representation of the acquisition cycle of a typical ADC. The turn-on time (the time the device takes to get ready to acquire a sample) is the first event. The acquisition time is next. This is the time the device takes to get to the point at which the output tracks the input sample, after the sample command or clock pulse. The aperture time delay is the time that elapses between the hold command and the point at which the sampling switch is completely open. The device then completes the hold cycle and the next acquisition is taken.

The process indicates that the real world of acquisition is not an ideal process at all, and the value sampled and converted could have some sources of error. Most of these errors increase with the sampling rate.

The difference between the original input and the digitized output, the quantization error, is denoted here by ε_q. For the characteristics of Figure 5.14a, ε_q varies as shown in Figure 5.14b, with the maximum occurring before each code transition. This error decreases as the resolution increases, and its effect can be viewed as additive noise (quantization noise) appearing at the output. Thus, even an "ideal" m-bit ADC introduces nonzero noise into the converted signal simply due to quantization.

We can formulate the impact of quantization noise on the performance as follows. For simplicity, consider a slightly different input/output characteristic, shown in Figure 5.14a, where code transitions occur at odd (rather than even) multiples of $\Delta/2$. A time domain waveform thus experiences both negative and positive quantization errors, as illustrated in Figure 5.14b. To calculate the power of the resulting noise, we assume that ε_q is (i) a random variable uniformly distributed between $-\Delta/2$ and $+\Delta/2$, and (ii) independent of the analog input. Although these assumptions are not strictly valid in the general case, they usually provide a reasonable approximation for resolutions greater than 4 bits. Razavi [10] provides details and the derivations of Equation 5.2 and Equation 5.3.

Full specification of the performance of ADCs requires a large number of parameters, some of which are defined differently by different manufacturers. Some important parameters frequently used in component data sheets and the like are described here. Figure 5.15 can be used to illustrate parameters

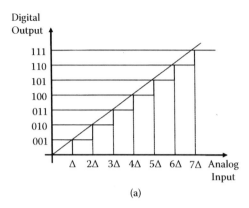

(a)

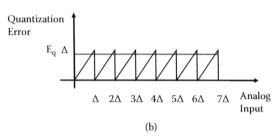

(b)

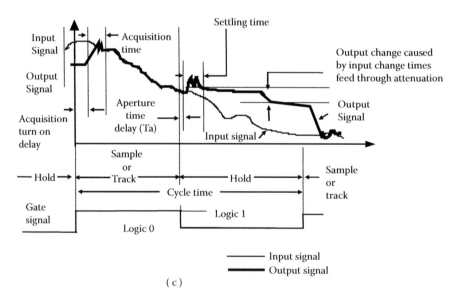

(c)

FIGURE 5.13

Quantization process, quantization error, and timing cycle: (a) basic input/output relationship of a 3-bit digitizer; (b) quantization error; (c) basic timing of the acquisition cycle.

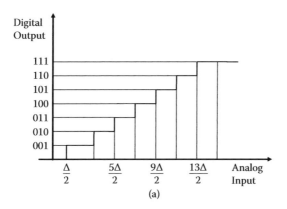

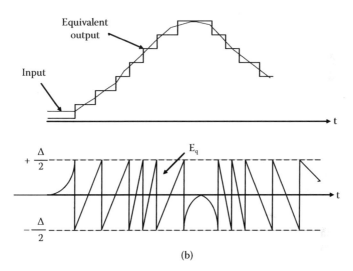

FIGURE 5.14
ADC characteristics and effect of amplitude quantization on a time domain waveform: (a) modified ADC characteristics; (b) effect of quantization.

such as differential nonlinearity (DNL), integral nonlinearity (INL), and off-set and gain errors—four important static parameters of the ADC process.

5.4.1 Differential Nonlinearity

Differential nonlinearity is the maximum deviation in the difference between two consecutive code transition points on the input axis from the ideal value of 1 LSB. The DNL is a measure of the deviation code widths from the ideal value of 1 LSB.

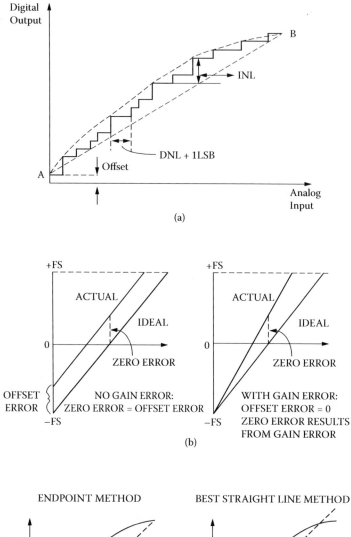

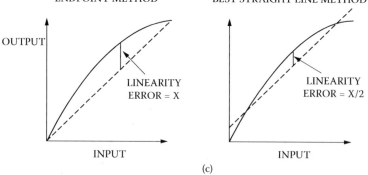

FIGURE 5.15
Static ADC metrics: (a) nonlinearities; (b) gain and offset errors for a case of bipolar coding; (c) two methods of curve fitting for linearity errors.

5.4.2 Integral Nonlinearity

Integral nonlinearity is the maximum deviation of the input/output characteristic from a straight line passed through its endpoints (line AB in Figure 5.15a). The overall difference plot is called the INL profile. The INL is the deviation of code centers from the converter's ideal transfer curve. The line used as the reference may be drawn through the endpoints (straight-line method, as in the left side of Figure 5.15c) or may be a best-fit line calculated from the data (right side of Figure 5.15c).

In the endpoint system, the deviation is measured from the straight line through the origin and the full-scale point (after gain adjustment). This is the most useful integral linearity measurement for measurement and control applications of data converters.

The best (or best-fit) straight line, however, does give a better prediction of distortion in AC applications and also gives a lower value of "linearity error" on a data sheet. The best-fit straight line is drawn through the transfer characteristic of the device using standard curve-filtering techniques, and the maximum deviation is measured from this line. In general, the integral linearity error measured in this way is only 50% of the value measured by endpoint methods. This makes the method good for producing impressive data sheets, but it is less useful for error budget analysis. For more details, see Kester [2].

The DNL and INL degrade as the input frequency approaches the Nyquist rate. The DNL shows up as an increase in quantization noise, which tends to elevate the converter's overall noise floor. Theoretical quantization noise for an ideal converter with the Nyquist bandwidth is

$$\text{ms quantization noise} = q/\sqrt{12}, \tag{5.4}$$

where q is the weight of the LSB.

At the same time, because the INL appears as a bend in the converter's transfer curve, it generates spurious frequencies (spurs) not in the original signal information. The testing of ADC linearity parameters is discussed in Shill [11].

5.4.3 Gain Error and Offset Error

The four DC errors in a data converter are offset error, gain error, differential of nonlinearity error, and integral nonlinearity error. Offset and gain errors are analogous to offset and gain errors in amplifiers, as shown in Figure 5.15c for a bipolar input range. (Unipolar and bipolar conversions are discussed further in section 5.8.2.) It is important to note that offset error and zero error, which are identical in amplifier and unipolar data converters, are not identical in bipolar converters and should be carefully distinguished.

The transfer characteristics of both DACs and ADCs may be expressed as a straight line given by $D = K + GA$, where D is the digital code, A is the analog signal, and K and G are constants in a unipolar converter. The ideal value of K is zero, and in an offset bipolar converter, it is −1 MSB. The offset error is the amount by which the actual value of K differs from its ideal value.

The gain error is the amount by which G differs from its ideal value and is generally expressed as the percentage difference between the two, although it may be defined as the gain error contribution (in millivolts or LSB) to the total error at full scale. These errors can usually be trimmed by the data converter user. Note, however, that amplifier offset is trimmed at zero input and then the gain is trimmed nearly to full scale. The trim algorithms for a bipolar data converter are not so straightforward. More discussion on this is found in Kester [2].

5.4.4   Testing of ADCs

A known periodic input is converted by an ADC under test at sampling times that are asynchronous relative to the input signal. The relative number of occurrences of the distinct digital output codes is termed the code density. For an ideal ADC, the code density is independent of the conversion rate and input frequency. These data are viewed in the form of a normalized histogram showing the frequency of occurrence of each code from zero to full scale. The code density data are used to compute all bit transition levels. Linearity, gain, and offset errors are readily calculated from the knowledge of the transition levels. This provides a complete characterization of the ADC in the amplitude domain. The effects of some of these static errors in the frequency domain for high-speed ADCs are discussed in Lauzon [12] and Lowenstein [13]. Doernberg et al. [14] provide ADC characterization methods based on code density test and spectral analysis using FFT.

5.5   Effects of Sample-and-Hold Circuits

The sample-and-hold amplifier (SHA) is a critical part of many data acquisition systems. It captures an analog signal and holds it during the analog-to-digital (A/D) conversion process. The circuitry involved is demanding, and unexpected properties of commonplace components such as capacitors and printed circuit boards may degrade SHA performance.

When an SHA is in the sample mode, the output follows the input with only a small voltage offset. In some SHAs, the output during the sample mode does not follow the input accurately and the output is accurate only during the hold period.

Today, high-density IC processes allow the manufacture of ADCs containing an integral SHA. Wherever possible, ADCs with an integral SHA (often known as sampling ADCs) should be used in preference to separate ADCs and SHAs. The advantage of such a sampling ADC, apart from the obvious ones of smaller size, lower cost, and fewer external components, is that the overall performance is specified. The designer need not spend time ensuring that no specification, interface, or timing issues are involved in combining a discrete ADC and a discrete SHA.

5.6 Basic SHA Operation

Regardless of the circuit details or type of SHA in question, all such devices have four major components. The input amplifier, energy storage device (capacitor), output buffer, and switching circuits are common to all SHAs, as shown in the typical configuration of Figure 5.16a.

The energy storage device, the heart of the SHA, is a capacitor. The input amplifier buffers the input by presenting a high impedance to the signal source and providing current gain to charge the hold capacitor. In the track mode, the voltage on the hold capacitor follows (or tracks) the input signal (with some delay and bandwidth limiting). Figure 5.13c depicts this process. In the hold mode, the switch is opened and the capacitor retains the voltage present before it was disconnected from the input buffer. The output buffer offers a high impedance to the hold capacitor to keep the held voltage from discharging prematurely. The switching circuit and its driver form the mechanism by which the SHA is alternately switched between track and hold.

Four groups of specifications describe basic SHA operation: track mode, track-to-hold transition, hold mode, and hold-to-track transition. These specifications are summarized in Table 5.2, and some of the SHA error sources are shown in Figure 5.16b. Because of both DC and AC performance implications for each of the four modes, properly specifying an SHA and understanding its operation in a system are complex matters.

5.6.1 Track Mode Specifications

Because an SHA in the sample (or track) mode is simply an amplifier, both the static and dynamic specifications in this mode are similar to those of any amplifier. The principal track mode specifications are offset, gain, nonlinearity, bandwidth, slew rate, settling time, distortion, and noise; however, distortion and noise in the track mode often are of less interest than in the hold mode.

5.6.2 Track-to-Hold Mode Specifications

When the SHA switches from track to hold, generally a small amount of charge is dumped on the hold capacitor because of nonideal switches. This results in a hold-mode DC offset voltage called the pedestal error. If the SHA is driving an ADC, the pedestal error appears as a DC offset voltage that

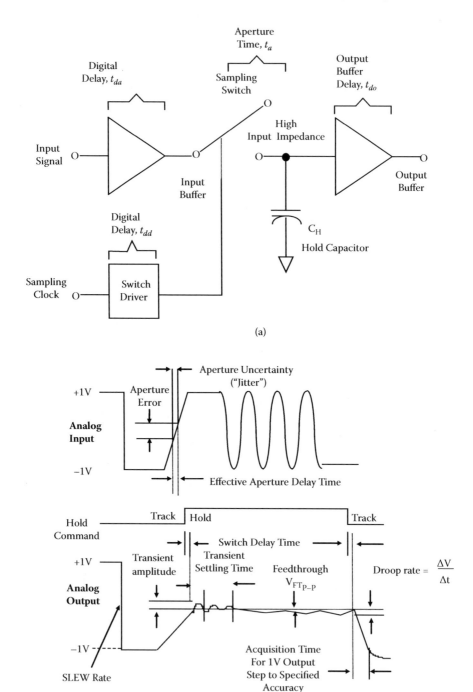

FIGURE 5.16
Sample-and-hold circuit and error sources: (a) basic sample-and-hold circuit; (b) some sources of error.

TABLE 5.2

Sample-and-Hold Specifications

	Track Mode	Track-to-Hold Transition	Hold Mode	Hold-to-Sample Transition
Static	Offset	Pedestal	Droop	
	Gain error nonlinearity	Pedestal nonlinearity	Dielectric absorption	
Dynamic	Settling time	Aperture delay time	Feedthrough	Acquisition time
	Bandwidth		Distortion	Switching transient
	Slew rate	Aperture jitter	Noise	
	Distortion	Switching transient		
	Noise	Settling time		

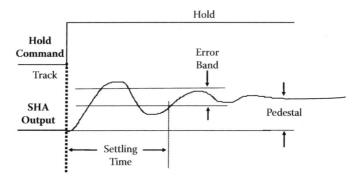

FIGURE 5.17
Hold-mode settling time.

may be removed by performing a system calibration. If the pedestal error is a function of the input signal level, the resulting nonlinearity contributes to hold-mode distortion. Pedestal errors may be reduced by increasing the value of the hold capacitor with a corresponding increase in acquisition time and a reduction in bandwidth and slew rate.

Switching from track to hold produces a transient, and the time required for the SHA output to settle to within a specified error band is called the hold-mode settling time. Occasionally, the peak amplitude of the switching transient is also specified (see Figure 5.17).

5.6.3 Aperture and Aperture Time

Perhaps the most misunderstood and misused SHA specifications are those that include the word aperture. The most essential dynamic property of an SHA is its ability to quickly disconnect the hold capacitor from the input buffer amplifier (see Figure 5.16a).

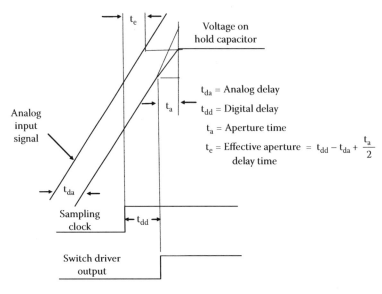

FIGURE 5.18
SHA waveform.

The short (but nonzero) interval required for this action is called the aperture time (t_a). The actual value of the voltage held at the end of this interval is a function of both the input signal and the errors introduced by the switching operation itself. Figure 5.18 shows what happens when the hold command is applied with an input signal of arbitrary slope. (For clarity, the sample-to-hold pedestal and switching transients are ignored.) The value finally held is a delayed version of the input signal, averaged over the aperture time of the switch, as shown in Figure 5.18. The first-order model assumes that the final value of the voltage on the hold capacitor is approximately equal to the average value of the signal applied to the switch over the interval during which the switch changes from a low to a high impedance (t_a).

The model shows that the finite time required for the switch to open (t_a) is equivalent to introducing a small delay in the sampling clock driving the SHA. This delay is constant and may be either positive or negative. Called the effective aperture delay time or simply the aperture delay (t_e), it is defined as the time difference between the analog propagation delay of the front-end buffer (t_{da}) and the switch digital delay (t_{dd}) plus half the aperture time ($t_a/2$). The effective aperture delay time is usually positive but may be negative if the sum of half the aperture time ($t_a/2$) and the switch digital delay (t_{dd}) is less than the propagation delay through the input buffer (t_{da}). The aperture delay specification thus establishes when the input signal is actually sampled with respect to the sampling clock edge.

The aperture delay time can be measured by applying a bipolar sine wave signal to the SHA and adjusting the synchronous sampling clock delay such that the output of the SHA is zero during the hold time. The relative delay

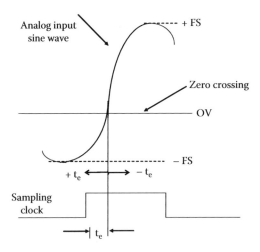

FIGURE 5.19
Measuring the effective aperture delay time.

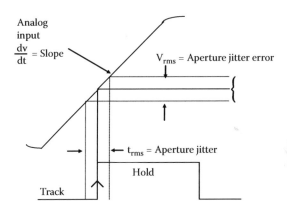

FIGURE 5.20
Effects of aperture jitter on SHA output.

between the input sampling clock edge and the actual zero crossing of the input sine wave is the aperture delay time (see Figure 5.19).

Aperture delay produces no errors but acts as a fixed delay in either the sampling clock input or the analog input (depending on its sign). If there is sample-to-sample variation in aperture delay (aperture jitter), then a corresponding voltage error is produced, as shown in Figure 5.20. This sample-to-sample variation in the instant that the switch opens, called aperture uncertainty or aperture jitter, usually is measured as an RMS value in picoseconds. The amplitude of the associated output error is related to the rate of change of the analog input. For any given value of aperture jitter, the aperture jitter error increases as the input dv/dt increases.

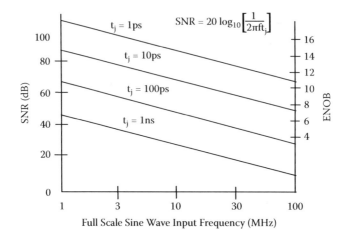

FIGURE 5.21

Effects of sampling clock jitter on the SNR. (Reproduced by permission of Analog Devices, Norwood, MA.)

Figure 5.21 shows the effects of total sampling clock jitter on the SNR of a sampled data system. The total RMS jitter is composed of a number of components, the actual SHA aperture jitter often being the least of them. See Reeder [15] for details.

5.6.4 Hold Mode Droop

During the hold mode, there are errors due to imperfections in the hold capacitor, switch, and output amplifier. If a leakage current flows in or out of the hold capacitor, it will slowly charge or discharge and its voltage will change, an effect known as droop in the SHA output (expressed in V/μs). Droop can be caused by leakage across a dirty PCB if an external capacitor is used or by a leaky capacitor, but most commonly droop is caused by leakage current in semiconductor switches and the bias current of the output buffer amplifier. An acceptable value of droop is found when the output of an SHA does not change by more than 0.5 LSB during the conversion time of the ADC it is driving (see Figure 5.22).

Droop can be reduced by increasing the value of the hold capacitor, but this will increase acquisition time and reduce the bandwidth in the track mode. Even quite small leakage currents can cause troublesome droop when SHAs use small hold capacitors. Leakage currents in PCBs can be minimized by the intelligent use of guard rings. Details of planning a guard ring are discussed in the *Linear Design Seminar* [5].

5.6.5 Dielectric Absorption

Hold capacitors for SHAs must have low leakage, but another characteristic is equally important: low dielectric absorption, discussed in Chapter 1.

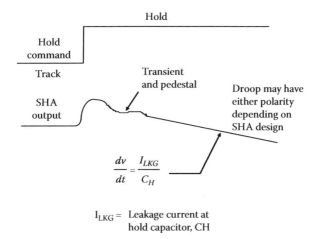

FIGURE 5.22
Hold mode droop.

Different capacitor materials have differing amounts of dielectric absorption: electrolytic capacitors are dreadful (and their leakage is high) and some high-potassium ceramic types are bad, whereas mica, polystyrene, and polypropylene generally are good. Unfortunately, dielectric absorption varies from batch to batch, and even occasional batches of polystyrene and polypropylene capacitors may be affected.

5.6.6 Hold-to-Track Transition Specification

When the SHA switches from hold to track, it must reacquire the input signal (which may have made a full-scale transition during the hold mode). Acquisition time is the interval of time required for the SHA to reacquire the signal to the desired accuracy when switching from hold to track. The interval starts at the 50% point of the sampling clock edge and ends when the SHA output voltage falls within the specified error band (usually, 0.1% and 0.01% times are given). Some SHAs also specify acquisition time with respect to the voltage on the hold capacitor, neglecting the delay and settling time of the output buffer. The hold capacitor acquisition time specification is applicable in high-speed applications, where the maximum possible time must be allocated for the hold mode. The output buffer settling time, of course, must be significantly smaller than the hold time.

5.6.7 SHA Architectures

There are numerous SHA architectures, and they are basically divided into open-loop and closed-loop types. For a discussion of SHA architectures, see Razavi [10].

5.7 ADC Architectures

In the 1990s and the latter part of the 1980s, many architectures for A/D conversion were implemented in monolithic form. Manufacturing process improvements achieved by mixed-signal product manufacturers led to this unprecedented development, which was fueled by the demand from product and system designers.

The most common ADC architectures in monolithic form are successive approximation, flash, integrating, pipeline, half-flash (or subranging), two-step, interpolative and folding, and sigma-delta (Σ-Δ). The following sections provide the basic operational and design details of several of these techniques. Whereas the Σ-Δ, successive approximation, and integrating types give very high resolution at lower speeds, flash architecture is the fastest but has high power consumption. However, recent architecture breakthroughs have allowed designers to achieve a higher conversion rate at low power consumption with integral track-and-hold circuitry on a chip [16].

5.7.1 Successive Approximation ADCs

The successive approximation register (SAR) ADC architecture is a popular and cost-effective form of converter for sampling frequencies up to a few megasamples per second (MSPS). A simplified block diagram of an SAR ADC is shown in Figure 5.23a. On the start conversion command, all the bits of the successive approximation register are reset to 0 except the most significant bit (MSB), which is set to 1. Bit 1 is tested in the following manner. If the ADC output is greater than the analog input, the MSB is reset; otherwise, it is left set. The next MSB is then tested by setting it to 1. If the DAC output is greater than the analog input, this bit is reset; otherwise, it is left set. The process is repeated with each bit in turn. When all the bits have been set, tested, and reset or not as appropriate, the contents of the SAR correspond to the digital value of the analog input, and the conversion is complete (Figure 5.23b).

An n-bit conversion takes n steps. On superficial examination, a 16-bit converter would seem to have a conversion time twice as long as an 8-bit one, but this is not the case. In an 8-bit converter, the DAC must settle to 8-bit accuracy before the bit decision is made, whereas in a 16-bit converter, it must settle to 16-bit accuracy, which takes much longer. In practice, 8-bit successive approximation ADCs can convert in a few hundred nanoseconds, whereas 16-bit ones generally take several microseconds.

The classic SAR ADC is only a quantizer—no sampling takes place—and for an accurate conversion, the input must remain constant for the entire conversion period. Most modern SAR ADCs are sampling types and have internal sample and hold so that they can process AC signals. They are specified for both AC and DC applications. An SHA is required in an SAR ADC because the signal must remain constant during the entire n-bit conversion cycle.

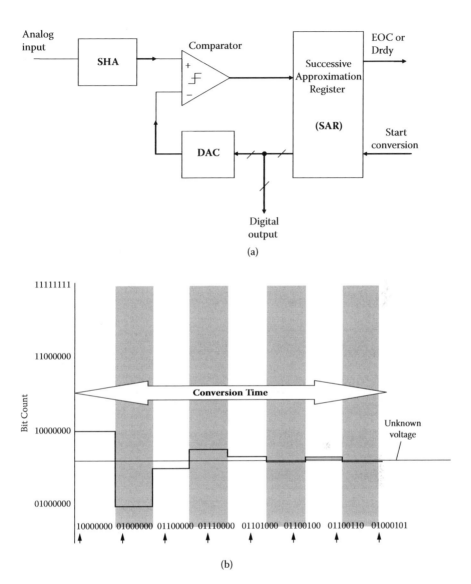

FIGURE 5.23
Successive approximation ADC: (a) simplified block diagram; (b) timing of an 8-bit SAR.

The accuracy of an SAR ADC depends primarily on the accuracy (differential and integral linearity, gain, and offset) of the internal DAC. Until recently, this accuracy was achieved using laser-trimmed thin-film resistors. Modern SAR ADCs utilize CMOS switched capacitor charge redistribution DACs. This type of DAC depends on the accurate ratio matching and stability of on-chip capacitors rather than thin-film resistors. For resolution greater than 12 bits, on-chip autocalibration techniques, using an additional calibration DAC and the accompanying logic, can accomplish the same thing as thin-film laser-trimmed resistors, at much less cost. Therefore, the entire ADC can be made on a standard submicron CMOS processor.

The successive approximation ADC has a very simple structure, low power, and reasonably fast conversion times. It is probably the most widely used ADC architecture and will continue to be used for medium-speed, medium-resolution applications. Current 16-bit SAR types can operate up to about 3 MSPS, whereas 18-bit versions run at more than 800 kSPS. Examples of typical state-of-the-art SAR ADCs are the AD7621 (16 bits at 3 MSPS) and the AD7674 (18 bits at 800 kSPS).

5.7.2   Flash Converters

Flash ADCs (sometimes called parallel ADCs) are the fastest ADCs and use large numbers of comparators. An n-bit flash ADC consists of 2^n resistors and $2^n - 1$ comparators arranged as in Figure 5.24. Each comparator has a reference voltage 1 LSB higher than that of the one below it in the chain. For a given input voltage, all the comparators below a certain point will have their input voltage larger than their reference voltage and a 1 logic output, and all the comparators above that point will have a reference voltage larger

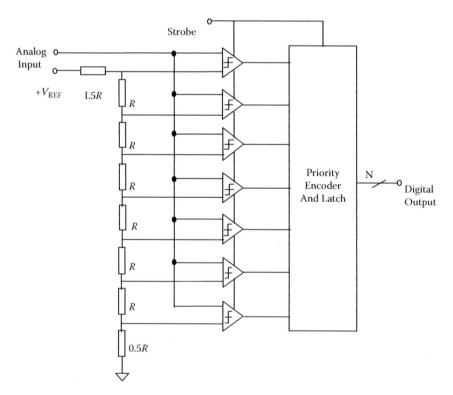

FIGURE 5.24
Flash or parallel ADC block diagram.

than the input voltage and a 0 logic output. The $2^n - 1$ comparator output thus behaves like a mercury thermometer, and the output code at this point is sometimes called a thermometer code. Since $2^n - 1$ data output is not really practical, these are processed by a decoder to an n-bit binary output.

The input signal is applied to all the comparators at once, so the thermometer output is delayed by only one comparator delay from the input and from the encoder n-bit output by only a few gate delays on top of that, so the process is very fast. However, the architecture uses large numbers of resistors and comparators and is limited to low resolutions; if it is to be fast, each comparator must run at relatively high power levels. Hence, the problems of flash ADCs include limited resolution, high power dissipation because of the large number of high-speed comparators (especially at sampling rates greater than 50 MSPS), and relatively large (and therefore expensive) chip sizes. In addition, the resistance of the reference resistor chain must be kept low to supply adequate bias current to the fast comparators, so the voltage reference has to source quite large currents (greater than 10 mA).

In practice, flash converters are available at up to 10 bits of resolution, but more commonly they have 8 bits of resolution. Maximum sampling rate of common parts can be as high as 500 MSPS, and input full-power bandwidths are in excess of 300 MHz. For special purposes such as storage oscilloscopes, etc., there are proprietary very high-speed versions.

However, as mentioned earlier, full-power bandwidths are not necessarily full-resolution bandwidths. Ideally the comparators in a flash converter are well matched both for DC and AC characteristics. Because the strobe is applied to all the comparators simultaneously, the flash converter is inherently a sampling converter. In practice, delay variations between the comparators and other AC mismatches cause a degradation in ENOB at high input frequencies. This is because the inputs are slewing at a rate comparable to the comparator conversion time.

The input to a flash ADC is applied in parallel to a large number of comparators. Each has a voltage-variable junction capacitance, and this signal-dependent capacitance results in all flash ADCs having reduced ENOB and higher distortion at high input frequencies. For more details, see Analog Devices [17], Swager [18], and Kester [19–21].

5.7.3 Integrating ADCs

The integrating ADC is a very popular architecture in applications where a very slow conversion rate is acceptable. A classic example is the digital multimeter.

All the converters discussed so far can digitize analog inputs at speeds of at least 10 kSPS. A typical integrating converter is slow relative to these high-speed converters. Useful for precisely measuring slowly varying signals, the integrating converter finds applications in low-frequency and DC measurement applications.

Integrating converters are based on an indirect conversion method. Here the analog input voltage is converted to a time period and later to a digital

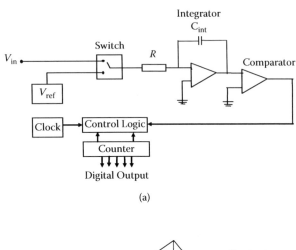

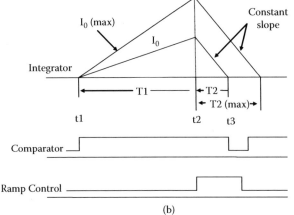

FIGURE 5.25
A typical integrating converter.

number using a counter. The integration eliminates the need for a sample-and-hold circuit to "capture" the input signal during the measurement period. The two common variations of the integrating converter are the dual-slope type and the charge balance or multislope type. The dual-slope technique is very popular among instrument manufacturers because of its simplicity, low price, and better noise rejection. The multislope technique is an improvement on the dual-slope method.

Figure 5.25a shows a typical integrating converter. It consists of an analog integrator, a comparator, a counter, a clock, and control logic. Figure 5.25b shows the circuit's charge (T_1) and discharge (T_2) waveforms. The conversion is started by closing the switch, thereby connecting the capacitor, C_{int}, to the unknown input voltage, V_{in}, through the resistor, R. This results in a linear ramp at the integrator output for a fixed period, T_1, controlled by the counter. The control circuit then switches the integrator input to the known reference voltage, V_{ref}, and the capacitor discharges until the comparator detects that the integrator has reached the original starting point. The counter measures the amount of time it takes for the capacitor to discharge.

Because the values of the resistor, the integrating capacitor, and the frequency of the clock remain the same for both the charge and discharge cycles, the ratio of the charge time to the discharge time is equal to the ratio of the reference voltage to the unknown input voltage. The absolute values of the resistor, capacitor, and clock frequency thus do not affect the conversion accuracy. Furthermore, any noise on the input signal is integrated over the entire sampling period, which imparts a high level of noise rejection to the converter. By making the signal integration period an integral multiple of the line frequency period, the user can obtain excellent line frequency noise rejection.

A charge balance integrating converter incorporates many of the elements of the dual-slope converter but uses a free-running integrator in a feedback loop. The converter continuously attempts to null its input by subtracting precise charge packets when the accumulated charge exceeds a reference value. The frequency of the charge packets (the number of packets per second) the converter needs to balance the input is proportional to that input. Clock-controlled synchronous logic delivers a serial output that a counter converts to a digital word in the circuit. There can be many variations of this technique as applied to digital multimeters; see Kularatna [22] for details.

5.7.4 Subranging ADCs (Half-Flash ADCs)

Although it is not practical to make them with high resolution, flash ADCs often are used as subsystems in "subranging" ADCs (sometimes known as half-flash ADCs), which are capable of much higher resolutions (up to 16 bits). A block diagram of an 8-bit subranging ADC based on two 4-bit flash converters is shown in Figure 5.26. Although 8-bit flash converters are readily available at high sampling rates, this case will be used to illustrate the theory.

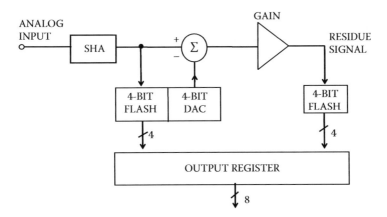

FIGURE 5.26
An 8-bit subranging ADC.

The conversion process is done in two steps. The four most significant bits are digitized by the first flash (to better than 8-bit accuracy) and the 4-bit binary output is applied to 4-bit DAC (again, better than 8-bit accuracy). The DAC output is subtracted from the held analog input, and the resulting residue signal is amplified and applied to the second 4-bit flash. The outputs of the two flash converters are combined into a single 8-bit binary output word. If the residue signal range does not exactly fill the range of the second flash converter, nonlinearities and perhaps missing codes will result.

Modern subranging ADCs use digital error correction techniques to eliminate problems associated with the architecture of Figure 5.26. A simplified block diagram of a 12-bit digitally corrected subranging ADC is shown in Figure 5.27. An example of such a practical ADC is the AD9042 from Analog Devices, a 12-bit, 41-MSPS device.

Note that a 6-bit and 7-bit ADC have been used to achieve an overall 12-bit output. These are not flash ADCs but utilize a magnitude-amplifier (MagAmp™) architecture.

If there are no errors in the first-stage conversion, the 6-bit "residue" signal applied to the 7-bit ADC by the summing amplifier will never exceed one-half the range of the 7-bit ADC. The extra range in the second ADC is used in conjunction with the error correction logic (usually just a full adder) to correct the output data for most of the errors inherent in the traditional uncorrected subranging converter architecture. It is important to note that the 6-bit DAC must be better than 12-bit accuracy because the digital error correction does not correct for DAC errors. In practice, "thermometer" or "fully decoded" DACs using one current switch per level (63 switches in the case

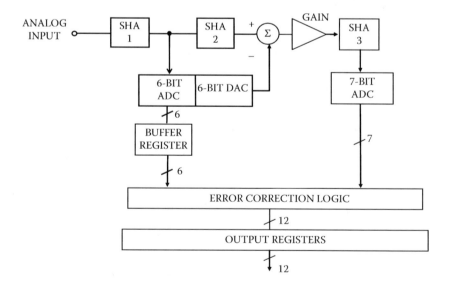

FIGURE 5.27
Pipeline-subranging ADC with a digital error correction. (Reproduced by permission of Analog Devices, Norwood, MA.)

of a 6-bit DAC) are often used instead of a "binary" DAC to ensure excellent differential and integral linearity and minimum switching transients [17].

The second SHA delays the held output of the first SHA while the first-stage conversion occurs, thereby maximizing throughput. The third SHA "deglitches" the residue output signal, allowing a full conversion cycle for the 7-bit ADC to make its decision. (The 6- and 7-bit ADCs in the AD9042 are bit-serial MagAmp ADCs, which require more settling time than a flash converter.) Additional shift registers in series with the digital output of the first-stage ADC ensure that its output ultimately is time-aligned with the last 7 bits from the second ADC when their outputs recombine in the error correction logic. A pipeline ADC thus has a specified number of clock cycles of latency—pipeline delay—associated with the output data. The leading edge of the sampling clock (for sample N) is used to clock the output register, but the data that appear as a result of that clock edge correspond to sample $N - L$, where L is the number of clock cycles of latency—in the case of the AD9042, two clock cycles of latency.

The error-correction scheme described previously is designed to correct for errors made in the first conversion. Internal ADC gain, offset, and linearity errors are corrected as long as the residue signal falls within the range of the second-stage ADC. These errors will not affect the linearity of the overall ADC transfer characteristic. Errors made in the final conversion, however, translate directly as errors in the overall transfer function. Also, linearity errors or gain errors either in the DAC or the residue amplifier will not be corrected and will show up as nonlinearities or nonmonotonic behavior in the overall ADC transfer function.

So far we have considered only two-stage subranging ADCs, as these are the easiest to analyze. There is no reason to stop at two stages, however. Three-pass and four-pass subranging pipeline ADCs are quite common and can be made in many different ways, usually with digital error correction. For details, see Analog Devices [17].

5.7.5 Two-Step Architectures

In a two-step ADC, first a coarse analog estimate of the input is obtained to yield a small voltage range around the input level. Subsequently, the input level is determined with higher precision within this range. Figure 5.28a illustrates a two-step architecture consisting of a front-end SHA, a coarse flash ADC stage, a DAC, a subtractor, and a fine flash ADC stage. The timing diagram is shown in Figure 5.28b.

For $t < t_1$, the SHA tracks the analog input. At $t = t_1$, the SHA enters a hold mode and the first flash stage is strobed to perform the coarse conversion. The first stage then provides a digital estimate of the signal held by the SHA (V_A), and the DAC converts this estimate to an analog signal (V_B), which is a coarse approximation of the SHA output. Next, the subtractor generates an output (V_C, called the residue) equal to the difference between V_A and V_B, which is subsequently digitized by the fine ADC. Comparison of timing in

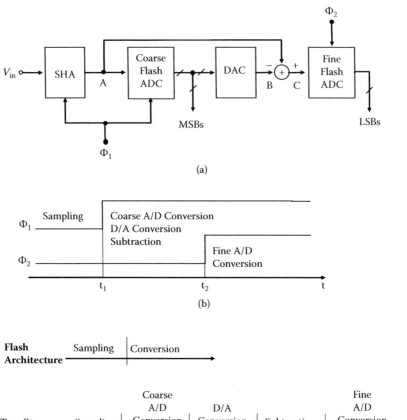

FIGURE 5.28
Two-step architecture: (a) block diagram; (b) timing; (c) comparison of timing in flash and two-step architecture.

flash and two-step architectures is shown in Figure 5.28c. For more details, see Kularatna [22] and Ushani [23–25].

A two-step ADC need not employ two separate flash stages to perform the coarse and fine conversions. One stage can be used for both; such *n*-architecture, shown in Figure 5.29, is called recycling architecture.

Here, during the coarse conversion, the flash stage senses the front-end SHA output, V_A, and generates the coarse digital output. This digital output is then converted to analog by the DAC and subtracted from V_A by the subtractor. During fine conversion, the subtractor output is digitized by the flash stage. Note that in this phase, the ADC full-scale voltage must be equal to that of the subtractor output. Therefore, for proper fine conversion, either the ADC reference voltage must be reduced or the residue must be amplified.

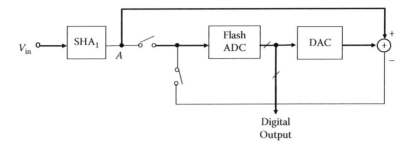

FIGURE 5.29
Recycling ADC architecture.

While reducing area and power dissipation by roughly a factor of 2 relative to two-stage ADCs, recycling converters suffer from other limitations. The converter must now employ either low-offset comparators (if the subtractor has a gain of 1), inevitably slowing down the coarse conversion, or a high-gain subtractor, increasing the interstage delay. This is in contrast to two-stage ADCs, where the coarse-stage comparators need not have a high resolution and hence can operate faster. Details on subranging ADCs are available in Kularatna [22] and Ushani [23–25].

5.7.6 Sigma-Delta Converters

Sigma-delta ADCs (Σ-Δ ADCs) have been in existence for nearly 40 years, but only after 1980s high-density VLSIs were the commercially practical. They are used in many applications where a low-cost, low-bandwidth, high-resolution ADC is required.

The literature contains innumerable descriptions of the architecture and theory of Σ-Δ ADCs [25–27]. As a text of this nature is not appropriate for describing their mathematical analysis and background, this section has been written to classify the subject. A practical monolithic Σ-Δ ADC contains very simple analog circuit blocks (a comparator, a switch, and one or more integrators and analog summing circuits) and quite complex digital computational circuitry. The circuitry consists of a digital signal processor that acts as a filter (generally, but not invariably, a low-pass filter). It is not necessary to know how the filter works to appreciate what it does. To understand how a Σ-Δ ADC works, one should be familiar with the concepts of oversampling, noise shaping, digital filtering, and decimation. We briefly discuss these concepts in Σ-Δ converters.

5.7.6.1 Key Concepts behind Σ-Δ ADCs

Figure 5.30 shows the transfer characteristics of a 3-bit unipolar Σ-Δ ADC. The input to an ADC is analog and is not quantized, but its output is quantized. The transfer characteristics therefore consist of eight horizontal steps. (When considering the offset, gain, and linearity of an ADC, we consider the

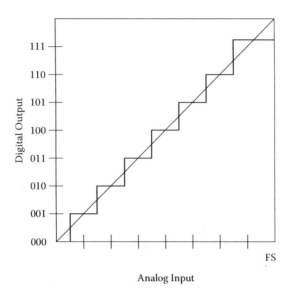

FIGURE 5.30
Transfer characteristics of a 3-bit unipolar ADC.

line joining the midpoints of these steps.) Digital full scale (all ones) corresponds to 1 LSB below the analog full scale (the reference or some multiple of it). This is because, as mentioned previously, the digital code represents the normalized ratio of the analog signal to the reference, and if this were unity, the digital code would be all zeros and ones in the bit above the MSB.

The (ideal) ADC transitions take place at 0.5 LSB above zero and thereafter every LSB, up to 1.5 LSB below analog full scale. This process generates the quantization error or the quantization uncertainty. In AC (sampling) applications, this quantization error gives rise to quantization noise. If we apply a fixed input to an ideal ADC, we will always obtain the same output and the resolution will be limited by the quantization error.

Suppose, however, that we add some AC (dither) to the fixed signal, take a large number of samples, and prepare a histogram of the results. We will obtain something like the result in Figure 5.31. If we calculate the mean value of a large number of samples, we will find that we can measure the fixed signal with greater resolution than that of the ADC we are using. This procedure is known as oversampling.

The AC (dither) that we add may be a sine wave, a triangular wave, or Gaussian noise (but not a square wave); with some types of sampling ADCs (including Σ-Δ ADCs), an external dither signal is unnecessary, because the ADC generates its own. Analysis of the effects of differing dither waveforms and amplitudes is complex and for the purpose of this section unnecessary. What we need to know is that with the simple oversampling described here, the number of samples must be doubled for each bit of increase in resolution.

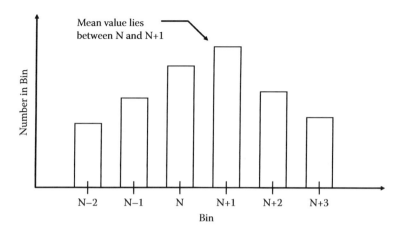

FIGURE 5.31
Oversampling.

If, instead of a fixed DC signal, the signal that we are oversampling is AC, then it is not necessary to add a dither signal to it to oversample, because the signal is moving, anyway. (If the AC signal is a single tone, harmonically related to the sampling frequency, dither may be necessary, but this is a special case.)

Consider the technique of oversampling with an analysis in the frequency domain. Where a DC conversion has a quantization error of up to 0.5 LSB, a sampled data system has quantization noise. As we already have seen, a perfect classical n-bit sampling ADC has an RMS quantization noise of $q/\sqrt{12}$ uniformly distributed within the Nyquist band of DC to $f_s/2$ (where q is the value of an LSB and f_s is the sample rate), giving us an SNR of $6.02N + 1.76$ dB with full-scale sine wave inputs (e.g., Equation 5.2) (see Figure 5.32). If the

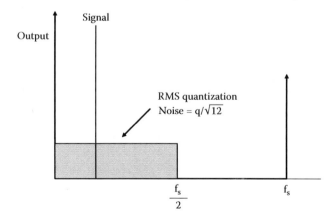

FIGURE 5.32
Sampling ADC quantization noise.

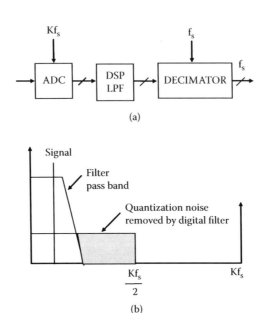

(a)

(b)

FIGURE 5.33
Oversampling and digital filtering: (a) block diagram; (b) output versus frequency.

ADC is less than perfect and its noise is greater than its theoretical minimum quantization noise, then its effective resolution will be less than n bits, given by the ENOB defined by Equation 5.3.

If we choose a much higher sampling rate ($K \times f_s$), as in Figure 5.33a, the quantization noise is distributed over a wider bandwidth, as shown in Figure 5.33b. If we then apply a digital low-pass filter to the output, we remove much of the quantization noise but do not affect the wanted signal, so the ENOB is improved. We have performed a high-resolution A/D conversion with a low-resolution ADC.

Because the bandwidth is reduced by the digital output filter, the output data rate may be lower than the original sampling rate and still satisfy the Nyquist criteria. This may be achieved by passing every Mth result to the output and discarding the remainder, a process is known as decimation by a factor of M. Here, M can have any integer value, provided that the output data rate is more than twice the signal bandwidth. Decimation causes no loss of information (Figure 5.34). As shown in Figure 5.34, after sampling at f_s and filtering, the output data rate may be reduced to f_s/M with no loss of information.

5.7.6.2   *Block Diagram of a Σ-Δ ADC*

In recent years, oversampled Σ-Δ ADCs have become more prevalent for high-accuracy, 12-bit to more than 22-bit A/D conversion of DC through moderately high (hundreds of kilohertz) AC signals. Their greatest advantage

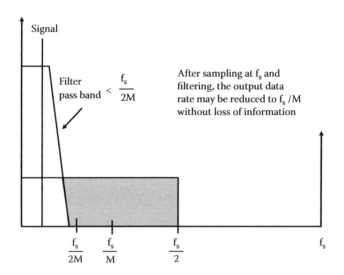

FIGURE 5.34
Decimation process.

is that they trade greatly reduced analog circuit accuracy requirements for increased digital circuit complexity. This is a distinct advantage of VLSI circuit technologies. VLSI circuit techniques can achieve circuit densities of hundreds of thousands of gates, allowing complex digital filters to be integrated on the chip. The result is high precision at low cost.

The basic oversampled Σ-Δ ADC (Figure 5.35) is an integrating ADC. The single-bit feedback DAC output is subtracted from the analog input signal, V_{in}, in the summing amplifier. The resulting error signal from the summing amplifier output is low-pass filtered by the integrator, and the integrated error signal polarity is detected by the single comparator. This comparator is effectively a 1-bit ADC. The output of the comparator drives the 1-bit DAC to a 1 or 0 (a 1, if during the previous sample time the integrator output was detected by the comparator as being too low, that is, below 0 V; a 0, if the difference detected during the previous sample was too high, that is, above 0 V reference of the comparator).

The 1-bit DAC, as in successive approximation ADCs, provides the negative feedback. This negative feedback for a 1 in the DAC is always in a direction to drive the integrator output toward 0 V. The DAC output for a 1 input is the reference voltage. The reference voltage is equal to or exceeds the expected full-scale analog input signal voltage. Then, for a small value of V_{in}, the integrator takes many clock pulses to cross 0 V, after a single 1 is generated, during which time the comparator is sending 0s to the digital filter. If V_{in} is at full scale, the integrator crosses 0 every clock time and the comparator output will be a string of alternate 1s and 0s. The digital filter's function is to determine a digital number at its output that is proportional to the number of 1s in the previous bit stream from the comparator. Various types of digital filters are used to perform this computational function, which is the most complex

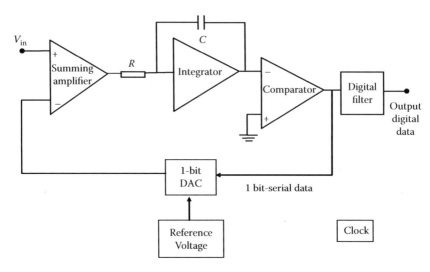

FIGURE 5.35
Oversampled Σ-Δ ADC.

function in this type of DAC. However, complex digital computations can be performed readily in a VLSI circuit.

Oversampled converters sample at much higher rates than the Nyquist rate. The oversampling ratio is equal to the actual sampling rate divided by the Nyquist rate. The oversampling rate can be hundreds to thousands of times the analog input signal frequency bandwidth. Because each sample is in a 1-bit low-accuracy conversion, sampling rates can be very high.

In cases where antialiasing filtering is required on the input analog signal, the filter does not require the sharp cutoff characteristics as are required to limit broadband signals prior to a successive approximation-type ADC operating at or near a small multiple of the Nyquist sampling rate. The reason is that the oversampling rate is many times the Nyquist rate. Therefore, a simple RC filter is adequate to prevent aliasing. The input filter can pass frequencies many times higher than the frequencies of interest before filter cutoff is required. The bandwidth of signals converted can be increased significantly by a Σ-Δ ADC at a given clock sampling rate by using a multibit ADC and DAC rather than a single-bit ADC and DAC. Digital filter design is another variable affecting the bandwidth of signals that can be accurately converted for a given oversampling rate.

However, the greatest advantage of an oversampled Σ-Δ converter is that it requires only a single-bit ADC and DAC with a relatively inaccurate differential summing amplifier and integrator (low-pass filter). These analog circuits are much easier to implement in a digital VLSI circuit than the accurate analog circuits required in parallel and successive approximation ADCs that require precision resistors or capacitors. This is especially true when accuracies exceed 12 bits.

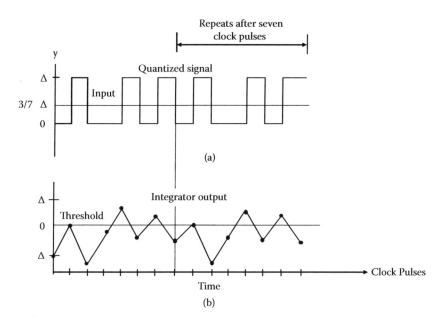

FIGURE 5.36
Waveforms in a Σ-Δ circuit for a constant input situated at 3Δ/7 above a quantization level: (a) quantized signal and input; (b) integrator output.

Based on the preceding description, we can show that the quantized signal bounces between two levels, keeping its mean equal to the input, when the input to the modulator is a DC signal. Figure 5.36 shows the quantized signal and the integrator output when the input signal is 3Δ/7 above 0 for a quantization level of Δ. The figure shows that the oscillation may be repetitive (it returns to its starting condition after seven clock periods). The frequency of repetition depends on the input level.

By using more than one integration and summing stage in the Σ-Δ modulator, higher orders of quantization noise shaping and better ENOB for a given oversampling ratio can be achieved. A block diagram of a second-order Σ-Δ ADC is shown in Figure 5.37.

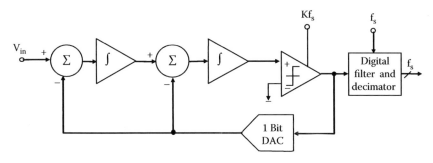

FIGURE 5.37
Second-order Σ-Δ ADC.

5.7.7 Self-Calibration Techniques

The integral of data converters usually depends on the matching and linearity of integrated resistors, capacitors, or current sources, and it is typically limited to approximately 10 bits with no calibration. For higher resolutions, means must be sought that can reliably correct nonlinearity errors. This is often accomplished by either improving the effective matching of individual devices or correcting the overall transfer characteristics. Since high-resolution ADCs typically employ a multistep architecture, they often impose two stringent requirements: small integral nonlinearity in their interstage DACs and precise gain (usually a power of 2) in their interstage subtractors/amplifiers. These constraints, in turn, demand correction for device mismatches if resolutions greater than 10 bits are required.

The ADC calibration techniques can be in two forms: use of analog processing techniques for correction of nonidealities and digital calibration techniques. A description of these techniques is beyond the scope of this chapter; for details, see Razavi [10], Stewart and Pfann [28], Jonston [29], and O'Leary [30].

5.7.8 Figure of Merit for ADCs

The demand for lower power-dissipating electronic systems has become a challenge to IC designers, including designers of ADCs. As a result, a figure of merit (FOM) was devised by the International Solid State Circuits Conference (ISSCC) Program Committee to compare available and future sampling-type ADCs. The FOM is derived by dividing the device's power dissipation (in watts) by the product of its resolution (in 2^n bits) and its sampling rate (in hertz). The result is multiplied by 10^{12}. This is expressed by the equation

$$\text{FOM} = \frac{PD}{R * SR} 10^{12}, \tag{5.5}$$

where PD is the power dissipation (in watts), R is the resolution (in 2^n bits), and SR is the sampling rate (in hertz). Therefore, a 12-bit ADC sampling at 1 MHz and dissipating 10 mW has an FOM rounded off to 2.5. This FOM is expressed in picojoules of energy per unit conversion (pJ/conversion). For details and a comparison of performance of some monolithic ICs, see O'Leary [31].

5.8 DACs

Digital-to-analog (D/A) conversion is an essential function in data-processing systems. DACs provide an interface between the digital output of signal processes and the analog world. Moreover, as discussed previously, multi-step ADCs employ interstage DACs to reconstruct analog estimates of the input signal. Each of these applications imposes certain speed, precision, and power dissipation requirements on the DAC, mandating a good understanding of various D/A conversion techniques and their tradeoffs.

5.8.1 General Considerations

A DAC produces an analog output, A, proportional to the digital input D:

$$A = \alpha D, \tag{5.6}$$

where α is a proportionality factor. Because D is a dimensionless quantity, α sets both the dimension and the full-scale range of A. For example, if α is a current quantity, I_{ref}, then the output can be expressed as

$$A = I_{ref}D. \tag{5.7}$$

In some cases, it is more practical to normalize D with respect to its full-scale value, $2n$, where m is the resolution. For example, if α is a voltage quantity, V_{ref}, then

$$A = V_{ref}\frac{D}{2^m}. \tag{5.8}$$

From Equation 5.7 and Equation 5.8, we can see that, in a DAC, each code at the digital input generates a certain multiple or fraction of a reference at the analog output. In practical monolithic DACs, conversion can be viewed as a reference multiplication or division function, where the reference may be one of the three electrical quantities: voltage, current, or charge.

The accuracy of this function determines the linearity of the DAC, whereas the speed at which each multiple or fraction of the reference can be selected and established at the output gives the conversion rate of the DAC. Figure 5.38

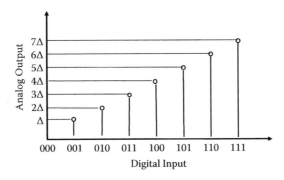

FIGURE 5.38
Input/output characteristic of an ideal 3-bit DAC.

TABLE 5.3

Binary, Thermometer, and One-of-n Codes

Decimal	0	1	2	3
Binary	00	01	10	11
Thermometer	0	0	0	0
	0	0	0	1
	0	0	1	1
	0	1	1	1
One-of-n	0	0	0	0
	0	0	0	1
	0	0	1	0
	0	1	0	0

shows the input/output characteristics of an ideal 3-bit DAC. The analog levels generated at the output follow a straight line passed through the origin and the full-scale point. We should mention that in some applications, such as "companding" (compressing and expanding) DACs, the desired relationship between D and A is nonlinear [31], but in this chapter we discuss only "linear" or "uniform" DACs, that is, those that ideally behave according to Equation 5.7 or Equation 5.8.

The digital input to a DAC can assume any predefined format but eventually must be of a form easily convertible to analog. Table 5.3 shows three formats often used in DACs: binary, thermometer, and one-of-n codes. The latter two are shown in column form to make visualization easier.

5.8.2 Performance Parameters and Data Sheet Terminology

In manufacturers' data books, many terms are used to characterize DACs. The following is a basic guideline only, and the reader is referred to manufacturers' data sheet guidelines for more application-oriented descriptions. Figure 5.41 illustrates some of the metrics that are listed in Table 5.4.

Converters can have three different formats, as indicated in Figure 5.39, namely, unipolar, offset bipolar, and sign magnitude bipolar. In unipolar converters, the analog port has only single polarity. These are the simplest type, but bipolar converters are generally more useful in real-world applications. An "offset bipolar" converter is merely a unipolar converter with an accurate 1 MSB of negative offset. Many converters are designed so that this offset can be switched in and out for operating as unipolar or bipolar converters. A sign magnitude converter is more complex and has n bits of magnitude information and an additional bit that corresponds to the sign of the analog signal. Sign magnitude DACs are not very common, and sign magnitude ADCs are used in digital multimeters.

Nonmonotonicity and missing codes are two important parameters of converters. As indicated in Figure 5.40a, nonideal transfer functions for a

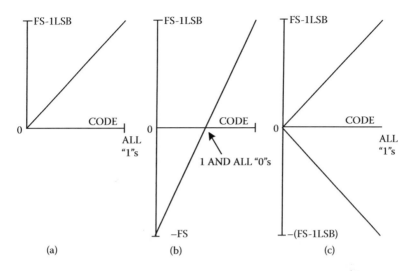

FIGURE 5.39
Unipolar and bipolar converters: (a) unipolar; (b) offset bipolar; (c) sign magnitude bipolar.

TABLE 5.4

DAC Performance Parameters

Parameter	Description
Differential nonlinearity (DNL)	Maximum deviation in the output step size from the ideal value of 1 LSB.
Integral nonlinearity (INL)	Maximum deviation of the input/output characteristic from a straight line passed through its endpoints. The difference between the ideal and actual characteristics is called the INL profile.
Offset	Vertical intercept of the straight line passed through the endpoints.
Gain error	Deviation of the slope of the line passed through the endpoints from its ideal value (usually unity).
Settling time	Time required for the output to experience a full-scale transition and settle within a specified error band around its final value.
Glitch impulse area	Maximum area under any extraneous glitch that appears at the output after the input code changes; also called glitch energy in the literature, even though it has no energy dimension.
Latency	Total delay from the time the digital input changes to the time the analog output has settled within a specified error band around its final value. Latency may include multiples of the clock period if the digital logic in the DAC is in a pipeline.
Signal-to-noise (+ distortion) ratio (SNDR or SINAD)	Ratio of the signal power to the total noise and harmonic distortion at the output when the input is a digital sine wave.

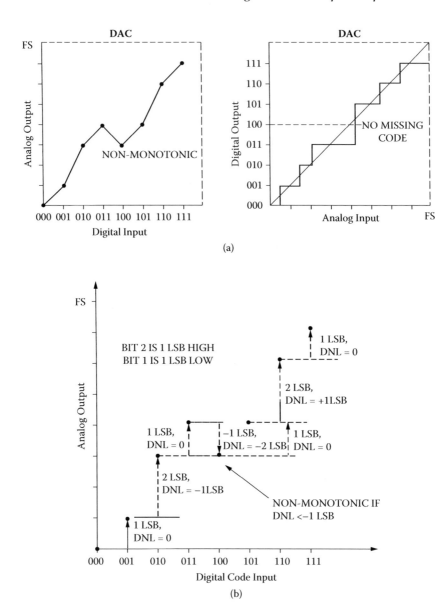

FIGURE 5.40
Nonmonotonicity and missing codes in converters: (a) transfer functions of 3-bit converters; (b) details of DAC differential nonlinearity; (c) details of ADC differential nonlinearity;
(d) nonmonotonic ADC with missing codes. (Courtesy of Analog Devices, Norwood, MA.)

DAC and ADC are due to the effects of the DNL error. The DNL of a DAC
is examined more closely in Figure 5.40b. If the DNL of a DAC is less than
−1 LSB at any transition, the DAC is nonmonotonic, where the transfer characteristic contains one or more localized maxima or minima. A DNL greater
than +1 LSB does not cause nonmonotonicity, but it is still undesirable. In

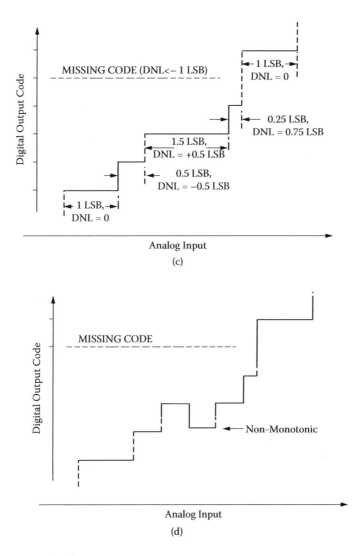

FIGURE 5.40 (continued)

many DAC applications, it is critically important that DACs be monotonic. For example, in closed-loop systems, this problem can change negative feedback to positive feedback. In Figure 5.40c, the DNL of an ADC is examined in more detail. ADCs can be nonmonotonic, but a more common result of excess DNL is missing codes. Missing codes in an ADC are as objectionable as nonmonotonicity in a DAC, which also results from DNL being less than −1 LSB. As shown in Figure 5.40d, not only can ADCs have missing codes, but they can also be nonmonotonic. These effects can present major problems in some applications, such as servo systems. In a DAC, there can be no missing codes, as each digital input word produces a corresponding analog output.

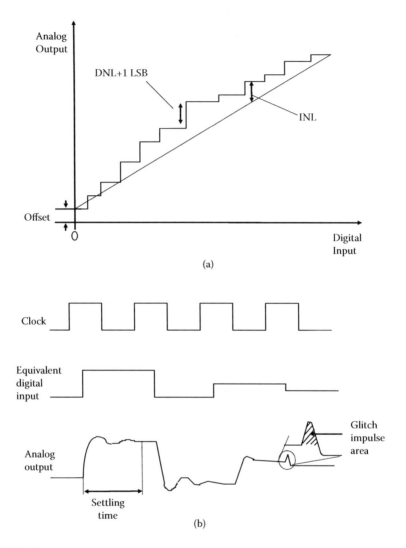

FIGURE 5.41
Parameters of DACs: (a) static parameters; (b) dynamic parameters.

A successive approximation ADC with an internal nonmonotonic DAC will generally produce missing codes but remain monotonic. ADCs that use subranging architecture may exhibit nonmonotonicity and missing codes, depending on the design [2]. Some of these metrics are listed in Table 5.4, and Figure 5.41 illustrates selected ones of these. Among these parameters, DNL and INL are usually determined by the accuracy of reference multiplication or division, settling time and delay are functions of output loading and switching speed, and glitch impulse depends on the DAC architecture and design.

5.9 Principles Used in DACs

The digital code transferred to the input of a DAC is ultimately converted into an analog signal based on three possibilities: voltage division, current division, and charge division. These are briefly discussed in the following sections.

5.9.1 Voltage Division

A given reference voltage, V_{REF}, can be divided into n equal segments using a ladder composed of n identical resistors $R_1 = R_2 = \ldots = R_n$ (n typically is a power of 2) (Figure 5.42a). An m-bit DAC requires a ladder with 2^m resistors manifesting the exponential growth of the number of resistors as a function of resolution. An important aspect of resistor ladders is the differential and integral nonlinearity they introduce when used in DACs. These errors result from mismatches in the resistors composing the ladder.

Most commonly used simple DAC structures are binary weighted DACs or ladder networks, but, although simple in structure, these require quite complex analysis. The simplest structure of all is the Kelvin divider, shown in Figure 5.42b. An n-bit version of this DAC simply consists of 2^n equal resistors in series. The output is taken from the appropriate tap by closing one of the 2^n switches.

5.9.2 Current Division

Instead of using voltage division, current division techniques can be used in DACs. Figure 5.43a shows how a reference current, I_{REF}, can be divided into n equal currents using n identical (bipolar or MOS) transistors. These currents can be combined to provide binary weighting, as depicted in Figure 5.43b, using a 3-bit case as the example. In this simple implementation, an m-bit DAC requires $2^m - 1$ transistors, resulting in a large number of devices for $m > 7$.

Although conceptually simple, the implementation of Figure 5.43a has two drawbacks: the stack of current division transistors on top of I_{REF} limits output voltage range, and I_{REF} must be n times each of the output currents. This requires a high-current device for the I_{REF} source transistor. There are techniques for alleviating these problems [10]. DACs that employ current division suffer from three sources of nonlinearity: current source mismatch, finite output impedance of current sources, and voltage dependence of the load resistor that converts the output current to voltage.

5.9.3 Charge Division

A reference charge, Q_{ref}, can be divided into n equal packets using n identical capacitors configured as in Figure 5.44. In this circuit, before S_1 turns on, C_1

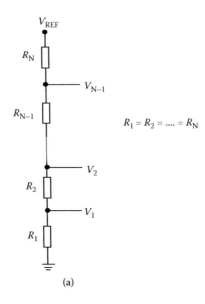

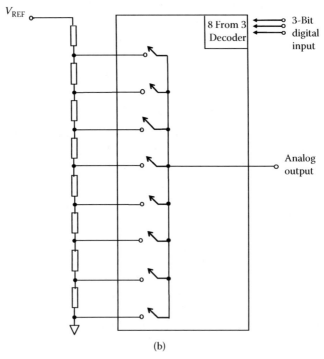

FIGURE 5.42
DAC using a voltage division technique: (a) basic resistor ladder; (b) Kelvin divider (3-bit DAC example).

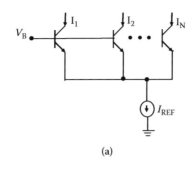

(a)

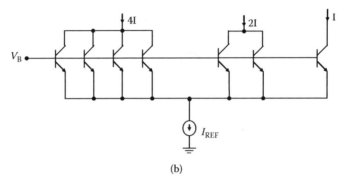

(b)

FIGURE 5.43
Current division: (a) uniform division; (b) binary division.

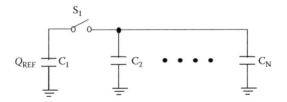

FIGURE 5.44
Simple charge division.

has a charge equal to Q_{ref}, whereas C_2 to C_n have no charge. When S_1 turns on, Q_{ref} is distributed equally among C_1 to C_n, yielding a charge of Q_{ref}/n on each. Further subdivision can be accomplished by disconnecting one of the capacitors from the array and redistributing its charge among some other capacitors.

Although the circuit of Figure 5.44 can operate as a DAC if a separate array is employed for each bit of the digital input, the resulting complexity prohibits its use for resolutions greater than 6 bits. A modified version of this circuit is shown in Figure 5.45a. Here, identical capacitors $C_1 = \ldots = C_n = C$ share the same top plate, and their bottom plates can be switched from ground to a reference voltage, V_{ref}, according to the input thermometer code. In other

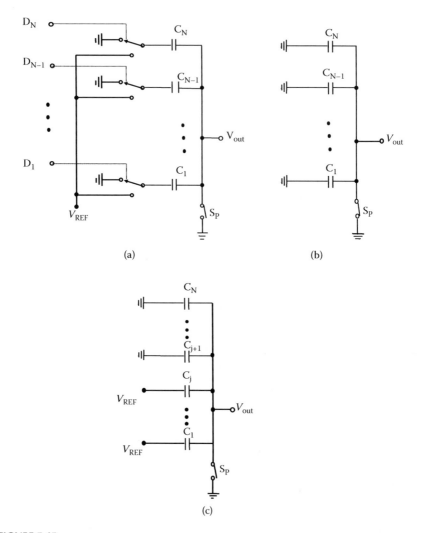

FIGURE 5.45
Modified charge division: (a) configuration; (b) circuit of (a) in discharge mode; (c) circuit of (a) in evaluate mode.

words, each capacitor can inject a charge equal to CV_{ref} onto the output node, producing an output voltage proportional to the height of the thermometer code. The circuit operates as follows. First, S_p is on and the bottom plates of C_1 to C_n are grounded, discharging the array to 0 (Figure 5.45b). Next, S_p turns off, and a thermometer code with height j is applied at D_1 to D_n, connecting the bottom plate of C_1 to C_j to V_{ref} and generating an output equal to jV_{ref}/n (Figure 5.45c). This circuit, in a strict sense, is a voltage divider rather than a charge divider. In fact, the expression relating its output voltage to V_{REF} and the value of the capacitors is quite similar to that of resistor ladders.

Nonetheless, in considering nonlinearity and loading effects, it is helpful to remember that the circuit's operation is based on charge injection and redistribution.

The nonlinearity of capacitor DACs arises from three sources: capacitor mismatch, capacitor nonlinearity, and the nonlinearity of the junction capacitance of any switches connected to the output code. For details and implementation of capacitor DACs, see Razavi [10].

5.10 DAC Architectures

With the basic principles of D/A conversion explained, we can study this function from an architectural perspective. This section describes DAC architecture based on resistor ladders and current-steering arrays, with an emphasis on stand-alone applications. Although capacitor DACs frequently are used in ADCs, they have not been popular as stand-alone circuits.

5.10.1 Resistor Ladder DAC Architectures

The simplicity of resistor ladder DACs using MOS switches makes these architectures attractive for many applications. Simple ladder networks with simple voltage division, as in Section 5.8.1, have several drawbacks: they require a large number of resistors and switches (2^m, where m is the resolution) and exhibit a long delay at the output. Consequently, alternative ladder topologies have been devised to improve the speed and resolution.

5.10.2 Ladder Architecture with Switched Subdividers

In high-resolution applications, the number of devices in a DAC can be prohibitively large. It therefore is plausible to decompose the converter into a coarse section and a fine section so that the number of devices becomes proportional to approximately $2^m/2$ rather than 2^m, where m is the overall resolution. Such an architecture is shown in Figure 5.46a. In this circuit, a primary ladder divides the main reference voltage, generating 2^j equal voltage segments. One of these segments is selected by the j most significant bits of $k + j = m$. If $k = j$, the number of devices in this architecture is proportional to $2^m/2$. It also is possible to utilize more than two ladders to further reduce the number of devices at high resolutions.

Figure 5.46b depicts a simple implementation of this architecture using MOS switches that are driven by one-of-n codes in both stages [32]. Depending on the environment, these codes are generated from binary or thermometer code inputs. The details and drawbacks of this implementation are discussed in Razavi [10].

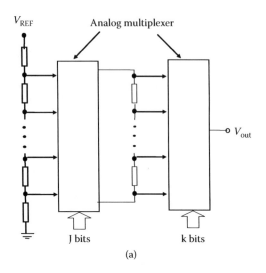

(a)

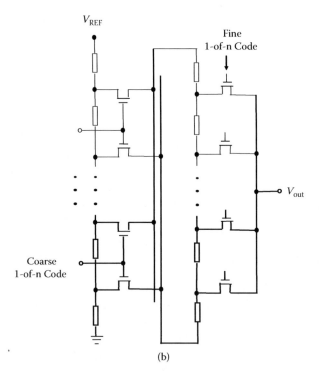

(b)

FIGURE 5.46
Resistor ladder DAC with a switched subdivider: (a) block diagram; (b) implementation.

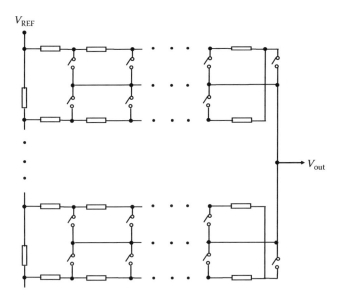

FIGURE 5.47
Intermeshed ladder architecture with one level multiplexing.

5.10.3 Intermeshed Ladder Architecture

Some of the drawbacks of ladder DACs can be alleviated through the use of intermeshed ladder architectures [10]. In these architectures, a primary ladder divides the main reference voltage into equal segments, each of which is sub-divided by a separate, fixed secondary ladder. Figure 5.47 illustrates such an arrangement [17], where all the switches are controlled by a one-of-n code.

The intermeshed ladder has several advantages over single-ladder or switched-ladder architectures. This configuration can have smaller equivalent resistance at each tap than a single-ladder DAC having the same resolution, allowing faster recovery. Also, because the secondary ladders do not switch, their loading on the primary ladder is constant and uniform. Furthermore, the DNL resulting from finite on-resistance of switches does not exist here.

5.10.4 Current-Steering Architecture

Most high-speed DACs are based on a current-steering architecture. Because these architectures can drive resistive loads directly, they require no high-speed amplifiers at the output and potentially are faster than other types of DACs. Although the high-speed switching of bipolar transistors makes them the natural choice for current-steering DACs, many designs have been recently reported in CMOS technology as well.

5.10.5   R-2R Network-Based Architectures

To realize binary weighting in a current-steering DAC, an R-2R ladder can be incorporated to relax device scaling requirements. Figure 5.48a illustrates an architecture that employs an R-2R ladder in the emitter network. A network with an R-2R ladder in collector networks is shown in Figure 5.48b. For details, see Razavi [10] and Hoeschele [32].

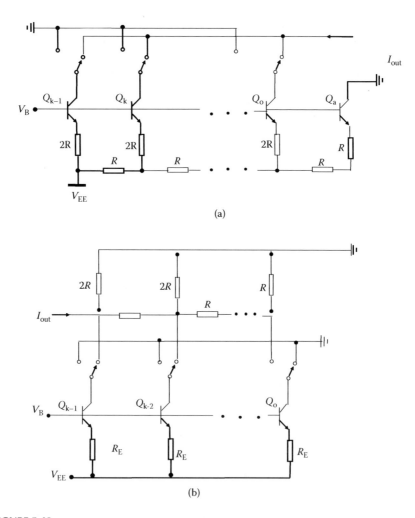

(a)

(b)

FIGURE 5.48
Current-steering DAC with an R-2R ladder: (a) R-2R ladder in the emitter; (b) R-2R ladder in the collector.

5.10.6   Other Architectures

Other architectures for DACs include segmented current-steering versions, multiplying DACs, and Σ-Δ types. For details, see the *Linear Design Seminar* [5], Razavi [10], and Hoeschele [32].

5.11   Data Acquisition System Interfaces

5.11.1   Signal Source and Acquisition Time

Continued demand for lower-power, lower-cost systems increases the likelihood that a mixed-signal design will operate from a single 3.3-V or 5-V power supply. Doing away with traditional ±15 V analog power supplies can help one meet power and cost goals, but it also will eliminate some options.

Most low-voltage ADC and data acquisition system chips are designed for easy analog and digital interfaces. The ICs' digital interfaces generally are compatible with popular microcontrollers, and the devices almost always can accept analog input signals that range from ground to the positive supply voltage; the span is set by an internal or external band-gap voltage reference. Virtually all ADCs that operate from 5 V or less are CMOS devices that use arrays of switches and capacitors to perform their conversions. Although the architectural details vary from design to design, the input stage of this type of converter usually includes a switch and a capacitor that present a transient load to the input signal source. The simplified schematic of Figure 5.49 shows how these input stages affect the circuits that drive them.

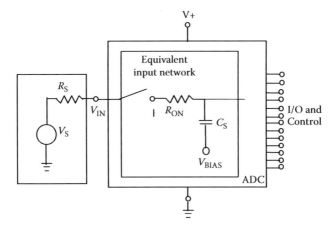

FIGURE 5.49
Simplified interface between a low-voltage ADC and a signal source.

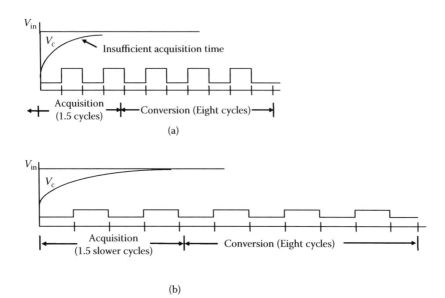

FIGURE 5.50
The effect of source resistance: (a) insufficient acquisition time; (b) slowing the clock to increase acquisition time.

R_{on} is not a separate component; it is the on-resistance of the internal analog switch. Sampling capacitor C_s connects to an internal bias voltage, whose value depends on the ADC's architecture. In a sampling ADC, the switch closes once per conversion during the acquisition (sampling) time. The on-resistance of the sampling switches ranges from about 5 to 10 kΩ in many low-resolution successive approximation ADCs to 70 Ω in some multistep or half-flash converters. The capacitors can be as small as 10 pF or more in higher-resolution devices.

When the sampling switch closes, the capacitor begins to charge through the switch and source resistance. After a time interval that is usually controlled by counters or timers within the ADC, the switch opens and the capacitor stops charging. The acquisition time described in Figure 5.13c is actually the time during which the switch is closed and the capacitor charges. As long as the source impedance is low enough, the capacitor has time to charge fully during the sampling period and can work properly at its rated speed with a reasonable source resistance (1 kΩ is common). Larger source impedance slows the charging of the sampling capacitor and can cause significant errors unless one takes steps to avoid them. Figure 5.50 illustrates this. Figure 5.50a illustrates the case of insufficient acquisition time. Figure 5.50b illustrates a case in which the problem can be solved by slowing the clock. For further details, see Lancanelle [33], Sheer [34] and Swager [35].

5.11.2 The Amplifier–ADC Interface

Operational amplifiers nearly always are present in data acquisition systems, performing basic signal conditioning ahead of the ADC. Their interactions with ADCs affect system performance. Although many amplifiers are good at driving a variety of static loads, the switched nature of the ADC input stage can introduce problems with some amplifiers, especially the low-power, low-speed devices most likely to be used in 3-V and 5-V systems. Using the simple model in Figure 5.51a, the load presented to the amplifier by the ADC input keeps switching abruptly between an open circuit and a series RC network connected

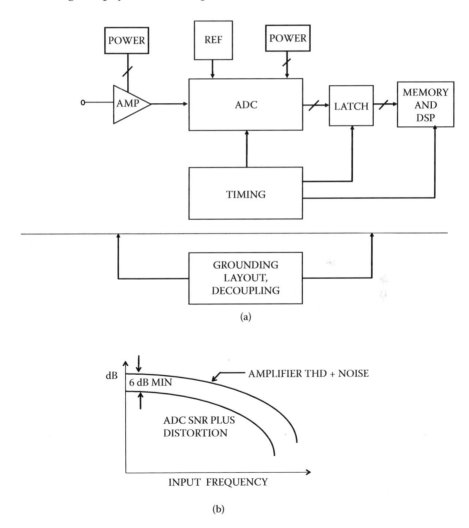

(a)

(b)

FIGURE 5.51
ADC–amplifier interface: (a) basic elements; (b) performance expected from ADC and amplifier. (Source: Analog Devices, Norwood, MA.)

TABLE 5.5

ADC Drive Amplifier Considerations

Performance Requirements	Parameter
AC performance	Bandwidth, settling time, harmonic distortion, total harmonic distortion, noise, THD + noise
DC performance	Gain, offset, drift, gain nonlinearity
General	As a general principle, select first for AC performance, then evaluate DC performance. Always consult the data sheet for recommendations.

to an internal voltage source. The op amp's response to the sudden load current and impedance change depends on several parameters. Among them are the device's gain-bandwidth product, slew rate, and output impedance.

Selecting the appropriate drive amplifier for an ADC involves many considerations. Because the ADC drive amplifier is in the signal path, its error sources (both DC and AC) must be considered in calculating the total error budget. Ideally, the AC and DC performance of the amplifier should be such that there is no degradation of the ADC performance. Achieving this is rarely possible, however; therefore, the effects of each amplifier error source on system performance should be evaluated individually.

Evaluating and selecting op amps based on the DC requirements of the system is a relatively straightforward matter. For many applications, however, it is more desirable first to select an amplifier on the basis of AC performance (bandwidth, THD, noise, etc.). The AC characteristics of ADCs are specified in terms of SNR, ENOB, and distortion. The drive amplifier should have performance better than that of the ADC so that maximum dynamic performance is obtained (see Figure 5.51b). If the amplifier AC performance is adequate, the DC specifications should be examined in terms of system performance. Table 5.5 summarizes the ADC drive amplifier considerations. Further details can be found in the *Linear Design Seminar* [5].

Other considerations in interfacing are input clamping and protection, drive amplifier noise configurations, ADC reference voltage considerations, settling time considerations, and the like. These are beyond the scope of this chapter, and the reader is referred to other sources for more details [2,5,7,17,33,34]. Interfaces between data converters and digital signal processors are discussed in Chapter 7. A good practical reference for designers is [37].

References

1. Garrett, P., *High Performance Instrumentation and Automation*, CRC Press, Boca Raton, FL, 2005.
2. Kester, W., ed., *Analog to Digital Conversion*, Analog Devices, Norwood, MA, 2004.

3. Hill, G., The benefits of undersampling, *Electronic Design*, July 11, 69, 1994.
4. Fonte, G.C.A., Apply fundamentals to avoid surprises with sample systems, *EDN*, June 24, 137, 1993.
5. *Linear Design Seminar*, Analog Devices, Norwood, MA, 1995.
6. Steer, R.W., Antialiasing filters reduce errors in A/D converters, *EDN*, March 30, 171, 1989.
7. *1992 Amplifier Application Guide*, Analog Devices, Norwood, MA, 1992.
8. Digital oscilloscope concepts, Engineering Note 37W-6136, Tektronix, Beaverton, OR, 1986.
9. Demler, M.J., Time domain techniques enhance testing of high speed ADCs, *EDN*, March 30, 115, 1992.
10. Razavi, B., *Data Conversion System Design*, IEEE Press, Piscataway, NJ, 1995.
11. Shill, M.A., Servo loop speeds tests of 16-bit ADCs, *Electronic Design*, February 6, 93, 1995.
12. Lauzon, P., Decipher high-sample rate ADC specs, *Electronic Design*, March 20, 91, 1995.
13. Lowenstein, E.B., Test analog-to-digital converters for DNL, *Test and Measurement World*, December, 15, 1994.
14. Doernberg, J., Lee, H.S., and Hodges, D., Full-speed testing of A/D converters, *IEEE Journal of Solid State Circuits*, 19, 820, 1984.
15. Reeder, R., Compare aperture delay between ADCs, *Test and Measurement World*, May, 31, 2002.
16. McGoldrick, P., Architectural breakthrough moves conversion into mainstream, *Electronic Design*, January 20, 67, 1997.
17. *High Speed Design Techniques*, Analog Devices, Norwood, MA, 1996.
18. Swager, A.W., Flash ADCs push speed and bandwidth limits, *EDN*, May 29, 93, 1989.
19. Kester, W., Flash ADCs provide the basis for high-speed conversion, *EDN*, January 4, 101, 1990.
20. Kester, W., DSP test techniques keep flash ADCs in check, *EDN*, January 18, 133, 1990.
21. Kester, W., Measure flash ADC performance for trouble free operation, *EDN*, February 1, 103, 1990.
22. Kularatna, N., *Digital and Analogue Instrumentation Testing and Measurement*, IEE, London, 2003.
23. Ushani, R.K., Subranging ADCs operate at high speed with high resolution, *EDN*, April 11, 139, 1991.
24. Ushani, R.K., Dynamic specifications describe performance of subranging ADCs, *EDN*, April 25, 155, 1991.
25. Ushani, R.K., Classical tests are inadequate for modern high-speed converters, *EDN*, May 9, 155, 1991.
26. Candy, J.C. and Temes, G.C., *Oversampling Delta-Sigma Data Converters: Theory, Design and Simulation*, IEEE Press, Piscataway, NJ, 1992.
27. Hares, D., Delta-sigma analog to analog converter solves tough design problems, *EDN*, April 27, 111, 1995.
28. Stewart, R.W. and Pfann, E., Oversampling and sigma-delta strategies for data conversion, *Electronics and Communication Engineering Journal*, February, 37, 1998.
29. Jonston, J.E., Techniques and benefits of self calibration in A/D converters, *Electronic Engineering*, November, 41, 1993.
30. O'Leary, S., Self-calibrating ADCs offer accuracy, flexibility, *EDN*, June 22, 93, 1995.

31. Goodenough, F., ADCs move to cut power dissipation, *Electronic Design*, January 9, 69, 1995.
32. Hoeschele, D.F., *Analog-to-Digital and Digital-to-Analog Conversion Techniques*, 2nd ed., John Wiley Interscience, New York, 1994.
33. Lacanette, K., To build data acquisition systems that run from 5 to 3.3V, know your ICs, *EDN*, September 29, 89, 1994.
34. Sheer, D., Monolithic high-resolution ADCs, *EDN*, May 12, 116, 1988.
35. Swager, A.W., Evolving ADCs demand more from drive amplifiers, *EDN*, September 29, 53, 1994.
36. Holdaway, M., Designing antialias filters for ADCs, *EDN*, November 26, 65, 2006.
37. Baker, B., *A Baker's Dozen: Real Analog Solutions for Digital Designers*, Newnes, London, 2005.

6

Configurable Logic Blocks for Digital Systems Design

Morteza Biglari-Abhari

CONTENTS

6.1   Introduction..343
6.2   Programmable Logic Devices 344
6.3   Field Programmable Gate Arrays..346
        6.3.1   FPGA Types 349
6.4   Design Tools for PLDs..351
6.5   Overview of a Hardware Description Language.....................352
6.6   A Design Example: Pulse-Width Measurement.....................357
6.7   FPGAs in Practice 361
References 361

6.1   Introduction

Recent significant advancements in semiconductor technology and high demands in consumer electronic products have changed the digital system design methodology. The reduced time to market, increased complexity of digital systems, and feasibility of integrating several hundred million transistors on a single chip have led to the employment of more sophisticated design automation tools. Complex programmable logic devices (CPLDs) and field programmable gate arrays (FPGAs), which previously were suitable for only small digital systems or rapid prototyping, have now become the target implementation platform. This chapter discusses the basics of CPLDs and FPGAs and presents an introduction to design methodology for digital systems based on these.

6.2   Programmable Logic Devices

The concept of programmable logic device (PLD) structures is based on realizing Boolean equations as sums of products. Consider the simple example of a combinatorial digital circuit shown in Figure 6.1. The logic implementation (without simplifying the Boolean equations) can be represented as shown in Figure 6.1c. In principle, any combinatorial circuit can be implemented using AND-OR gates in two logic levels. Input signals are connected to the AND gates, and the outputs of AND gates are used as inputs for OR gates where the final output signals are generated. Therefore, sum-of-products representation of the circuit can be implemented using an AND array followed by an OR array, as illustrated in Figure 6.2, which is the basis for PLDs.

PLDs are represented in three general forms: programmable read-only memories (PROMs), programmable logic arrays (PLAs), and programmable array logics (PALs). The main difference among these three forms is the way the programmability can be achieved. In PROMs, only OR arrays are programmable and AND arrays are fixed. In PLAs, OR arrays are fixed whereas AND arrays are programmable. PALs provide programmability for both OR and AND arrays, so they are more flexible.

PLAs and PALs have been used as glue logic in systems. The type of programmability (using fuses and antifuses, for example) differentiates different devices to be used for a specific application. CPLDs are a group of simple PLDs with programmable communication channels in a single chip, as shown in Figure 6.3. Storage elements such as latches and flip-flops can be added to the output of OR gates to implement sequential circuits.

A	B	C	D	Z
0	0	0	0	1
0	0	0	1	0
0	0	1	0	0
0	0	1	1	1
0	1	0	0	1
0	1	0	1	0
0	1	1	0	0
0	1	1	1	1
1	0	0	0	1
1	0	0	1	0
1	0	1	0	0
1	0	1	1	1
1	1	0	0	1
1	1	0	1	0
1	1	1	0	0
1	1	1	1	1

(a) Truth Table

$$Z = \overline{ABCD} + \overline{AB}C\overline{D} + \overline{A}B\overline{CD} + \overline{A}BCD + A\overline{BCD} +$$
$$A\overline{B}CD + AB\overline{CD} + ABCD$$

(b) Boolean Equation

$$Z = \Sigma m\ (0, 3, 4, 7, 8, 11, 12, 15)$$

(c) Sum-of-Products Realization

FIGURE 6.1
An example of a combinatorial circuit.

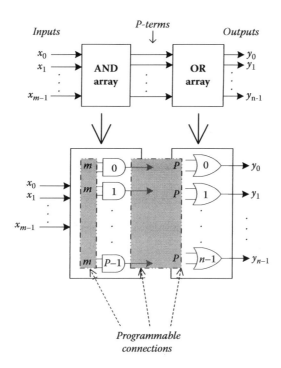

FIGURE 6.2
General structure of a PLD.

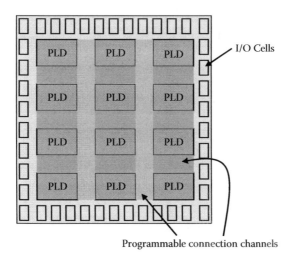

FIGURE 6.3
General block diagram of a CPLD.

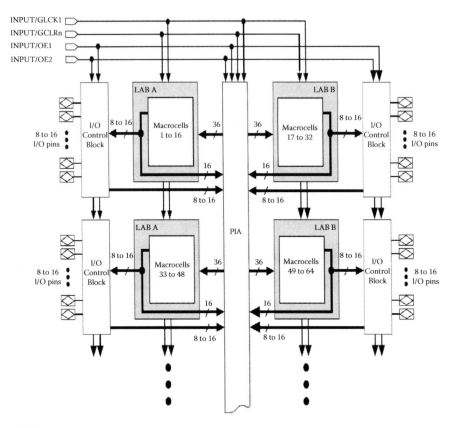

FIGURE 6.4
Architecture of an Altera CPLD device [16]. (Source: Altera.)

In commercial CPLDs, such as Altera's MAX 7000s devices, a group of PLDs (eight PLDs in this case) is organized as a bigger block, which is called a logic array block (LAB), as shown in Figure 6.4. Faster interconnection channels for communication among these PLDs are provided in an LAB. External routing channels provide interconnection resources among LABs. All communication channels are programmable and are properly programmed automatically for each application through the use of an electronic design automation (EDA) tool. Altera's MAX 7000s devices are based on electrically erasable PROM (EEPROM) technology that can be programmed in-system through the Joint Test Action Group (JTAG) interface.

6.3   Field Programmable Gate Arrays

To provide more flexibility and reconfigurability, FPGAs are preferred to CPLDs for most applications. FPGA architectures are designed based on

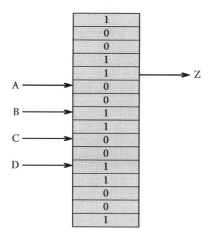

FIGURE 6.5
LUT-based implementation of $z =$ $\Sigma m(0,3,4,7,8,11,12,15)$.

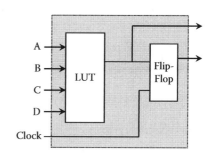

FIGURE 6.6
Basic structure of a logic element.

look-up tables (LUTs). Any N-input combinatorial circuit can be implemented using an N-input LUT, which is a small random access memory (RAM) block, and each memory location stores only one bit. The N inputs of an LUT are used to address different locations of the memory that store either a logical 0 or 1, depending on the Boolean equation representing the combinatorial circuit. Figure 6.5 shows the LUT implementation for the example of Figure 6.1. Latches and flip-flops can be added to the output of the LUT to provide a programmable element to implement sequential circuits. Figure 6.6 shows the basic structure of a logic cell or logic element that is the basic programmable primitive in most commercial FPGA families. A multiplexer (MUX) determines if the output of the LUT should be registered or not. Most early commercial FPGAs from Altera [1] and Xilinx [2] used a four-input LUT in each logic element. Recent FPGA architectures employ adaptive LUT structures in each logic element that can be programmed as an up-to-seven-input LUT or two four-input LUTs and two registers [3]. Different FPGA architectures employ LUT and storage elements differently to provide a generic reprogrammable resource referred to as a logic cell, logic element (LE), or slice.

Figure 6.7 illustrates the internal architecture of an LE for one Altera FPGA device family referred to as Cyclone [4]. Dotted lines show the extra programmable logic included to configure the LUT and the flip-flop. As addition/subtraction is a frequently used operation, additional logic such as carry chains is provided in each LE to implement fast adders. Also, more control signals are available to initialize flip-flops synchronously or asynchronously, as shown in Figure 6.7. To optimize the interconnection network among the neighboring logic elements, multiple LEs are grouped as a bigger block called a logic array block (LAB), which uses local faster connection

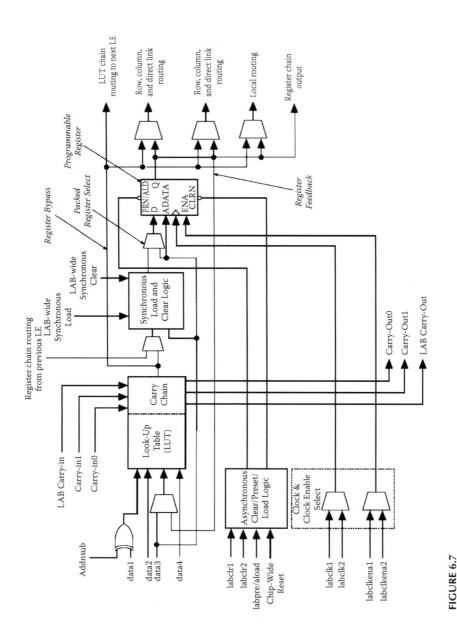

FIGURE 6.7

Architecture of an LE in Altera Cyclone FPGA devices [4]. (Source: Altera.)

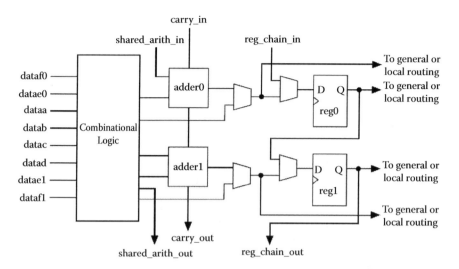

FIGURE 6.8
Simplified block diagram of a Stratix II ALM [3]. (Source: Altera.)

wires. Connections among LABs are provided through routing channels distributed throughout the FPGA chip.

State-of-the-art FPGA devices can provide design resources equivalent to several million gates to implement complex digital circuits. The number of inputs for LUTs and the number of flip-flops in each LE have been increased in recent FPGA architectures. For example, the Altera Stratix II family [3] provides adaptive logic modules (ALMs) that have an LUT that can be configured either as two four-input LUTs or one LUT with up to seven inputs, as indicated in Figure 6.8 and Figure 6.9.

In addition to LEs, embedded memory blocks in different sizes and speeds, special-purpose blocks such as multipliers, and multiply-and-accumulate (MAC) units are included in an FPGA for more flexible and efficient implementation of digital circuits. Some FPGA manufacturers provide hard-core blocks such as processor cores, which are fabricated in silicon on the FPGA chip, in addition to LEs and other special-purpose blocks. Figure 6.10 shows the chip layout of an Altera Stratix device, which has three different types of embedded memory blocks and digital signal processing (DSP) blocks (that have MAC units) as special-purpose blocks.

6.3.1 FPGA Types

The FPGAs can be categorized based on their programmable capabilities. The above FPGA device families are the so-called static RAM (SRAM)-based devices. Configuration bits for LEs and interconnections are stored in RAM. In this way, these FPGAs are considered to be reprogrammable. In some devices from Xilinx [2], it is possible to reconfigure part of the FPGA during run time. FPGA devices from Altera can be reconfigured over an Ethernet

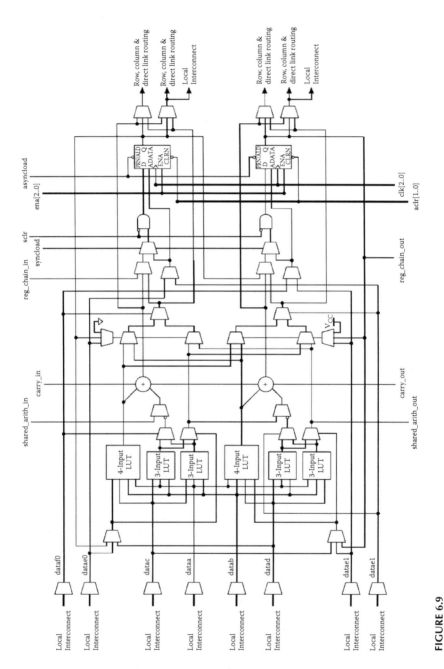

FIGURE 6.9

Architecture of an ALM in a Stratix II device [3]. (Source: Altera.)

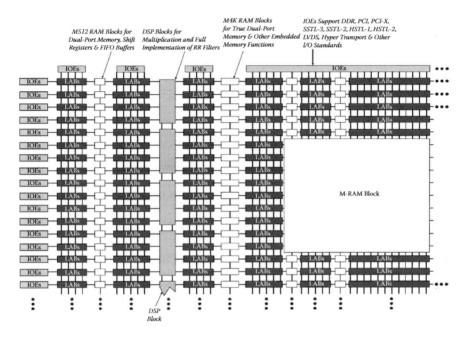

FIGURE 6.10
Block diagram of an Altera Stratix device [17]. (Source: Altera.)

connection. To reprogram the SRAM-based FPGAs, a nonvolatile storage device is present on the board, and during the power-up the device is configured through the configuration bits stored in the nonvolatile storage.

There are other types of FPGAs on the market. For example, Actel [5] and QuickLogic [6] FPGAs can be configured only once, as they use antifuse to set up connections and program the FPGA. Antifuse-based FPGAs consume less power and do not need extra storage for device configuration at power-on. However, SRAM-based devices are more flexible, as they can be reconfigured many times. Flash-based devices are also available to employ the benefit of antifuse- and SRAM-based devices.

6.4 Design Tools for PLDs

Designing to target PLDs (CPLDs and FPGAs) begins with capturing the system's specifications and its functional description using either a hardware description language (such as very high speed integrated circuit [VHSIC] hardware description language [VHDL] or Verilog) to model the functionality, or predesigned components (or intellectual properties [IPs]) to design the system. The synthesis tool, which is a part of the EDA tool, extracts the Boolean equations and state transition information from the functional

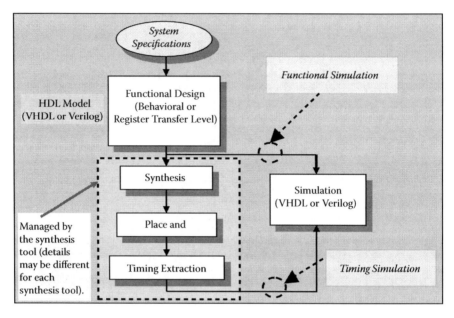

FIGURE 6.11
Design flow for CPLDs and FPGAs.

description. This logic information is mapped to the resources available on the target device (e.g., LEs for FPGAs) and proper connections are configured to satisfy the design constraints. Figure 6.11 gives the general steps that are followed in a typical design.

To cope with the design complexity, sophisticated EDA tools have become available to capture the system functional description, perform verification, and provide final implementation. Quartus II from Altera [7] is one of these tools. To clarify how these tools can be employed, a quick overview of a hardware description language is presented in the next section, followed by a design example.

6.5　Overview of a Hardware Description Language

One of the two most popular hardware description languages used for application-specific integrated circuits (ASICs) and FPGA/CPLD-based digital system design is VHDL. This section cannot cover all features of VHDL, and interested readers are referred to books covering this language in greater detail [8–10]. A quick introduction to VHDL is presented here to show how digital systems can be described for implementation on CPLDs and FPGAs. Verilog is the other popular hardware description language [11,12].

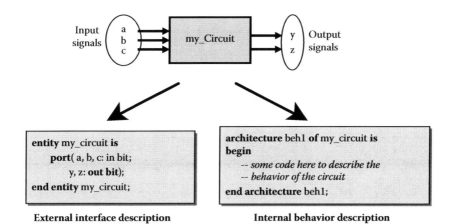

| entity my_circuit is port(a, b, c: in bit; y, z: out bit); end entity my_circuit; | architecture beh1 of my_circuit is begin -- some code here to describe the -- behavior of the circuit end architecture beh1; |

External interface description **Internal behavior description**

FIGURE 6.12
General structure of a VHDL description of digital systems.

Generally, a digital system can be modeled by describing its input/output (I/O) interface and functional behavior. In VHDL, the I/O interface description is separated from the functional behavior. This provides a mechanism for several functional descriptions at different levels of detail for the same system. For example, the system can be described at the behavior or algorithmic level. Later, when more timing information or predesigned components are available, a more appropriate description can be used.

Figure 6.12 illustrates the general structure of how I/O interfaces and the behavior of the circuit can be modeled. Highlighted words are VHDL keywords. VHDL is not a case-sensitive language. The keywords *entity* and *end entity* encapsulate the I/O interfacing of the circuits where I/O signals are pointed through the keyword *port* with their direction (as *in*, *out*, or *inout*) and their types (as **bit**, group of bits represented as **bit_vector** or a specific type such as **integer**, etc.). Table 6.1 and Table 6.2 show the most frequently used data types and operators to model the behavior.

To model the behavior of a digital system, the hardware description language must provide mechanisms to describe concurrency of operations and the structure of the system. This makes a hardware description language

TABLE 6.1

Some of the Built-In Data Types in VHDL

Type	Range of Values	Example
bit	0, 1	signal x: bit;
bit_vector	Array with each element of type bit	signal y: bit_vector(0 to 7);
integer	Implementation defined (32 or 64 bits)	signal a: integer := -1;
real	Implementation defined	signal z1: real := 1.3;
boolean	(TRUE, FALSE)	signal x: boolean

TABLE 6.2

Some of the Operators in VHDL for Behavior Description

Logical operators	and	or	nand	nor	xor	xnor
Relational operators	=	/= (not equal)	<	<=	>	>=
Addition operators	+	-	& (concatenation)			
Multiply/ division operators	*	/	mod (modulus)	rem (remainder)		
Miscellaneous operators	** power	abs (absolute value)	not (invert)			

different compared to a high-level programming language like C/C++ or Pascal. Structure represents how different components are connected. To model wires, the keyword *signal* is used. Concurrency is modeled by using *process* statements and *concurrent signal assignments*. The *process* statement uses sequential statements, which are mostly similar to what is provided in a high-level programming language, to describe the behavior of a part of the circuit. Each process can be suspended until occurrence of a specific event, which represents the result of changes on the signals, and then resumed. For example, if a *process* statement is used to model a combinatorial circuit, it must be invoked once in the beginning and then suspended until a value change or event occurs on one of the input signals where the process is sensitive to it. In this way, the behavior of a combinatorial circuit is modeled correctly. Figure 6.13 indicates the general structure of the process statement. An example of how it can be used is shown in Figure 6.14, which models a 1-bit full-adder circuit. Simpler circuits can be modeled using concurrent signal assignment as well. However, a process statement is general enough to model more complex circuits. The purpose of the example in Figure 6.14 is to show how a digital circuit can be modeled in VHDL; it does not mean it is the best description for this circuit. For example, a simpler description for the same 1-bit full-adder circuit uses concurrent signal assignment statements.

Variables and signals are two different concepts in VHDL. Variables keep the intermediate results temporarily similar to variables in a high-level programming language. However, signals represent wires or logical primitives and storage elements. Signal assignment is done using the <= operator, whereas variables are assigned using :=. Values are assigned to variables immediately so the following statements can use the new assigned values. However, this is not the case for signal assignments. In this case, new values are not assigned until the process reaches the last statement or is suspended using a wait statement. This is usually a source of confusion and introduces bugs for beginners in VHDL. Figure 6.15 shows an example to indicate the differences between signal and variable assignments. In Figure 6.15a,

```
[process-label:] process [ (sensitivity-list) ] [is]
  [process-item-declarations( such as variables) ]
begin
-- sequential-statements are
        variable-assignment-statement
        signal-assignment-statement
        wait-statement
        if-statement
        case-statement
        loop-statement
        null-statement
        exit-statement
        next-statement
        procedure-call-statement
        return-statement
end process [process-label];
```

Items in brackets are optional. Any number of sequential statements may be in a process.

FIGURE 6.13
General form of a process statement.

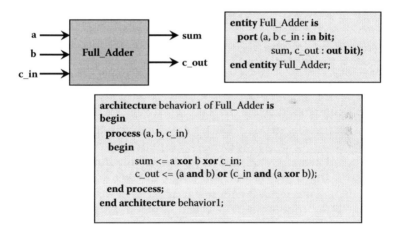

```
entity Full_Adder is
  port (a, b c_in : in bit;
            sum, c_out : out bit);
end entity Full_Adder;
```

```
architecture behavior1 of Full_Adder is
begin
  process (a, b, c_in)
  begin
        sum <= a xor b xor c_in;
        c_out <= (a and b) or (c_in and (a xor b));
  end process;
end architecture behavior1;
```

FIGURE 6.14
The VHDL description of a 1-bit full adder using process statement.

the variable **v1** is assigned immediately; so its new value is available to be assigned to signal **s2**. Figure 6.15b shows the waveform for Figure 6.15a. However, in Figure 6.15c, the current value of **x** is not assigned to signal **s1** immediately; so the old value of signal **s1** is assigned to signal **s2**. This results in the waveform shown in Figure 6.15d.

Signal assignments outside a process statement are called concurrent signal assignments and are executed concurrently. However, when signal assignment statements are used inside a process, they are considered as sequential statements in the process.

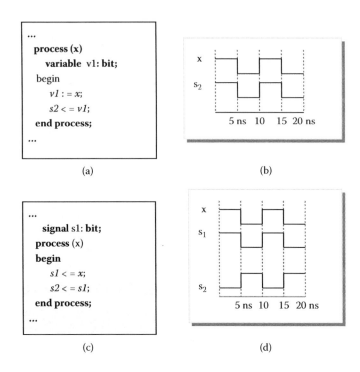

(a)   (b)

(c)   (d)

FIGURE 6.15
Pieces of VHDL code demonstrate the differences between signal and variable assignments.

To conclude this section, the VHDL code for a parameterized modulo-10 counter is shown in Figure 6.16. The term parameterized means that a default value is given to specific parameters at design time. When these systems are used as a component in another design, the default value of the parameters can be overridden to suit the current design. The keyword *generic* in VHDL is used to define and initialize these parameters. In this example, the maximum value of the counter is represented as $N - 1$, and N is declared as an integer set to 10.

Attributes in VHDL are used to determine some types of information about an object (such as signal). For example, *event* is an attribute for signals that returns TRUE if an event occurs on the signal and otherwise FALSE. This attribute is used to detect the active edge of the *clk* signal in this example. *If* statements in VHDL are used to determine only a part of the code to be evaluated based on a Boolean condition. As shown in Figure 6.15, the first *if* statement determines the occurrence of the active clock edge. The statements in the second *if* or its *else* parts are evaluated depending on the activity of the *reset* signal, or *enable* signal for normal operation of the counter.

It should be noted that signals that are declared as output port cannot be read like other signals. Therefore, a variable of the same type as the output port signal (variable **v_Q** in this example) is defined to be used for behavioral

```
entity counter is
    generic (N : integer: = 10);
    port (clk, reset, enable : in bit;
            Q_out : out integer range 0 to N-1);
end entity counter;
```

```
architecture behavior of counter is
  begin
    process (clk)
        variable v_Q: integer range 0 to N-1;
    begin
      if (clk'event and clk = '1') then
        if (reset = '1') then
            v_Q := 0;
        elsif (enable = '1') then
            v_Q := v_Q + 1;
        end if;
        Q_out <= v_Q;
      end if;
    end process;
end architecture behavior;
```

FIGURE 6.16
The VHDL code for a modulo-*N* binary counter.

description. Then its value is assigned to the output port signal **Q_out**. Similarly, input port signals can only be read.

VHDL can also be used to write test bench code to verify the correctness of the functionality of the system described in VHDL. Interested readers are referred to various references [8–10] for more details about VHDL features to model digital circuits and write test benches.

6.6 A Design Example: Pulse-Width Measurement

This section presents an example of a digital system to demonstrate the design flow for FPGA-based system design. A pulse-width measurement system is to be designed to measure the pulse width in the range of 1 ms to 300 ms with a resolution of 5 ns. An input signal *start* indicates when the system should start the measurement. Once it is done, an output signal *done* is activated and a number appears in the output as *count* to represent the pulse width.

In order to measure the pulse width at the given resolution (i.e., 5 ns), a counter with period of 5 ns can be used to count up during the active pulse. This number indicates the pulse width, an error of at most ±5 ns. A higher-input clock frequency can be used to improve this resolution when necessary. Current state-of-the-art FPGAs can work at up to 500 MHz, depending on the circuit complexity and critical timing paths.

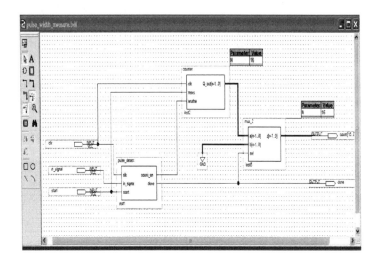

FIGURE 6.17
Schematic diagram of a pulse-width measurement system.

Figure 6.17 shows the block diagram of the circuit, which was designed using Altera Quartus II schematic diagram entry. Each component is already described in VHDL. The **pulse_detect** component determines the active edge of the input pulse and then asserts the *count_en* signal to enable the counter. Once the pulse width is measured, the *done* signal selects the right input of multiplexer **mux_2** to display the count number as the pulse width.

Figure 6.18, Figure 6.19, and Figure 6.20 show the VHDL code for components **counter**, **mux_2**, and **pulse_detect**, respectively. The VHDL code for the top level, which is captured as a schematic diagram, can be generated automatically in Quartus II. Figure 6.21 shows the simulation result performed using Quartus II. This digital system can be implemented on an Altera cyclone FPGA device (EP1C3T100C6) using 35 LEs.

VHDL provides a mechanism for using predefined functions, procedures, data types, and components encapsulated in a library. One of the popular standard libraries is IEEE 1164. This library provides different packages that define new data types, utilities, and conversion functions to be used in VHDL descriptions of digital systems. The first three lines of code in Figure 6.18 indicate that two packages from the IEEE library were used in this design. Data types *std_logic* and *std_logic_vector* are extended version of types *bit* and *bit_vector*, respectively, which define new values such as **X** (as do not care) and **U** (as undefined), in addition to **0** and **1** available in bit type. Also, the + and − operators are overloaded to be used with the *std_logic_vector* type.

As the purpose of this chapter is not to teach VHDL, only a very brief overview of VHDL statements used for the components in this example has been provided. Details about VHDL and the Altera Quartus II tool can be found in the references cited earlier.

```
library IEEE;
use IEEE.std_logic_1164.all;
use IEEE.std_logic_unsigned.all;

entity counter is
    generic (N : integer := 16);
    port (clk, reset, enable : in std_logic;
        Q_out : out std_logic_vector(N-1 downto 0));
end entity counter;

architecture behavior of counter is
  begin
    process (clk)
        variable v_Q: std_logic_vector(N-1 downto 0);
    begin
      if (clk'event and clk = '1') then
          if (reset = '1') then
              -- (others => '0') means use as many '0' as needed
              v_Q := (others => '0');
          elsif (enable = '1') then
              v_Q := v_Q + 1;
          end if;
          Q_out <= v_Q;
      end if;
    end process;
end architecture behavior;
```

FIGURE 6.18
The VHDL code for an *N*-bit binary counter.

```
library IEEE;
use IEEE.std_logic_1164.all;

entity mux_2 is
  generic (N : integer : = 16);
  port (a, b : in std_logic_vector(N-1 downto 0);
    sel : in std_logic;
    z : out std_logic_vector(N-1 downto 0));
end entity mux_2;

architecture arc1 of mux_2 is
begin
  z <= a when (sel = '1') else b;
end architecture arc1;
```

FIGURE 6.19
The VHDL code for an N bit, two-input multiplexer.

```vhdl
library IEEE;
use IEEE.std_logic_1164.all;

entity pulse_detect is
  port (clk, in_signal, start : in std_logic;
        count_en, done : out std_logic);
end entity pulse_detect;

architecture arc1 of pulse_detect is
  signal edge_detected : std_logic;
begin
  process (clk)
  begin
    if (clk'event and clk = '1') then
      if (start = '1') then
            edge_detected <= '0';
            done <= '0';
        else
          if (in_signal = '1' and edge_detected = '0') then
            edge_detected <= '1';
            count_en <= '1';
          elsif (in_signal = '0' and edge_detected = '1') then
            edge_detected <= '0';
            count_en <= '0';
            done <= '1';
          end if;
        end if;
    end if;
  end process;
end architecture arc1;
```

FIGURE 6.20
The VHDL code for a pulse_detect component.

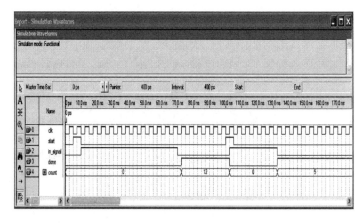

FIGURE 6.21
Simulation result for the pulse-width measurement system.

6.7   FPGAs in Practice

Today, FPGAs are a popular target platform in automotive and consumer electronic products [13] as well as medical applications, as low-cost development boards are now available from FPGA manufacturers. For example, a radar-based adaptive cruise control system has been implemented using FPGA to perform as a driver assistant [14]. Another example is the NASA Mars Rover, in which FPGA was used in some parts of the system because of its "on-the-fly" reprogrammability [15]. As FPGA technology advances, it is expected that FPGA use will increase as cost-effective application-specific digital systems.

References

1. Altera Corporation home page, http://www.altera.com.
2. Xilinx, Inc. home page, http://www.xilinx.com.
3. Literature: Stratix II Devices; available at http://www.altera.com/literature/lit-stx2.jsp.
4. Literature: Cyclone Devices; available at http://www.altera.com/literature/lit-cyc.jsp.
5. Actel Corporation home page, http://www.actel.com.
6. QuickLogic Corporation home page, http://www.quicklogic.com.
7. Quartus II Development Software Literature; available at http://www.altera.com/literature/lit-qts.jsp.
8. Ashenden, P.J., *The Designer's Guide to VHDL*, 2nd ed., Morgan Kaufmann, San Francisco, 2002.
9. Navabi, Z., *VHDL Analysis and Modeling of Digital Systems*, McGraw-Hill, New York, 1993.
10. Salcic, Z. and Smailagic, A., *Digital Systems Design and Prototyping Using Field-Programmable Logic and Hardware Description Languages*, Kluwer Academic, New York, 2000.
11. Navabi, Z., *Embedded Core Design with FPGAs*, McGraw-Hill, New York, 2007.
12. Mano, M.M. and Ciletti, M.D., *Digital Design*, 4th ed., Prentice Hall, Upper Saddle River, NJ, 2007.
13. Marsh, D., Low-cost kits: the new FPGA designer trend, *EDN*, November 26, 49, 2006.
14. Driver Assistance; available at http://www.altera.com/end-markets/auto/driver/aut-assist.html.
15. Xilinx Chips Land on Mars, Press Release 0412; available at http://www.xilinx.com/prs_rls/design_win/0412_marsrover.htm.
16. Literature: MAX 7000 Devices; available at http://www.altera.com/literature/lit-m7k.jsp.
17. Literature: Stratix Device Handbook; available at http://www.altera.com/literature/lit-stx.jsp.

7

Digital Signal Processors

CONTENTS

7.1   Introduction...364
7.2   What Is a DSP? 365
7.3   Comparison of a Microprocessor and a DSP...........................366
    7.3.1   The Importance of the Sample Period and Latency in the
           DSP World..367
    7.3.2   The Merging of Microprocessors and DSPs.............................369
7.4   Filtering Applications and the Evolution of DSP Architecture..........369
    7.4.1   Digital Filters...370
           7.4.1.1   FIR Filter..370
           7.4.1.2   IIR Filter...371
    7.4.2   Filter Implementation in DSPs.....................................373
    7.4.3   DSP Architecture 375
           7.4.3.1   Basic Harvard Architecture...........................376
           7.4.3.2   SISC Architectures..377
           7.4.3.3   Multiple-Access Memory-Based Architectures..........378
    7.4.4   Modifications to Harvard Architecture........................379
7.5   Special Addressing Modes 379
    7.5.1   Circular Addressing..379
    7.5.2   Bit-Reversed Addressing 381
7.6   Important Architectural Elements in a DSP............................381
    7.6.1   Multiplier/Accumulator..383
    7.6.2   Address Generation Units 385
           7.6.2.1   Data Address Units of ADSP-21xx Family: An
                    Example 385
    7.6.3   Shifters..387
    7.6.4   Loop Mechanisms 391
7.7   Instruction Set 393
    7.7.1   Computation Instructions: A Summary of the ADSP-21xx
           Family..394
           7.7.1.1   MAC Functions 394
           7.7.1.2   ALU Group Functions....................................397
           7.7.1.3   Shifter Group Functions................................397
    7.7.2   Other Instructions 398

7.8   Interface between DSPs and Data Converters 398
         7.8.1   Interface between ADCs and DSPs 399
                  7.8.1.1   Parallel Interfaces with ADCs 399
                  7.8.1.2   Interface between Serial Output ADCs 400
         7.8.2   Interfaces with DACs 401
                  7.8.2.1   Parallel Input DACs 401
                  7.8.2.2   Serial Input DACs 403
7.9   A Few Simple Examples of DSP Applications 403
7.10   Practical Components and Developments in the Late 1990s 407
7.11   Recent Technology Trends in DSP Design 408
Acknowledgment 410
References 410

7.1   Introduction

During the period from 1975 to 1985, many designers recognized the value of microprocessors in the design of electronic systems. Early generations of 4- and 8-bit microprocessors have evolved into 16-, 32-, and 64-bit components with complex instruction set computer (CISC) or reduced instruction set computer (RISC) architectures. Digital signal processors (DSPs) can be considered as special cases of RISC architectures or sometimes parallel developments of CISC systems to tackle real-time signal processing needs. Over the past quarter century, the field of digital signal-processing has grown from a theoretical infancy to a powerful practical tool and matured into an economical yet successful technology. At early stages, audio and the many other familiar signals in the same frequency band have appeared as a magnet for DSP-based systems development. The 1970s saw the implementation of signal-processing algorithms, especially for filters and fast Fourier transforms (FFTs), by means of digital hardware developed for the purpose. Early sequential program DSPs are described in Jones and Watson [1]. In the late 1990s, the market for DSPs was mostly generated by wireless, multimedia, and several other applications. By 2001, the market for DSPs was expected to grow to about US$9.1 billion [2]. Currently, communications represents more than half of the applications for DSPs [2].

With the rapid developments related to data converter ICs and other mixed-signal components allowing the conversion of analog real-time signals into digital, instrument designers saw the advantage of using DSP chips inside communication products and instruments. Cellular products and wireless local loop transceivers are classic examples of common DSP-based communication products. FFT analyzers, the latest families of digital storage oscilloscopes (DSOs), and new spectrum analyzers have made use of these low-cost parts to improve the real-time processing of digitized signals. In many early cases, both DSP and microprocessor chips were used inside the instruments, whereas some of the new signal processor chips have mixed DSP/microcontroller capabilities. This chapter provides an essential guide for designers to understand the DSPs and briefly compares microprocessors and DSPs.

7.2 What Is a DSP?

A digital signal processor (DSP) accepts one or more discrete time inputs, $x_i[n]$, and produces one or more items of output, $y_i[n]$, for $n = ..., -1, 0, 1, 2, ...,$ and $I = 1, ... , N$, as depicted in Figure 7.1a. The input can represent appropriately sampled (and analog-to-digital [A/D] converted) values of continuous time signals of interest, which are then processed in the discrete time domain to produce outputs in discrete time that can then be converted to continuous time, if necessary. The operation of the DSP on the input samples can be linear or nonlinear and time invariant or time varying, depending on the application of interest. The samples of the signal are quantized to a finite number of bits, and this word length can be either fixed or variable within the processor. Signal processors operate on millions of samples per second, require large memory bandwidth, and are computationally very demanding, often requiring as many as a few hundred operations on each sample processed. These real-time capabilities are beyond the capabilities of conventional microprocessors and microcontrollers. A practical example of voice processing by DSP is shown in Figure 7.1b.

Signal processors can be either programmable or of a dedicated nature. Programmable signal processors allow flexibility of implementation of a variety of algorithms that can use the same computational kernel, whereas

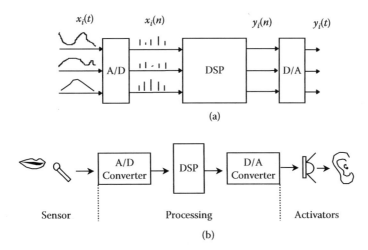

FIGURE 7.1
A digital signal processing system: (a) mathematical representation; (b) a practical example.

dedicated signal processors are hardwired to a specific algorithm or to a specific class of algorithms. Dedicated processors are often faster than or dissipate less power than general-purpose programmable processors, although this is not always the case.

Digital signal processors have traditionally been optimized to compute finite impulse response convolutions (sum of products), infinite impulse response recursive filtering, and FFT-type (butterfly) operations that typically characterize most signal-processing algorithms. They also include interfaces to external data ports for real-time operation. It is interesting to note that one of the earliest digital computers, ENIAC, had characteristics of a DSP [3].

7.3   Comparison of a Microprocessor and a DSP

General architectures for computers and single-chip microcomputers fall into two categories. The architectures for the first significant electromechanical computer had separate memory spaces for the program and the data, so that each could be accessed simultaneously. This is known as a Harvard architecture, having been developed in the late 1930s by Howard Aiken, a physicist at Harvard University. The Harvard Mark 1 computer became operational in 1944. A quick guide to basic microprocessor architecture is available in Kularatna [4].

The first general-purpose electronic computer was probably the ENIAC (Electronic Numerical Integrator and Calculator), built from 1943 to 1946 at the University of Pennsylvania. The architecture was similar to that of the Harvard Mark 1, with separate program and data memories. Because of the complexity of two separate memory systems, the Harvard architecture has not proved popular in general-purpose computer and microcomputer design.

One of the consultants to the ENIAC project, John von Neumann, a Hungarian-born mathematician, is widely recognized as the creator of a different and very significant architecture, published by Burks, Goldstine, and von Neumann [5]. The so-called von Neumann architecture set the standard for developments in computer systems over the next 40 years and more. The idea was very simple and based on two main premises: there is no intrinsic difference between instructions and data, and instructions can be partitioned into two major fields containing the operation command and the address of the operand (the data to be operated upon); therefore, a single memory space can contain both instructions and data.

Common general-purpose microprocessors share what is now known as the von Neumann architecture. These and other general-purpose microprocessors also have other characteristics typical of most computers over the past 40 years. The basic computational blocks are an arithmetic logic unit (ALU) and a shifter. Operations such as add, move, and subtract are easily performed in a very few clock cycles. Complex instructions such as multiply and divide are built up from a series of simple shift, add, and subtract operations. Devices of this type are known as CISCs. CISC devices have multiply instructions, but this simply executes a series of microcode instructions that are hard coded in on-chip read-only memory (ROM). The microcoded multiply operation therefore takes many clock cycles.

Figure 7.2 compares the basic differences between traditional microprocessor architecture and typical DSP architecture. Real-time digital signal-processing applications require many calculations of the form

$$A = BC + D. \tag{7.1}$$

This simple equation involves a multiply operation and an add operation. Because of its slow multiply instruction, a CISC microcomputer is not very efficient at calculating it. We need a machine that can multiply and add in just one clock cycle. For this, we need a different approach to computer architecture.

Many embedded applications are well defined in scope and require only a few calculations to be performed, but they require very fast processing. Examples of such applications are digital compression of images, compact disc players, and digital telephones. In addition to these computation-intensive functions demanding continuous processing, the processor has to perform comparatively simple functions such as menu control for satellite TV, selection of tracks for CD players, and number processing in a digital private branch exchange (PBX), all of which require significantly less processing power.

In such applications, computation-intensive functions such as digital filtering and data compression require continuous signal processing, which requires multiplication, addition, subtraction, and other mathematical functions. Whereas RISC processor architectures can be optimized to handle these situations by incorporating cache memory and direct access internal registers, DSP systems provide more computing-intensive functions such as FFT, convolution, and digital filters. Particularly in a DSP-based system, such tasks should be performed on a real-time basis, as in Figure 7.3. This indicates that the sample period and computational latency are becoming key parameters.

7.3.1 The Importance of the Sample Period and Latency in the DSP World

The sample period (the time interval between the arrival of successive samples of the input signal) depends on the technology employed in the processor. The time interval between the arrival of the input and the departure of the corresponding output sample is the computational latency of the processor. To ensure the stability of the input ports, the output samples have to depart at the same sample period as the input samples. In signal-processing applications, the minimum sample period that can be achieved is often more important than the latency of the circuit. Once the first output sample emerges,

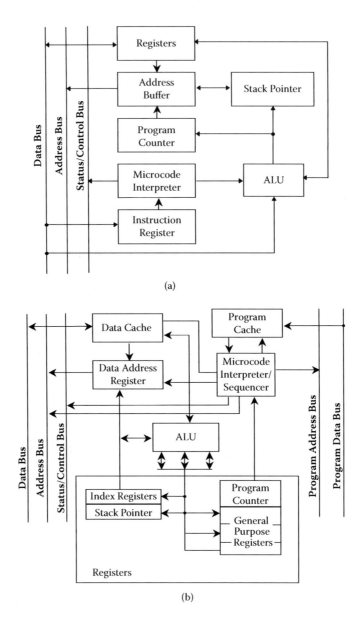

FIGURE 7.2
Comparison of microprocessor and DSP architectures: (a) traditional microprocessor architecture; (b) typical DSP architecture.

successive samples will also emerge at the sample period rate, hiding the effects of a large latency of circuit operation. This makes sense because typical signal-processing applications deal with a few million samples of data in every second of operation. For details on the relationship between these two parameters, see Madisetti [6].

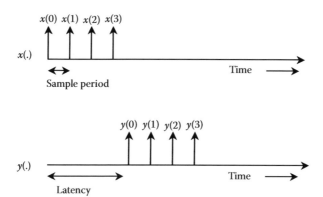

FIGURE 7.3
Sample period and latency.

Other important measures are the area of the VLSI implementation and its power dissipation. These directly contribute to the cost of a DSP chip. One or more of these measures usually is optimized at the cost of others. These trade-offs again depend on the application. For instance, signal processors for portable communications require low power consumption combined with small size, usually at the cost of an increased sample period and latency.

7.3.2 The Merging of Microprocessors and DSPs

Diverse, high-volume applications such as cell phones, disc drives, antilock brakes, modems, and fax machines require both microprocessor and DSP capability. This requirement has led many microprocessor vendors to build in DSP functionality. In some cases, such as in Siemens' Tricore architecture [7], the functional merging is so complete that it is difficult to determine whether one should call the device a DSP or a microprocessor. At the other extreme, some vendors claim that their microprocessors have high-performance DSP capability when, in fact, they have added only a simple 16 × 16-bit multiplication instruction.

7.4 Filtering Applications and the Evolution of DSP Architecture

Digital signal processing techniques are based on mathematical concepts that are familiar to most engineers. From these basic ideas spring the myriad applications of DSPs, including FFT, linear prediction, nonlinear filtering, decimation and interpolation, and many more (see Figure 7.4). One of the most common signal-processing functions is linear filtering. High-pass, low-pass, and band-pass filters, which traditionally are analog designs, can be

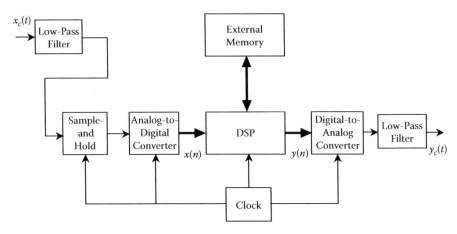

FIGURE 7.4
A DSP-based filter implementation.

constructed with DSP techniques. To build a linear filter with digital methods, a continuous time input signal, $x_c(t)$, is sampled to produce a sequence of numbers, $x[n] = x_c(nT)$. This sequence is transformed by a discrete time system, that is, a computational algorithm, into an output sequence of numbers, $y[n]$. Finally, a continuous time output signal, $y_c(t)$, is reconstructed from the sequence $y[n]$. Essentials of filtering and sampling concepts as applied to the world of DSPs are discussed in Chapter 5.

7.4.1 Digital Filters

Digital filters have, for many years, been the most common application of DSPs. Digital design of any kind ensures repeatability. Two other significant advantages accrue with respect to filters. First, it is possible to reprogram the DSP and drastically alter the filter's gain or phase response. For example, we can reprogram a system from low pass to high pass without throwing away the existing hardware. Second, we can update the filter coefficients while the program is running; that is, we can build "adaptive" filters. The two basic forms of digital filter, the finite impulse response (FIR) filter and the infinite impulse response (IIR) filter, are explained next. The initial descriptions are based on a low-pass filter. It is very easy to change a low-pass filter to any other type: high pass, band pass, etc. Parks and Burrus [8] and Oppenheim and Schafer [9] cover this in detail.

7.4.1.1 FIR Filter

The mechanics of the basic FIR filter algorithm are straightforward. The blocks labeled z^{-1} in Figure 7.5 are unit delay operators; their output is a copy of the input sample delayed by one sample period. A series of storage elements (usually memory locations) is used to simulate a series of these delay elements (called a delay line). The FIR filter is constructed from a series

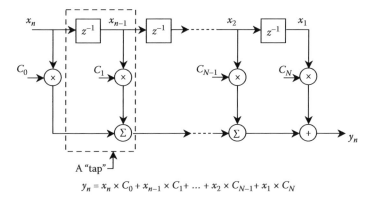

$$y_n = x_n \times C_0 + x_{n-1} \times C_1 + \dots + x_2 \times C_{N-1} + x_1 \times C_N$$

FIGURE 7.5
A finite impulse response filter.

of taps. Each tap includes a multiplication and an accumulation operation. At any given time, $n - 1$ of the most recent input samples reside in the delay line, where n is the number of taps in the filter. Input samples are designated x_k; the first input sample is x_1, the next is x_2, and so on. Each time a new input sample arrives, the previously stored sample is shifted one place to the right along the delay line, and a new output sample is computed by multiplying the newly arrived sample and each of the previously stored input samples by the corresponding coefficient. In Figure 7.5, coefficients are represented as C_N, where N is the coefficient number. The results of each multiplication are summed to form the new output sample, y_n. Later we discuss how DSPs are designed to help implement these.

7.4.1.2 IIR Filter

The other basic form of digital filter is the IIR filter. A simple form of this is shown in Figure 7.6. Using the same notations that we just used for the FIR, we can see that

$$y(n) = x(n) + a_1 y(n-1) + a_2 y(n-2) \qquad (7.2)$$

$$= x(n) + [a_1 z^{-1} + a_2 z^{-2}] y(n)$$

$$= x(n) \frac{1}{1 - a_1 z^{-1} - a_2 z^{-2}}. \qquad (7.3)$$

Take the math for granted—it is just a relatively simple substitution. Therefore, the transfer function is given by

$$H(n) = \frac{y(n)}{x(n)} = \frac{1}{1 - a_1 z^{-1} - a_2 z^{-2}}. \qquad (7.4)$$

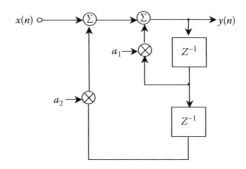

FIGURE 7.6
A simple IIR filter.

From Equation 7.2 we can see that each output, $y(n)$, is dependent on the input value, $x(n)$, and two previous outputs, $y(n-1)$ and $y(n-2)$. Taking this one step at a time, let us assume that there were no previous input samples before $n = 0$; then

$$y(0) = x(0).$$

At the next sample instant,

$$y(1) = x(1) + a_1y(0)$$

$$= x(1) + a_1x(0).$$

Then at $n = 2$,

$$y(2) = x(2) + a_1y(1) + a_2y(0)$$

$$= x(2) + a_1[x(1) + a_1x(0)] + a_2x(0).$$

Then at $n = 3$,

$$y(3) = x(3) + a_1y(2) + a_2y(1)$$

$$= x(3) + a_1[x(2) + a_1[x(1) + a_1x(0)] + a_2x(0)] + a_2[x(1) + a_1x(0)].$$

We already can see that any output depends on all the previous inputs, and we could go on, but the equation just gets longer. An alternative way of expressing this is to say that each output is dependent on an infinite number of inputs. This is why this filter type is called an infinite impulse response.

If we look again at Figure 7.6, the filter is actually a series of feedback loops, and as with any such design, we know that under certain conditions it may become unstable. Although instability is possible with an IIR design, it has the advantage that, for the same roll-off rate, it requires fewer taps than FIR filters. This means that if we are limited in the processor resources available to perform our desired function, we may find ourselves having to use an IIR. We just have to be careful to design a stable filter. More advanced forms of these filters are discussed with simple explanations in Marven and Ewers [3].

7.4.2 Filter Implementation in DSPs

To explain the filter implementation, let us take the case of a first-order recursive filter. A signal flow graph (SFG) or signal flow diagram is a convenient representation of a signal-processing algorithm. Consider the first-order recursive filter shown in Figure 7.7a. The sequential computations involved are not clearly evident in the SFG, because it appears as if all the operations can be evaluated at the same time. However, operations have to follow a certain precedence to preserve correct operation. It is also not clear where the data operands and coefficients are stored prior to their utilization in the computation. A more convenient mode of description would be the one in Figure 7.7b, which shows the storage locations for each operand and the

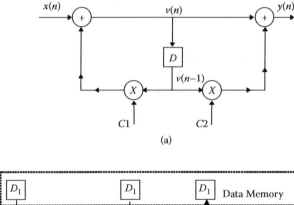

(a)

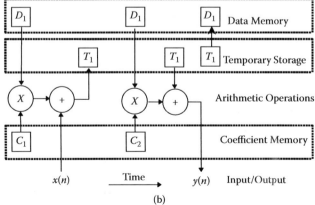

(b)

FIGURE 7.7
Filter implementation by DSP techniques: (a) first-order IIR filter; (b) assembler instructions at the RTL.

sequence of computations in terms of micro-operations at the register transfer level (RTL) ordered in time from left to right.

We assume that the state variable $v(n - 1)$ is stored in the data memory (DM) at location D_1, whereas the coefficient C_1 is stored in a coefficient memory (CM) at location C_1. Both these operands are fetched and multiplied and the result is added to the input sample, $x[n]$, and the sum is stored in a temporary location T_1. Then another multiplication is performed using coefficient C_2, and the product is added to the contents of T_1. The final result is output $y[n]$. Then the new variable $v[n]$ is stored in memory location D_1. One may wonder why temporary location T_1 has been used. Temporary locations such as T_1 often provide a longer word length (or precision) than the word length of the memory. Repeated sums of products as required in this example can quickly exceed the dynamic range provided by the word length. Temporary locations provide the additional bits required to offset the deleterious effects of overflow. One also can observe that, in this example, the multiplier and adder operate in tandem, and the second coefficient multiplication can utilize the same multiplier when the input sample is being added. Thus, only one multiplier and one adder are required as arithmetic units. One data memory location, two coefficient memory locations, and one temporary storage register are required for correct operation of the filter. The specification of the sequence of micro-operations required to perform the computation is called programming in assembler.

From the preceding discussion, any candidate signal processor architecture for the IIR filter must have a coefficient memory, a data memory, temporary registers for storage, a multiplier, an adder, and interconnection. In addition, addresses must be calculated for the memories as well as interpretation (or decoding) of the instruction (obtained from the program memory). The coefficient memory and the program memory can be combined into one memory (the program memory). Nothing can be written into this ROM. Data can be written and read from the random access memory (RAM). The architecture shown in Figure 7.8 is suitable for this application.

The program counter and the index registers are used in computing the addresses of the next instruction and the coefficients. The instruction is decoded by the instruction register (IR), where the address of the data is calculated using the adder and the base index register provided with the data memory. The program bus and the data bus are separate from each other, as are the program and data memories. This separation of data and program memories and buses characterizes the so-called Harvard architecture. The shifter is provided to allow incorporation of multiple word lengths within the data path (the multiplier and the adder) and the data and program buses. The T_1 register is configured as a higher-precision accumulator. Input samples are read in from the input buffer and written into the output buffer. The DSP can interface to a host computer via the external interface. In Figure 7.8, the integers represent the number of bits carried on each bus. For a detailed discussion of digital filters, see Chapter 7 of Jones and Watson [1].

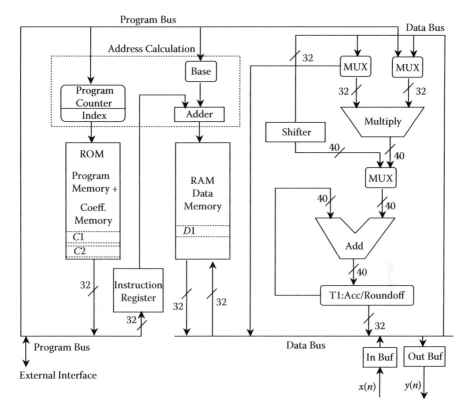

FIGURE 7.8
A candidate DSP architecture for IIR/FIR-type filtering. (Source: Oppenheim and Schafer [9].)

The inherent advantages of digital filters include:

- They can be made to have no insertion loss.
- Linear phase characteristics are possible.
- Filter coefficients are easily changed to enable adaptive performance.
- Frequency response characteristics can be made to approximate closely to the ideal.
- They do not drift.
- Performance accuracy can be controlled by the designer.
- They can handle very low frequency signals.

7.4.3 DSP Architecture

The simplest processor memory structure is a single bank of memory, which the processor accesses through a single set of address and data lines, as shown in Figure 7.9. This structure, which is common among non-DSP

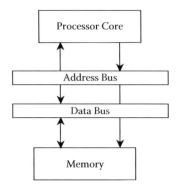

FIGURE 7.9
Von Neumann architecture for non-DSP processors.

processors, is often considered a von Neumann architecture. Both program instructions and data are stored in the single memory. In the simplest (and most common) case, the processor can make one access (either read or write) to memory during each instruction cycle.

If we consider programming a simple von Neumann architecture machine to implement our example FIR filter algorithm, the shortcomings of the architecture become immediately apparent. Even if the processor's data path is capable of completing a multiply-and-accumulate operation in one instruction cycle, it will take four instruction cycles for the processor to actually perform the multiply-and-accumulate operation, because the four memory accesses outlined above must proceed sequentially, with each memory access taking one instruction cycle. This is one reason why conventional processors often do not perform well in DSP-intensive applications and why designers of DSP processors have developed a wide range of alternatives to the von Neumann architecture, which we explore next.

The previous discussions indicate that parallel memories are preferred in DSP applications. In most DSPs, Harvard architectures coexist with pipelined data and instruction processors in a very efficient manner. These systems, with specific addressing modes for signal-processing applications, can be best described as special instruction set computers (SISCs). The SISC architecture is characterized by a memory-oriented special-purpose instruction set.

7.4.3.1 Basic Harvard Architecture

Harvard architecture refers to a memory structure in which the processor is connected to two independent memory banks via two independent sets of buses. In the original Harvard architecture, one memory bank holds program instructions and the other holds data. Commonly, this concept is extended slightly to allow one bank to hold program instructions and data while the other bank holds data only. This "modified" Harvard architecture is shown in Figure 7.10. The key advantage of the Harvard architecture is that two

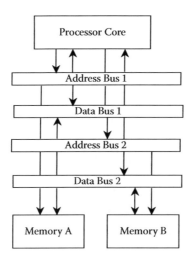

FIGURE 7.10
Harvard architecture.

memory accesses can be made during any one instruction cycle. Thus, the four memory accesses required for the example FIR filter can be completed in two instruction cycles. This type of memory architecture is used in many DSP families, including Analog Devices ADSP-21xx.

7.4.3.2 SISC Architectures

Whereas microprocessors are based on register-oriented architectures, signal processors are memory-oriented architectures. Multiple memories for both program and data have been present even in the first-generation DSPs, such as the TMS320C10. Modern DSPs have as many as six parallel memories for the use of the instruction and data processors. External memory is as easily accessible as internal memory. In addition, a rich set of addressing modes tailored for signal-processing applications is also provided. We describe these architectures as representative of SISCs and expect that future generations of SISCs will have communication primitives as part of the standard instruction set.

The basic instruction cycle is a unit of time measurement in the context of signal-processing architectures, in some sense it is the average time required to execute an ALU instruction. The basic instruction cycle is further divided into subcycles (usually two to four). The memory cycle time is that time required to access one operand from the memory. The high-memory bandwidth requirement in SISCs can be met by providing for either memories with very low memory cycle times or multiple memories with relatively slower cycle times. Typically, an instruction cycle is twice as long as a memory cycle for on-chip memory (and equal to the memory cycle for external memory). Clearly, this facilitates the use of operand fetch-and-execution pipelines of two-operand instructions with on-chip data memories. If parallel data memories are provided, then the total number of memory cycles per instruction cycle is increased. The total number of memory cycles possible within a single basic instruction cycle is defined as the demand ratio [10] for an SISC machine. Higher demand ratios lead to a higher throughput of instructions:

$$\text{Demand ratio} = \frac{(\text{basic instruction cycle time}) \times (\text{number of memories})}{\text{memory cycle time}}. \quad (7.5)$$

7.4.3.3 Multiple-Access Memory-Based Architectures

As discussed earlier, the Harvard architecture achieves multiple memory accesses per instruction cycle by using multiple, independent memory banks connected to the processor data path via independent buses. Although a number of DSP processors use this approach, there are also other ways to achieve multiple memory accesses per instruction cycle. These include using fast memories that support multiple, sequential accesses per instruction cycle over a single set of buses, and using "multiported" memories that allow multiple concurrent memory accesses over two or more independent sets of buses.

Achieving increased memory access capacity by the use of multiported memory is becoming popular with the development of memory technology. A multiported memory has multiple independent sets of address and data connections, allowing multiple independent memory accesses to proceed in parallel. The most common type of multiported memory is the dual-ported variety, which provides two simultaneous accesses. However, triple- and even quadruple-ported varieties are sometimes used. Multiported memories dispense with the need to arrange data among multiple, independent memory banks to achieve maximum performance. The key disadvantage of multiported memories is that they are much more costly (in terms of chip area) to implement than standard, single-ported memories. The memory architecture shown in Figure 7.11, for example, includes a single-ported program memory with a dual-ported data memory. This arrangement provides one program memory access and two data memory accesses per instruction word and is

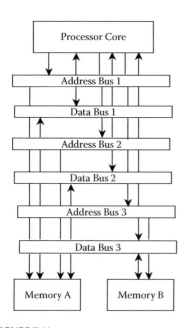

FIGURE 7.11
Modified Harvard architecture with dual-ported memory.

used in the Motorola DSP561xx processors. For more detailed discussion on these techniques, see Lapsley et al. [11].

7.4.4 Modifications to Harvard Architecture

The basic Harvard architecture can be modified into six different types. This discussion is beyond the scope of the chapter; for details, see Lee [12,13].

7.5 Special Addressing Modes

In addition to general addressing modes used in microprocessor systems, several special addressing modes are used in DSPs, including circular addressing and bit-reversed addressing. For a comprehensive discussion on addressing modes, see Lapsley et al. [11], as only circular addressing and bit-reversed addressing are discussed here.

7.5.1 Circular Addressing

Many DSP applications need to manage data buffers. A data buffer is a section of memory that is used to store data that arrive from an off-chip source or a previous computation until the processor is ready to process the data. In real-time systems, where dynamic memory allocation is prohibitively expensive, the programmer usually must determine the maximum amount of data that a given buffer will need to hold and then set aside a portion of memory for that buffer. The buffers generally use a first-in-first-out (FIFO) protocol, meaning that data values are read out of the buffer in the order in which they arrive.

In managing the movement of data into and out of the buffer, the programmer maintains two pointers, which are stored in registers or in memory: a read pointer and a write pointer. The read pointer points to (i.e., contains the address of) the memory location containing the next data value to arrive, as illustrated in Figure 7.12. Each time a read or write operation is performed, the read or write pointer is advanced and the programmer must check to see whether the pointer has reached the last location in the buffer. The action of checking after each buffer operation whether the pointer has reached the end of the butter, and resetting it if it has, is time consuming. For systems that use buffers extensively, this linear addressing can cause a significant performance bottleneck.

To address this bottleneck, many DSPs have a special addressing capability that allows them, after each buffer address calculation, to automatically check whether the pointer has reached the end of the buffer and reset it at the buffer start location, if necessary. This capability is called modulo addressing or circular addressing.

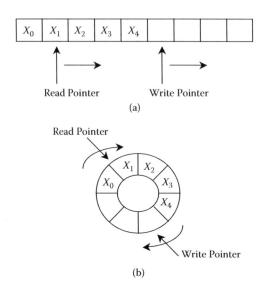

FIGURE 7.12
Comparison of linear and circular addressing: (a) FIFO buffer with linear addressing; (b) the same data in a FIFO buffer with circular addressing.

The term modulo refers to modulo arithmetic, wherein numbers are limited to a specific range. This is similar to the arithmetic used in a clock, which is based on a 12-hour cycle. When the result of a calculation exceeds the maximum value, it is adjusted by repeatedly subtracting from it the maximum representable value until the result lies within the specified range. For example, 4 hours after 10 o'clock is 2 o'clock (14 modulo 12).

When modulo address arithmetic is in effect, read and write pointers (address registers) are updated using pre- or postincrement register-indirect addressing [11]. The processor's address generation unit performs modulo arithmetic when new address values are computed, creating the appearance of a circular memory layout, as illustrated in Figure 7.12b. Modulo address arithmetic eliminates the need for the programmer to check the read and write pointers to see whether they have reached the end of the buffer and reset them once they have reached the end. This results in much faster buffer operations and makes modulo addressing a valuable capability for many applications.

In most real-time signal-processing applications, such as those found in filtering, the input is an infinite stream of data samples. These samples are "windowed" and used in filtering applications. For instance, a sliding window of N data samples is used by an FIR filter with N taps. The data samples simulate a tapped delay line, and the oldest sample is written over by the most recent sample. The filter coefficients and the data samples are written into two circular buffers. Then they are multiplied and accumulated to form the output sample result, which is stored. The address pointer for the data buffer is then updated and the samples appear shifted by one sample period, the oldest data being written out and the most recent data being written into that location.

7.5.2 Bit-Reversed Addressing

Perhaps the most unusual of addressing modes, bit-reversed addressing is used only in very specialized circumstances. Some DSP applications make heavy use of the FFT algorithm. The FFT is a fast algorithm for transforming a time-domain signal into its frequency-domain representation and vice versa [9,14]. However, the FFT has the disadvantage that it either takes its input or leaves its output in a scrambled order. This dictates that the data be rearranged to or from the natural order at some point.

The scrambling required depends on the particular variation of the FFT. The radix-2 implementation of an FFT, a very common form, requires reordering of a particularly simple nature, bit-reversed ordering. The term bit-reversed refers to the observation that if the output values from a binary counter are written in reverse order (i.e., the least significant bit first), the resulting sequence of counter output values will match the scrambled sequence of the FFT output data. This phenomenon is illustrated in Figure 7.13.

Because the FFT is an important algorithm in many DSP applications, many DSP processors include special hardware in their address generation units to facilitate generating bit-reversed address sequences for unscrambling FFT results. For example, the Analog Devices ADSP-210xx provides a bit-reversed mode, which is enabled by setting a bit in a control register. When the processor is in bit-reversed mode, the output of one of its address registers is bit-reversed before being applied to the memory address bus.

An alternative approach to implementing bit-reversed addressing is the use of reverse-carry arithmetic. With reverse-carry arithmetic, the address generation unit reverses the direction in which carry bits propagate when an increment is added to the value in an address register. If reverse-carry arithmetic is enabled in the address generation unit (AGU), and the programmer supplies the base address and increment value in bit-reversed order, then the resulting addresses will be in bit-reversed order. Reverse-carry arithmetic is provided in the AT&T DSP32xx, for example.

7.6 Important Architectural Elements in a DSP

Compared to architectural elements of microprocessors (see Chapter 4 of Kularatna [4]), it may be relevant for us to discuss special-function blocks in a DSP chip. Performing efficient digital signal processing on a microprocessor is tricky business. Although the ability to support single-cycle multiply-and-accumulate operations is the most important function a DSP performs, many other functions are critical for real-time DSP applications. Executing

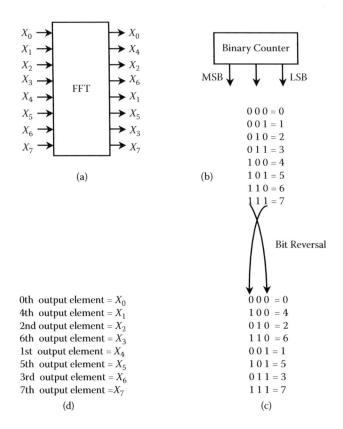

FIGURE 7.13
The output of an FFT algorithm and bit-reversed addressing: (a) FFT output and input relationships; (b) binary counter output; (c) bit reversal; (d) transformation of output into order.

a real-time DSP application requires an architecture that supports high-speed data flow to and from the computation units and memory through a multiport register file. This execution often involves the use of direct memory access (DMA) units and AGUs that operate in parallel with other chip resources. AGUs, which perform address calculations, allow the DSP to bring two pieces of data per clock, which is a critical need for real-time DSP algorithms.

It is important for DSPs to have an efficient looping mechanism, because most DSP code is highly repetitive. The architecture allows for zero-overhead looping, in which one uses no additional instructions to check the completion of loop iterations. Generally, DSPs take looping a step further by including the ability to handle nested loops.

DSPs typically handle extended precision and dynamic range to avoid overflow and minimize round-off errors. To accommodate this capability, DSPs typically include dedicated accumulators with registers wider than the nominal word size to preserve precision. DSPs also must support circular buffers to handle algorithmic functions, such as tapped delay lines and coefficient buffers. DSP hardware updates circular buffer pointers during every cycle in parallel with other chip resources. During each clock cycle, the circular buffer hardware performs an end-of-buffer comparison test and resets the pointer without overhead when it reaches the end of the buffer. FFTs and other DSP algorithms also require bit-reversed addressing.

7.6.1 Multiplier/Accumulator

The multiplier/accumulator (MAC) provides high-speed multiplication, multiplication with cumulative addition, multiplication with cumulative subtraction, saturation, and clear-to-zero functions. A feedback function allows part of the accumulator output to be directly used as one of the multiplicands of the next cycle. To explain MAC operation, consider a real-life example from the ADSP-21xx family (see Figure 7.14).

The multiplier has two 16-bit input ports, X and Y, and a 32-bit product output port, P. The 32-bit product is passed to a 40-bit adder/subtractor, which adds or subtracts the new product from the content of the multiplier result (MR) register or passes the new product directly to the MR register. The MR register is 40 bits wide. In this discussion we refer to the entire register as the MR register, which actually consists of three smaller registers: MR0 and MR1, which are 16 bits wide, and MR2, which is 8 bits wide.

The adder/subtractor is greater than 32 bits to allow for intermediate overflow in a series of multiply-and-accumulate operations. The multiply overflow (MV) status bit is set when the accumulator has overflowed beyond the 32-bit boundary, that is, when there are significant (nonsign) bits in the top 9 bits of the MR register (based on two's-complement arithmetic). The input/output registers of the MAC section are similar to the ALU. The X input port can accept data from either the MX register file or any register on the result (R) bus. The R bus connects the output registers of all the computational units, permitting them to be used directly as input operands. Two registers in the MX register file, MX0 and MX1, can be read and written from the DMD bus. The MX register file output is dual ported so that one register can provide input to the multiplier while either one drives the DMD bus.

The Y input port can accept data from either the MY register file or the MF register. The MY register file has two registers, MY0 and MY1, that can be read and written from the DMD bus and written from the program memory data (PMD) bus. The ADSP-2101 instruction set also provides for reading these registers over the PMD bus, but with no direct connection; this operation uses the DMD–PMD bus exchange unit. The MY register file output is also dual ported so that one register can provide input to the multiplier while either one drives the DMD bus.

The output of the adder/subtractor goes to either the MF register or the MR register. The MF register is a feedback register that allows bits 16 to 31 of the result to be used directly as the multiplier Y input on a subsequent cycle. The 40-bit adder/subtractor register (MR) is divided into three sections: MR2,

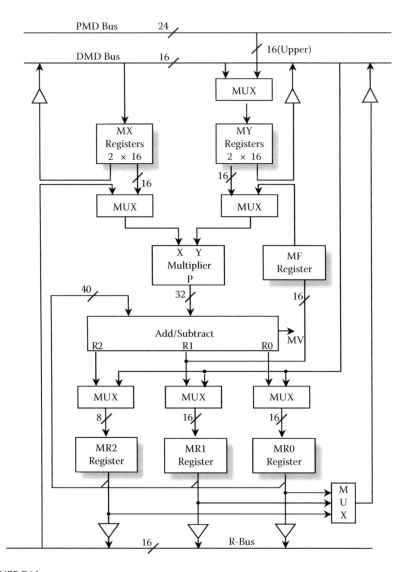

FIGURE 7.14
MAC block diagram of the ADSP-2104. (Reproduced by permission of Analog Devices, Norwood, MA.)

MR1, and MR0. Each of these registers can be loaded directly from the DMD bus and its output to either the DMD bus or the R bus.

Any register associated with the MAC can be both read and written in the same cycle. Registers are read at the beginning of the cycle and written at the end of the cycle. A register read instruction therefore reads the value loaded at the end of a previous cycle. A new value written to a register cannot be read out until the subsequent cycle. This allows an input register to provide an operand to the MAC at the beginning of the cycle and be updated with the next operand from memory at the end of the same cycle. It also allows a result register to be stored in memory and updated with a new result in the same cycle.

The MAC contains a duplicate bank of registers, shown in Figure 7.14, behind the primary registers. There are actually two sets of MR, MF, MX, and MY register files. Only one bank is accessible at a time. The additional bank of registers can be activated for extremely fast context switching. A new task, such as an interrupt service routine, can be executed without transferring current states to storage. The selection of the primary or alternate bank of registers is controlled by bit 0 in the processor mode states register (MSTAT). If this bit is a 0, the primary bank is selected; if it is 1, the secondary bank is selected. For details, see Ingle and Proakis [15] and New [16].

7.6.2 Address Generation Units

Most DSP processors include one or more special AGUs that are dedicated to calculating addresses. Manufacturers refer to these units by various names. For example, Analog Devices calls its AGU a data address generator, and AT&T calls its AGU a control arithmetic unit. An AGU can perform one or more complex address calculations per instruction cycle without using the processor's main data path. This allows address calculations to take place in parallel with arithmetic operations on data, improving processor performance. The differences among address generation units are manifested in the types of addressing modes provided and the capability and flexibility of each addressing mode. As an example, let us look at data address units in ADSP-21xx family.

7.6.2.1 Data Address Units of ADSP-21xx Family: An Example

Data address generator (DAG) units contain two independent address generators so that program and data memories can be accessed simultaneously. Let us discuss the operation of DAGs, taking the ADSP-2101 as an example. DAGs provide indirect addressing capabilities and perform automatic address modification. In the ADSP-2101, the two DAGs differ: DAG1 generates data memory addresses and provides an optional bit-reversal capability; DAG2 can generate both data memory and program memory addresses but has no bit reversal.

Figure 7.15 shows a block diagram of a single DAG. There are three register files: the modify (M) register file, the index (I) register file, and the length (L) register file. Each file contains four 14-bit registers that can be read from and written to via the DMD bus. The I registers (I0–I3 in DAG1, I4–I7 in DAG2) contain the actual addresses used to access memory. When the data are accessed in the indirect mode, the address stored in the selected I register becomes the memory address. With DAG1, the output address can be bit reversed by setting the appropriate mode bit in the MSTAT, as discussed next. Bit reversal facilitates FFT addressing.

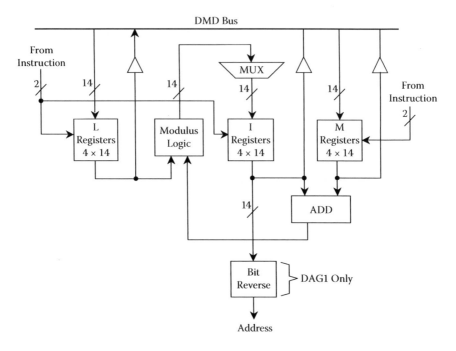

FIGURE 7.15
Data address generator block diagram of the ADSP-2101. (Courtesy of Analog Devices, Norwood, MA.)

The DAG employs a postmodification scheme. After an indirect data access, the specified M register (M0–M3 in DAG2) is added to the specified I register to generate the new I value. The choice of the I and M registers is independent within each DAG. In other words, any register in the I0–I3 set may be modified by any register in the M0–M3 set in any combination, but not by those in DAG2 (M4–M7). The modification values stored in the M register are signed numbers so that the next address can be either higher or lower. The address generators support both linear and circular addressing. The value of the L register determines which addressing scheme is used. For circular buffer addressing, the L register is initialized with the length of the buffer. For linear addressing, the modulus logic is disabled by setting the corresponding L register to zero. L registers and I registers are paired, and the selection of the L register (L0–L3 in DAG1, L4–L7 in DAG2) is determined by the I register used. Each time an I register is selected, the corresponding L register provides the modulus logic with the length information. If the sum of the M register content and the I register content crosses the buffer boundary, the modified I register value is calculated by the modulus logic using the L register value.

All DAG resisters (I, M, and L registers) are loadable and readable from the lower 14 bits of the DMD bus. Because the I and L register content is considered unsigned, the upper 2 bits of the DMD bus are padded with zeros when reading them. The M register content is signed; when reading an M register, the upper 2 bits of the DMD bus are sign extended. The modulus logic implements automatic pointer wraparound for accessing circular buffers. To calculate the next address, the modulus logic uses the following information:

- The current location, found in the I register (unsigned)
- The modify value, found in the M register (signed)
- The buffer length, found in the L register (unsigned)
- The buffer base address

From such input, the next address is calculated with the formula:

$$\text{Next address} = (I + M - B) \text{ modulo } (L) + B, \tag{7.6}$$

where I is the current address, M is the modify value (signed), B is the base address (generated by the linker), and L is the buffer length. Value of M is less than L which ensures that the next address cannot wrap around the buffer more than once in one operation.

7.6.3 Shifters

Shifting a binary number allows scaling. A shifter unit in a DSP provides a complete set of shifting functions, which can be divided into two categories: arithmetic and logical. A logical left shift by 1 bit inserts a 0 bit in the least significant bit, and a logical right shift by 1 bit inserts a 0 bit in the most significant bit. In contrast, an arithmetic right shift duplicates the sign bit (either a 1 or 0, depending on whether the number is negative or not) into the most significant bit. Although some use the term arithmetic left shift, arithmetic and logical left shifts are really identical: both shift the word left and insert a 0 in the least significant bit.

Arithmetic shifting provides a way of scaling data without using the processor's multiplier. Scaling is especially important in fixed-point processors, where proper scaling is required to obtain accurate results from mathematical operations.

Virtually all DSPs provide shift instructions of one form or another. Some processors provide the minimum, that is, instructions to do arithmetic left or right shifting by one bit. Some processors may additionally provide instructions for 2- or 4-bit shifts. These can be combined with single-bit shifts to synthesize n-bit shifts, although at a cost of several instruction cycles.

Increasingly, many DSPs feature a barrel shifter and instructions that use the barrel shifter to perform arithmetic or logical left or right shifts by any number of bits. Examples include the AT&T DSP16xx, the Analog Devices ADSP-21xx and ADSP-210xx, the DSP Group Oak DSP Core, the Motorola DSP563xx, the SGS-Thompson D950-CORE, and the Texas Instruments TMS320C5x and TMS320C54x. If one begins with a 16-bit input, a complete

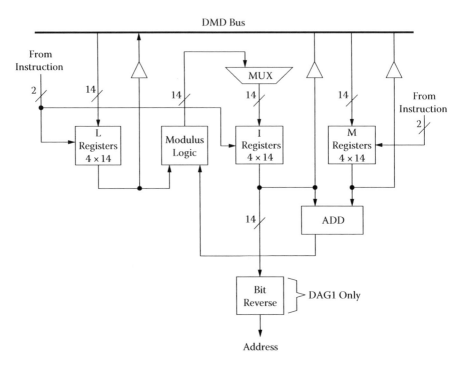

FIGURE 7.16
Block diagram of the ADSP-2101's shifter. (Courtesy of Analog Devices, Norwood, MA.)

set of shifting functions needs a 32-bit output. These functions include arithmetic shift, logical shift, and normalization. The shifter also performs derivation of exponent and derivation of common exponent for an entire block of numbers. These basic functions can be combined to efficiently implement any degree of numerical format control, including full floating point representation. Figure 7.16 shows a block diagram of the ADSP-2101 shifter.

A variable shifter section in the ADSP-2100 can be divided into a shifter array, an OR/PASS logic, an exponent detector, and the exponent compare logic. The shifter array is a 16 × 32 barrel shifter. It accepts a 16-bit input and can place it anywhere in the 32-bit output field, from off-scale right to off-scale left, in a single cycle. This gives 49 possible placements within the 32-bit field. The placement of the 16 input bits is determined by a control code (C) and a HI/LO reference signal.

The shifter array and its associated logic are surrounded by a set of registers. The shifter input (SI) register provides input to the shifter array and the exponent detector. The SI register is 16 bits wide and is readable and writable from the DMD bus. The shifter array and the exponent detector also take as inputs arithmetic, shifter, or multiplier results via the R bus. The shifter result (SR) register is 32 bits wide and is divided into two 16-bit sections, SR0 and SR1. The SR0 and SR1 registers can be loaded from the DMD bus and sent to either the DMD bus or the R bus. The SR register is also fed back to the OR/PASS logic to allow double-precision shift operations. The shifter exponent (SE) register is 8 bits wide and holds the exponent during the normalize and denormalize operations. The SE register is loadable and readable from the lower 8 bits of the DMD bus. It is a two's-complement integer value.

The shifter block (SB) register is important in block floating point operations, where it holds the block exponent value, that is, the value by which the block values must be shifted to normalize the largest value. The SB is 5 bits wide and holds the most recent block exponent value. The SB register is loadable and readable from the lower 5 bits of the DMD bus. It is a two's-complement integer value.

Whenever the SE or SB registers are loaded onto the DMD bus, they are sign extended to form a 16-bit value. Any of the SI, SE, or SR registers can be read and written in the same cycle. Registers are read at the beginning of the cycle and written at the end of the cycle. All register reads thus read values loaded at the end of a previous cycle. A new value written to a register cannot be read out until a subsequent cycle. This allows an input register to provide an operand to the shifter at the beginning of the cycle and be updated with the next operand at the end of that cycle. It also allows a result register to be stored in memory and updated with a new result in the same cycle.

The shifter section contains a duplicate bank of registers, shown in Figure 7.16, behind the primary registers. There actually are two sets of SE, SB, SI, SR1, and SR0 registers, and only one bank is accessible at a time. The additional bank of registers can be activated for extremely fast context switching. A new task, such as an interrupt service routine, can be executed without transferring current states to storage. The selection of the primary or alternate bank of registers is controlled by bit 0 in the processor MSTAT. If this bit is a 0, the primary bank is selected; if it is a 1, the secondary bank is selected.

Shifting of the input is determined by a control code (C) and a HI/LO reference signal. The control code is an 8-bit signed value that indicates the direction and number of places the input is to be shifted. Positive codes indicate a left shift (up-shift) and negative codes indicate a right shift (down-shift). The control code can come from three sources: the content of the SE register, the negated content of the SE register, or an immediate value from the instruction.

The HI/LO signal determines the reference point for the shifting. In the HI state, all shifts are referenced to SR1 (the upper half of the output field); in the LO state, all shifts are referenced to SR0 (the lower half). The HI/LO reference feature is useful when shifting 32-bit values, because it allows both halves of the number to be shifted with the same control code. The HI/LO reference signal is selectable each time the shifter is used.

The shifter fills any bits to the right of the input value in the output field with zeros, and bits to the left are filled with the extension bit (X). The extension bit can be fed by three possible sources, depending on the instruction being performed: the most significant bit (MSB) of the input, the AC bit from the arithmetic status register (ASTAT), or a zero.

The OR/PASS logic allows the shifted sections of a multiprecision number to be combined into a single quantity. When PASS is selected, the shifter array output is passed through and loaded into the shifter result (SR) register, unmodified. When OR is selected, the shifter array is bitwise ORed with the current contents of the SR register before being loaded there.

The exponent detector derives an exponent for the shifter input value. The exponent detector operates in one of three ways, which determine how the input value is interpreted. In the HI state, the input is interpreted as a single-precision number or the upper half of a double-precision number. The exponent detector determines the number of leading sign bits and produces a code that indicates how many places the input must be up-shifted to eliminate all but one of the sign bits. The code is negative so that it can become the effective exponent for the mantissa formed by removing the redundant sign bits.

In the HI-extend (HIX) state, the input is interpreted as the result of an add or subtract performed in the ALU section, which may have overflowed. Therefore, the exponent detector takes the arithmetic overflow (AV) status into consideration. If AV is set, then a +1 exponent becomes the output to indicate an extra bit is needed in the normalized mantissa (the ALU carry bit); if AV is not set, then HIX functions exactly like the HI state. When performing a derive exponent function in the HI or HIX modes, the exponent detector also outputs a shifter sign (SS) bit, which is loaded into the ASTAT. The sign bit is the same as the MSB of the shifter input, except when AV is set; when AV is set in the HIX state, the MSB is inverted to restore the sign bit of the overflow value. In the LO state, the input is interpreted as the lower half of a double-precision number. In the LO state, the exponent detector interprets the SS bit in the ASTAT as the sign bit of the number. The SE register is loaded with the output of the exponent detector only if SE contains P15. This occurs only when the upper half, which must be processed first, contains all sign bits. The exponent detector output is also offset by P16 to indicate that the input actually is the lower half of a 32-bit value.

The exponent compare logic is used to find the largest exponent value in an array of shifter input values. The exponent compare logic, in conjunction with the exponent detector, derives a block exponent. The comparator compares the exponent value derived by the exponent detector with the value stored in the SB exponent register and updates the SB register only when the derived exponent value is larger than the value in the SB register.

Shifters in different DSPs have different capabilities and architectures. For example, the TMS320C25 scaling shifter shifts to the left from 0 to 16 bits. Two other shifters can shift data coming from the multiplier left 1 bit or 4 bits, or they can shift data coming from the accumulator left from 0 to 7 bits. These two shifters add the advantage of being able to scale data during the data move instead of requiring an additional shifter operation.

7.6.4 Loop Mechanisms

The DSP algorithms frequently involve the repetitive execution of a small number of instructions, so-called inner loops or kernels. FIR and IIR filters, FFTs, matrix multiplications, and a host of other application kernels are performed by repeatedly executing the same instruction or sequence of instructions. DSPs have evolved to include features to efficiently handle this sort of repeated execution. To understand the evolution, we look at the problems associated with traditional approaches to related instruction execution. First, a natural approach to looping uses a branch instruction to jump back to the start of the loop. Second, because most loops execute a fixed number of times, the processor must use a register to maintain the loop index, that is, the count of the number of times the processor has been through the loop. The processor's data path must be used to increment or decrement the index and test to see if the loop condition has been met. If not, a conditional branch brings the processor back to the top of the loop. All of these steps add overhead to the loop and use precious registers.

DSPs have evolved to avoid these problems via hardware looping, also known as zero-overhead looping. Hardware loops are special hardware control constructs that repeat between hardware loops and software loops so that hardware loops lose no time incrementing or decrementing counters, checking to see if the loop is finished, or branching back to the top of the loop. This can result in considerable savings. In order to explain how a loop mechanism improves efficiency, we once again use the ADSP-2101 as an example (Figure 7.17).

The ADSP-2100A program sequencer supports zero-overhead DO UNTIL loops. Using the count stack, loop stack, and loop comparator, the processor can determine whether a loop should terminate and the address of the next instruction (either the top of the loop or the instruction after the loop) with no overhead cycle.

A DO UNTIL loop may be as large as program memory size permits. A loop may terminate when a 16-bit counter expires or when any other arithmetic condition occurs. The example below shows a three-instruction loop that is to be repeated 100 times:

```
CNTR = 100
Do Label UNTIL CE
          First instruction of loop
          Second instruction of loop
Label:    Last instruction of loop
          First instruction outside loop
```

The first instruction loads the counter with 100. The DO UNTIL instruction contains the address of the last instruction in the loop (in this case, the address represented by the identifier Label) and the termination condition (in this case, the count expiring, CE). The execution of the DO UNTIL

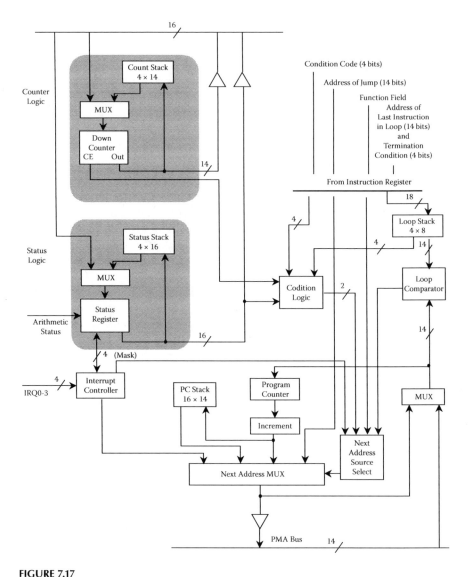

FIGURE 7.17
The ADSP-2100A program sequencer architecture. (Reproduced by permission of Analog Devices, Norwood, MA.)

instruction causes the address of the first instruction of the loop to be pushed on the program counter stack and the address of the last instruction of the loop to be pushed on the loop stack (see Figure 7.17).

As instruction addresses are sent to the program memory address bus and the instruction is retrieved, the loop comparator checks to see if the instruction is the last instruction of the loop. If it is, the program sequencer checks the status and condition logic to see if the termination condition is satisfied. The program sequencer then either takes the address from the program counter stack (to go back to the top of the loop) or simply increments the program counter (to go to the first instruction outside the loop).

The looping mechanism of the ADSP-2100A is automatic and transparent to the user. As long as the DO UNTIL instruction is specified, all stack and counter maintenance and program flow are handled by the sequencer logic with no overhead. This means that, in one cycle, the last instruction of the loop is being executed and, in the very next cycle, the first instruction of the loop is executed or the first instruction outside the loop is executed, depending upon whether the loop terminated or not. For further details of program sequencer and loop mechanisms of the ADSP-2100A, see Ingle and Proakis [15] and Fine [17].

7.7 Instruction Set

Generally, a DSP instruction set is tailored to the computation-intensive algorithms common to DSP applications. This is possible because the instruction set allows data movement between various computational units with minimum overhead. For example, sustained single-cycle multiplication/ accumulation operations are possible.

Again, we use the ADSP-2101 as an example. The instruction set provides full control of the ADSP-2101's three computation units: the ALU, MAC, and shifter. Arithmetic instructions can process single-precision 16-bit operands directly with provisions for multiprecision operations. The ADSP-2101 assembly language uses an algebraic syntax for arithmetic operations and for data moves. The sources and destinations of computations and data moves are written explicitly, eliminating cryptic assembler mnemonics. There is no performance penalty for this; each program statement assembles into one 24-bit instruction, which executes in one cycle. There are no multicycle instructions in the ADSP-2101 instruction set. Some 50 registers surrounding the computational units are dual purpose and are available for general-purpose on-chip storage when not used in computation. This saves many memory access cycles and provides excellent freedom in coding. The control instructions provide conditional execution of most calculations and, in addition to the usual JUMP and CALL, support a DO UNTIL looping instruction. Return from interrupt (RTI) and return from subroutine (RTS) are also provided. These services are made compact and speedy by the single-cycle context save. The contents of the primary register set are held constant while the alternate set is enabled for subroutine and interrupt services. This eliminates the cluster of PUSHes and POPs of stacks common in general-purpose microprocessors.

The ADSP-2101 also provides the IDLE instruction for idling the processor until an interrupt occurs. IDLE puts the processor into a low-power state while waiting for interrupts. Two addressing modes are supported for memory fetches. Direct addressing uses immediate values; indirect addressing uses the two data-addressing generators.

The 24-bit instruction word allows a high degree of parallelism in performing operations. The instruction set allows for a single-cycle execution of any of the following combinations:

- Any ALU, MAC, or shifter operation (may be conditional)
- Any register-to-register move
- Any data memory read or write
- A computation with any data register/data register move
- A computation with any memory read or write
- A computation with a read from two memories

The instruction set provides moves from any register to any other register or from most registers to and from either memory. For combining operations, almost any ALU, MAC, or shifter operation may be combined with any register-to-register moves or with a register move to or from either internal or external memory.

There are five basic categories of instructions: computational instructions, data move instructions, multifunction instructions, program flow control instructions, and miscellaneous instructions, all of which are described in the next several sections, with tables summarizing the syntax of each instruction category. The notation used in an instruction is shown in Table 7.1.

As it is beyond the scope of this chapter to explain the whole group of instructions, the computation instructions of ADSP-2101 are described in summary form. A more detailed instruction set overview can be found in Ingle and Proakis [15] and the ADSP literature.

7.7.1   Computation Instructions: A Summary of the ADSP-21xx Family

The computation group executes all ALU, MAC, and shifter instructions. There are two functional classes: standard instructions, which include the bulk of the computation operations, can be executed conditionally (IF condition...), test the ALU status register, and may be combined with a data transfer in single-cycle multifunction instructions; and special instructions, which form a small subset and must be executed individually. Table 7.2 indicates permissible conditions for computation instructions, and Table 7.3 describes the computational input/output registers.

7.7.1.1   *MAC Functions*

7.7.1.1.1   *Standard Functions*

Standard MAC instructions include multiply, multiply/accumulate, multiply/subtract, transfer AR conditionally, and clear. As an example, consider a MAC instruction for multiply/accumulate in the form

TABLE 7.1

Notation Used in Instruction Set of ADSP-21xx Family

Symbol	Meaning
+, -	Add, subtract
*	Multiply
a=b	Transfer into a the contents of b
,	Separates multifunction instruction
DM (addr)	The contents of data-memory at location "addr"
PM (addr)	The contents of program-memory at location "addr"
[option]	Anything within square brackets is an optional part of the instruction statement
|option a|	List of parameters enclosed by parallel vertical lines requires the choice of one parameter from among the available list
CAPITAL LETTERS	Capital letters denote reserved words. These are instruction words, register names, and operand selections
Lower-case letters	Parameters are shown in small letters and denote an operand in the instruction for which there are numerous choices
<data>	These angle brackets denote an immediate data value
<addr>	These angle brackets denote an immediate value of an address to be coded in the instruction
;	End of instruction

Courtesy of Analog Devices, Norwood, MA.

```
[IF Condition]  MR = MR + xop * yop  (SS) ;
                MF                     SU
                                       US
                                       UU
                                       RND
```

If the options MR and UU are chosen, if xop and yop are the contents of MXO and MYO, respectively, and if the MAC overflow condition is chosen, then a conditional instruction would read:

```
IF NOT MV MR = MR + MXO * MYO (UU) ;
```

The conditional expression, IF NOT MV, tests the MAC overflow bit. If the condition is not true, an NOP is executed. The expression MR = MR + MXO * MYO is the multiply/accumulate operation: the multiplier result register (MR) gets the value of itself plus the product of the X and Y input registers selected. The modifier selected in parentheses (UU) treats the operands as unsigned. Only one such modifier can be selected from the available set: SS means both are signed, US and SU mean that either the first or second operand is signed, and RND means to round the (implicitly signed) result.

Accumulator saturation is the only MAC special function:

```
IF MV SAT MR ;
```

TABLE 7.2

Permissible Conditions for Computation
Instructions of ADSP-2101

Condition	Keyword
ALU result is: equal to zero not equal to zero greater than zero greater than or equal to zero less than zero less than or equal to zero	EQNEGTGELTLE
ALU carry status: carry not carry	ACNOT AC
x-input sign: positive negative	POSNEG
ALU overflow status: overflow not overflow	AVNOT AV
MAC overflow status: overflow not overflow	MVNOT MV
Counter status: not expired	NOT CE

Reproduced by permission of Analog Devices, Norwood, MA.

TABLE 7.3

Computational Input/Output Registers

Source for X Input (xop)	Source for Y Input (yop)	Destination*
ALU		
AX0, AX1, AR	AY0, AY1	AR
MR0, MR1, MR2	AF	AF
SR0, SR1		
MAC		
MX0, MX1, AR	MY0, MY1	MR(MR2, MR1, MR0)
MR0, MR1, MR2	MF	MF
SR0, SR1		
Shifter		
SI, SR0, SR1		SR (SR1, SR0)
AR		
MR0, MR1, MR2		

* Destination for output port *R* for ALU and MAC or destination for shifter output.
Reproduced by permission of Analog Devices, Norwood, MA.

The instruction tests the MAC overflow bit (MV) and saturates the MR register (for only one cycle) if that bit is set.

7.7.1.2 ALU Group Functions

Standard ALU instructions include add, subtract, logic (AND, OR, NOT, exclusive-OR), pass, negate increment, decrement, clear, and absolute value. The − function does two's-complement subtraction, whereas NOT obtains a one's complement. The PASS function passes the listed operand but tests and stores status information for later sign/zero testing. As an example, consider an ALU addition instruction for add/add-with-carry in the form

```
[IF Condition] AR = xop + yop      ;
               AF        + c
                         + yop + c
```

Instructions are in a similar form for subtraction and logical operations. If the options AR and + yop + C are chosen, and if xop and yop are the contents of AXO and AYO, respectively, the unconditional instruction would read

```
AR = AXO + AYO + C;
```

This algebraic expression means that the ALU result register AR gets the value of the ALU x-input and y-input registers plus the value of the carry-in bit. This shortens the code and speeds execution by eliminating many separate register-move instructions.

When an optional IF condition is included, and if ALU carry-bit status is chosen, then the conditional instruction reads

```
IF AC AR = AXO + AYO + C ;
```

The conditional expression, IF AC, tests the ALU carry bit. If there is a carry from the previous instruction, this instruction executes; otherwise, an NOP occurs and execution continues with the next instruction.

Division is the only ALU special function. It is executed in two steps: DIVS computes the sign, and then DIVQ computes the quotient. A full divide of a signed 16-bit divisor into a signed 32-bit quotient requires a DIVS followed by 15 DIVQs.

7.7.1.3 Shifter Group Functions

Shifter standard functions include arithmetic and logical shifts as well as floating point and block floating point scaling operations, derive exponent, normalize, denormalize, and block exponent adjust. As an example, consider a shifter instruction for normalize:

```
IF NOT CE SR = SR OR NORM SI (HI) ;
```

TABLE 7.4

Instruction Set Groups (Using ADSP-21xx Family as an Example)

Instruction Type	Purpose
Data move instructions	Move data to and from data registers and external memory
Multifunction instructions	Exploit the inherent parallelism of a DSP by combinations of data moves and memory writes/reads in a single cycle
Program flow control instructions	Direct the program sequence. In normal order, the sequence automatically fetches the next contiguous instruction for execution. This flow can be altered by these instructions.
Miscellaneous instructions	Such as NOP [no operation] and PUSH/POP, etc. These are general instructions which do not fall into the above categories. Another example is Mode Control instructions.

The conditional expression, IF NOT CE, tests the "not counter expired" condition. If the condition is false, an NOP is executed. The destination of all shifting operations is the shifter result register, SR. (The destination of the exponent detection instructions is SE or SB.) In this example, SI, the shifter input register, is the operand. The amount and direction of the shift are controlled by the signed value in the SE register in all shift operations except an immediate shift. Positive values cause left shifts; negative values cause right shifts.

The "SR OR" modifier (which is optional) logically ORs the result with the current contents of the SR register; this allows the user to construct a 32-bit value in SR from two 16-bit pieces. NORM is the operator, and HI is the modifier that determines whether the shift is relative to the HI or LO (16-bit) half of SR. If SR OR is omitted, the result is passed directly into SR.

Shift-immediate is the only shifter special function. The number of places (exponents) to shift is specified in the instruction word.

7.7.2 Other Instructions

Other instructions in a DSP can be grouped as in Table 7.4. Details are dependent on the DSP family, and hence Table 7.4 should be considered only as a guideline.

7.8　Interface between DSPs and Data Converters

Advances in semiconductor technology have given DSPs fast processing capabilities, and data converter ICs have the conversion speeds to match the faster processing speeds. This section considers the hardware aspects of practical design.

7.8.1 Interface between ADCs and DSPs

Precision-sampling ADCs generally have either parallel data output or a single serial output data link. We consider these separately.

7.8.1.1 Parallel Interfaces with ADCs

Many parallel output sampling ADCs offer three-state output that can be enabled or disabled using an output enable pin on the IC. Although it may be tempting to connect these three-state outputs directly to a back plane data bus, severe performance-degrading noise problems will result. All ADCs have a small amount of internal stray capacitance between the digital output and the analog input (typically 0.1 to 0.5 pF). Every attempt is made during the design and layout of the ADC to keep this capacitance to a minimum. However, if there is excessive overshoot, ringing, and possibly other high-frequency noise on the digital output lines (as would probably be the case if the digital output were connected directly to a back plane bus), this digital noise will couple back into the analog input through the stray capacitance. The effect of this noise is to decrease the overall ADC signal-to-noise ratio (SNR) and effective number of bits (ENOB). Any code-dependent noise will also tend to increase the ADC harmonic distortion.

The best approach to eliminating this potential problem is to provide an intermediate three-state output buffer latch, which is located close to the ADC data outputs. This latch isolates the noisy signals on the data bus from the ADC data outputs, minimizing any coupling back into the ADC analog input. The ADC data sheet should be consulted regarding exactly how the ADC data should be clocked into the buffer latch. Usually, a signal called a conversion complete or busy from the ADC is provided for this purpose.

It is also a good idea not to access the data in the intermediate latch during the actual conversion time of the ADC. This practice will further reduce the possibility of corrupting the ADC analog input with noise. The manufacturer's data sheet timing information should indicate the most desirable time to access the output data.

Figure 7.18 shows a simplified parallel interface between the AD676—a 16-bit, 100-kSPS ADC—(or the AD7884) and the ADSP-2101 microcomputer. (Note that the actual device pins shown have been relabeled to simplify the following general discussion.) In a real-time DSP application (such as in digital filtering), the processor must complete its series of instructions within the ADC sampling interval. Note that the entire cycle is initiated by the sampling clock edge from the sampling clock generator. Even though some DSP chips offer the capability to generate lower-frequency clocks from the DSP master clock, the use of these signals as precision sampling clock sources is not recommended, due to the probability of timing jitter. It is preferable to generate the ADC sampling clock from a well-designed low-noise crystal oscillator circuit, as has been previously described.

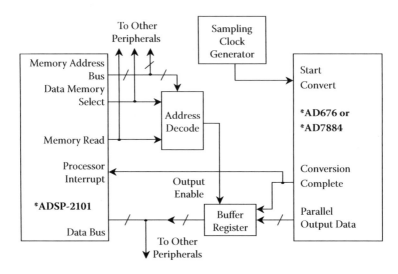

FIGURE 7.18
Generalized DSP-to-ADC parallel interface. (Reproduced by permission of Analog Devices, Norwood, MA.)

The sampling clock edge initiates the ADC conversion cycle. After the conversion is completed, the ADC conversion complete line is asserted, which in turn interrupts the DSP. The DSP places the address of the ADC that generated the interrupt on the data memory address bus and asserts the data memory select line. The read line of the DSP is then asserted. This enables the external three-state ADC buffer register outputs and places the ADC data on the data bus. The trailing edge of the read pulse latches the ADC data on the data bus into the DSP internal registers. At this time, the DSP is free to address other peripherals that may share the common data bus.

Because of the high-speed internal DSP clock (50 MHz for the ADSP-2101), the width of the read pulse may be too narrow to properly access the data in the buffer latch. If this is the case, adding the appropriate number of programmable software wait states in the DSP will both increase the width of the read pulse and cause the data memory select and the data memory address lines to remain asserted for a correspondingly longer period of time. In the case of the ADSP-2101, one wait state is one instruction cycle, or 80 ns.

7.8.1.2 Interface between Serial Output ADCs

ADCs that have a serial output (such as the AD677, AD776, and AD1879) have interfaces to the serial port of many DSP chips, as shown in Figure 7.19. The sampling clock is generated from the low-noise oscillator. The ADC output data are presented on the serial data line one bit at a time. The serial clock signal from the ADC is used to latch the individual bits into the serial input shift register of the DSP serial port. After all the serial data are transferred

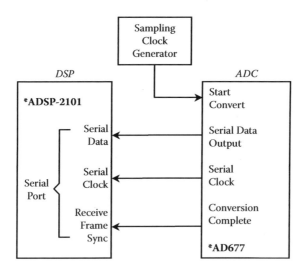

FIGURE 7.19

Generalized serial DSP-to-ADC interface. (Reproduced by permission of Analog Devices, Norwood, MA.)

into the serial input register, the serial port logic generates the required processor interrupt signal. The advantages of using serial output ADCs are the reduction in the number of interface connections as well as reduced noise because fewer noisy digital program counter tracks are close to the converter. In addition, SAR and Σ-Δ ADCs are inherently serial output devices. The number of peripheral serial devices permitted is limited by the number of serial ports available on the DSP chip.

7.8.2   Interfaces with DACs

7.8.2.1   Parallel Input DACs

Most of the principles previously discussed regarding interfaces with ADCs also apply to interfaces with DACs. A generalized block diagram of a parallel input DAC is shown in Figure 7.20a. Most high-performance DACs have an internal parallel DAC latch that drives the actual switches. This latch de-skews the data to minimize the output glitch. Some DACs designed for real-time sampling data DSP applications have an additional input latch so that the input data can be loaded asynchronously with respect to the DAC latch strobe. Some DACs have an internal reference voltage that can be either used or bypassed with a better external reference. Other DACs require an external reference.

The output of a DAC may be a current or a voltage. Fast video DACs generally are designed to supply sufficient output current to develop the required signal levels across resistive loads (generally 150 Ω, corresponding to a 75-Ω

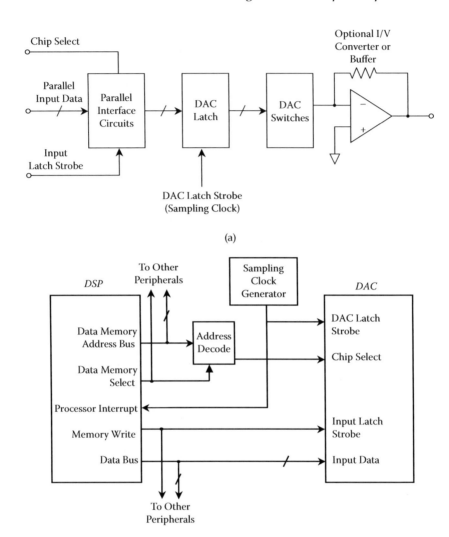

FIGURE 7.20
Interface between DSPs and parallel DACs: (a) parallel input DAC; (b) DSP and parallel DAC input.

source and load-terminated cable). Other DACs are designed to drive a current into a virtual ground and require a current-to-voltage converter (which may be internal or external). Some high-impedance voltage output DACs require an external buffer to drive reasonable values of load impedance.

A generalized parallel DSP-to-DAC interface is shown in Figure 7.20b. The operation is similar to that of a parallel DSP-to-ADC interface, described earlier. In most DSP applications, the DAC is operated continuously from a stable sampling clock generator external to the DSP. The DAC requires double buffering because of the asynchronous interface to the DSP. The sequence of events is as follows. Asserting the sampling clock generator line clocks the word contained in the DAC input latch into the DAC latch (the latch that drives the DAC switches). This causes the DAC output to change to the new value. The sampling clock edge also interrupts the DSP, which then addresses the DAC, enables the DAC chip select, and writes the next data into the DAC input latch using the memory write and data bus lines. The DAC is now ready to accept the next sampling clock edge.

7.8.2.2 Serial Input DACs

A block diagram of a typical serial input DAC is shown in Figure 7.21a. The digital input circuitry consists of a serial-to-parallel converter driven by a serial data line and a serial clock. After the serial data are loaded, the DAC latch strobe clocks the parallel DAC latch and updates the DAC switches with a new word. Interface between DSPs and serial DACs is quite easy using the DSP serial port (Figure 7.21b). The serial data transfer process is initiated by the assertion of the sampling clock generator line. This updates the DAC latch and causes the serial port of the DSP to transmit the next word to the DAC using the serial clock and the serial data line.

7.9   A Few Simple Examples of DSP Applications

In most microprocessor- or microcontroller-based systems, a continuously running assembler program may be interrupted and deviated to subroutines as and when interrupts occur. Compared to this process, a DSP-based system allows the processing of a continuously changing set of data samples, with no necessity to buffer and store the samples.

One simple example is to calculate the moving average of a set of data samples (Figure 7.22). To calculate a moving average, say, for the case of temperature measurements taken at 1 s intervals, one must first save several sequential measurements. When the current temperature is obtained, one that occurred 10 s ago can be discarded for the case of the average of 10 samples. By moving the 10-reading span for each average, one performs a moving average.

A moving average provides a useful tool that can uncover a trend and smooth sudden fluctuations. For example, in the case of temperature samples, one very high or very low temperature will not unduly influence the moving average calculated with 10 values. Any effect an odd temperature has on the average lasts for only 10 averages. The general formula for the temperature moving average contains one term per value:

$$\text{Average} = \frac{1}{10}\left(T_{now}\right) + \frac{1}{10}\left(T_{now-1}\right) + \ldots + \frac{1}{10}\left(T_{now-9}\right).$$

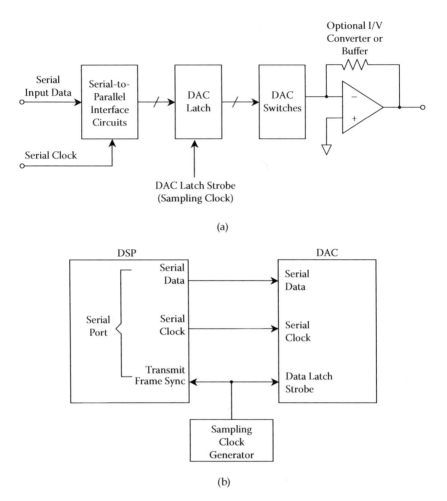

FIGURE 7.21
Interface between DSPs and serial DACs: (a) serial input DAC; (b) DSP and parallel DAC input.

One can expand or contract the equation to average more or fewer readings, depending on the specific averaging need. The general formula for a moving average incorporates as many values as needed:

$$\text{Average} = \frac{1}{n}\left(X_n\right) + \frac{1}{n}\left(X_{n-1}\right) + \ldots + \frac{1}{n}\left(X_1\right).$$

Averaging makes up a small portion of a larger class of common DSP operations, namely filtering. Whereas analog circuit engineers think of filter designs in terms of passive components and op amps, DSP system designers think of filters in terms of algorithms—ways to process information.

Most DSP filter algorithms fall into two general categories: IIR and FIR. Either type can be used to develop DSP equivalents of analog high-pass, band-pass, low-pass, and band-stop filters. Even if one has little DSP background, mastering the math behind an FIR filter takes little time. The FIR filter equation below practically duplicates the equation for a moving average:

$$F_1 = (C_n \times X_n) + (C_{n-1} \times X_{n-1}) + \ldots + (C_1 \times X_1).$$

Unlike the averaging equation, the FIR filter equation uses different coefficients for each multiplication term. Having discussed FIR and IIR filters earlier, let's discuss the case of an FIR filter compared to a moving average routine. Like a moving average routine, a FIR filter routine operates on the n most recent data values. (The number of coefficients, n, depends on a filter's characteristics.) Typically, a filter routine discards the oldest value as it acquires the newest value (Figure 7.23). After each new value arrives, the software multiplies each of the n values by its corresponding coefficient and sums the results.

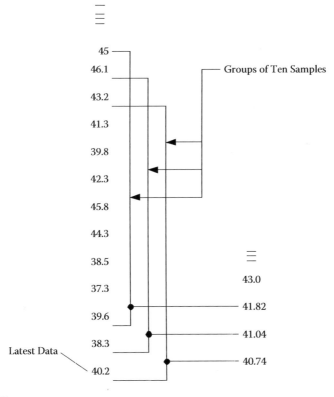

FIGURE 7.22
Moving average calculation as a simple example of DSP applications.

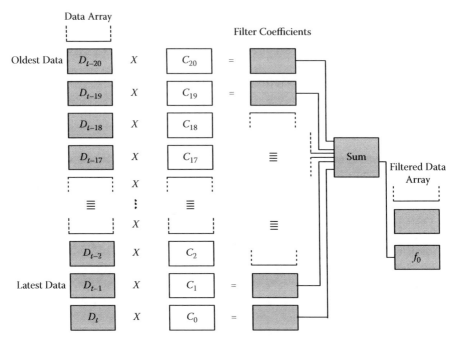

FIGURE 7.23
FIR filter implementation by multiplication and addition of pointed data.

The sum goes into a separate section of the system memory that holds the filtered information. If one uses a 20-term FIR filter routine to process a signal, the computer must perform 20 multiplications and 20 additions for each value. Some FIR filter routines may require even more coefficients. The heavy use of math operations differentiates DSP operations from other types of software tasks.

In spite of the amount of time a computer can spend working on math for DSP, processing signals as discrete values has its advantages. To acquire analog signals from 30 sensors for a strip chart recorder, for example, one may have to build, test, and adjust 30 analog front-end circuits. Also, there will be problems associated with thermal drift, voltage offsets, and component tolerances. If test conditions change, one must modify and retest all 30 circuits.

When one acquires signals with an ADC and processes them digitally, many problems disappear. Instead of needing 30 individual filters, one DSP filter routine can operate on all the data, one channel at a time. If one plans to apply the same type of filtering to each channel, one set of coefficients will suffice. To perform different types of filtering on some channels, one simply sets up an array of coefficients for each filter type. However, one still may have to provide antialiasing filters. For details, see Titus [18].

7.10   Practical Components and Developments in the Late 1990s

During the late 1990s, incredible developments took place in the DSP components world. Vendors focused on several key aspects of the DSP architectures. The most obvious architectural improvements were in increased "parallelism": the number of operations the DSP can perform in an instruction cycle. An extreme example of parallelism is Texas Instruments' C6x very long instruction word (VLIW) DSP, with eight parallel functional units. Although Analog Devices' super-Harvard architecture (SHARC) could perform as many as seven operations per cycle, the company and other vendors were working feverishly to develop their own "VLIW-ized" DSPs. In contrast to superscalar architectures, VLIW simplifies a DSP's control logic by providing independent control for each processing unit. During the late 1990s, the following important developments were achieved [19]:

- While announcing the first general-purpose VLIW DSP, Texas Instruments also announced the end of the road for the C8x DSP family. The company emphasized the importance of the compilers for DSPs with the purchase of DSP-compiler company Tartan.

- Analog Devices broke the $100 price barrier with its SARC floating point architecture.

- Lucent Technologies discontinued new designs incorporating its 32-bit floating point DSP. The company also focused its energy on application-specific rather than general-purpose DSPs. The application-specific products targeted modems and other communication devices.

- Motorola's DSP Division became the Wireless Signal Processing Division, although the company still supports many general-purpose DSP and audio applications.

Among the hottest architectural innovations during the same period was the move to dual MAC units. The architectures of these MACs perform twice the digital signal processing as before. Texas Instruments kicked off this evolution with its VLIW-based C6x. Meanwhile, engineers designing with DSPs need a simple method to compare processor performance. Unfortunately, as processor architectures diversify, traditional metrics such as MIPS and MOPS have become less relevant. Alternatively, Berkeley Design Technology (BDTI) has become well known in the DSP industry for providing DSP benchmarks. Instead of using full-application benchmarks, BDTI has adopted benchmark methodology based on DSP algorithm kernels, such as FFTs and FIR filters. BDTI implements its suite of 11 kernel-based benchmarks (the BDTI Benchmarks) on a variety of processors. One can find the results of these benchmarks in the company's *Buyer's Guide to DSP Processors* (available at http://www.bdti.com). Cushman [20] discusses the merging of microprocessor and DSP architectures, and Levy [19,21] discusses the maturing stages of DSP architectures.

7.11   Recent Technology Trends in DSP Design

Ever-increasing demands for higher performance, lower energy consumption, and lower cost in the consumer electronics market require optimizing these factors for different system components. As digital signal processing is one of the major computational demands in overall processing, DSP designers should employ state-of-the-art technological advancements in order to achieve rapid design turnaround as well as to satisfy performance, energy consumption, and cost demands.

DSP designers usually start the design process from a high-level algorithm description. Therefore, technology advances can be employed in different ways, such as using a sophisticated DSP or implementing an application-specific hardware accelerator, depending on the designer's experience and available intellectual properties (IPs).

DSPs are considered reprogrammable devices, which are designed to optimize the most frequent signal-processing operations [22]. DSPs, as the main processing components in most of today's signal-processing applications, have evolved architecturally. In addition, significant advances in semiconductor technology have helped them reach higher levels of performance at higher clock rates. However, this leads to significant increases in power consumption, as it is directly proportional to clock rates.

Recent DSPs, such as Texas Instrument's TMS320C64x+ [23] and Analog Devices' TigerSHARC [24], are designed based on VLIW architectural features. In this way, fine-grained instruction-level parallelism available in the application is extracted through a sophisticated compiler and exposed to the processor's multiple function units. In each cycle, multiple independent operations are executed concurrently.

Application-specific hardware accelerators can be designed to optimize the performance and energy consumption of a specific function. Their main disadvantage is their lack of programmability, which requires redesigning if further changes become necessary. The use of FPGA technology can overcome this problem, as FPGAs can be reconfigured to satisfy design constraints. In addition, FPGAs provide many resources for parallel implementation of the critical parts of the application. Soft-core processors can be implemented for general-purpose operations and control dominant parts of the application on the same FPGA chip.

Using FPGAs to design a hardware accelerator for digital signal-processing applications used to require hardware design skills that were not usually available to algorithm designers. MathWorks' MATLAB and Simulink [25] are very popular tools for designing and verifying DSP algorithms. FPGA manufacturers such as Altera and Xilinx provide tools to directly implement hardware accelerators from algorithmic and system-level descriptions in MATLAB and Simulink. Altera's DSP Builder [26] is an example that integrates hardware design tools based on a hardware description language (such as VHDL or Verilog) and algorithmic system-level design tools (MATLAB and Simulink) to synthesize an algorithmic description.

As shown in Figure 7.24, the system is first described in MATLAB or Simulink. Then the compiler (known as the signal compiler) analyzes the design and generates VHDL code for hardware synthesis of the design. This VHDL code is synthesized for the target FPGA device by the Quartus II synthesis tool. In order to achieve this, DSP Builder provides synthesizable blocks (as Simulink blocks), which can be used to model the system in Simulink targeted for hardware synthesis. This assists DSP algorithm designers in synthesizing their systems without the need to know a hardware description language. It is also possible to set up a cosimulation environment to use both Simulink (for system level) and a VHDL simulator (using ModelSim [27]) to

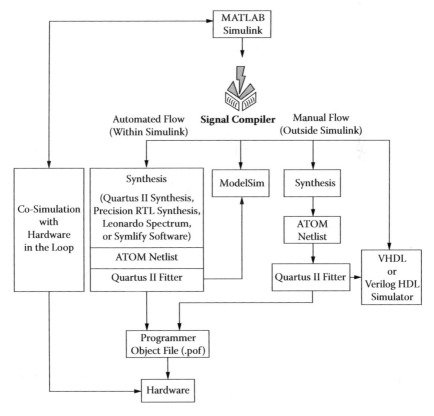

FIGURE 7.24
System-level design flow in DSP Builder [26]. (Source: Altera.)

verify the functionality of the system. It is expected that further advances in these system-level design tools will accelerate architectural explorations at the early stages of design for platform-based implementations.

Acknowledgment

This chapter is an edited version of Chapter 13 of the IEE Measurement Series, Volume 11, *Digital and Analog Instrumentation Testing and Measurement*, by Nihal Kularatna, reproduced with permission of the publisher. Additional contributions have come from Dr. Morteza Biglari-Abhari.

References

1. Jones, N.B. and Watson, J.D., *Digital Signal Processing: Principles, Devices and Applications*, IEE, London, 1990.
2. Schneiderman, R., Faster, more highly integrated DSPs: designed for designers, *Wireless Systems Designs*, November, 12, 1996.
3. Marven, C. and Ewers, G., *A Simple Approach to Digital Signal Processing*, Texas Instruments, Dallas, TX, 1994.
4. Kularatna, N., *Modern Component Families and Circuit Block Design*, Newnes, London, 2000.
5. Burks, A.W., Goldstine, H.H., and von Neumann, J., *Preliminary Discussion of the Logical Design of an Electronic Computing Instrument*, Institute for Advanced Study, Princeton, NJ, June 28, 1946; Reprinted in Bell, C.G. and Newell, A., eds., *Computer Structures: Readings and Examples*, McGraw-Hill, New York, 1971.
6. Madisetti, V.K., *VLSI Digital Signal Processors*, Butterworth Heinemann, Oxford, 1995.
7. Levy, M., Microprocessors and DSP technologies unite for embedded applications, *EDN*, March 2, 73, 1998.
8. Parks, T.W. and Burrus, C.S., *Digital Filter Design*, John Wiley & Sons, New York, 1987.
9. Oppenheim, A.V. and Schafer, R.W., *Digital Signal Processing*, Prentice Hall, Upper Saddle River, NJ, 1988.
10. Kogge, P.M., *The Architecture of Pipe-Lined Computers*, McGraw-Hill, New York, 1981.
11. Lapsley, P., Bier, J., Shoham, A., and Lee, E.A., *DSP Processor Fundamentals: Architecture and Features*, IEEE Press, New York, 1997.
12. Lee, E.A., Programmable DSP architectures: part I, *IEEE ASSP Magazine*, October, 4, 1988.
13. Lee, E.A., Programmable DSP architectures: part II, *IEEE ASSP Magazine*, January, 4, 1989.
14. Kularatna, N., *Modern Electronic Test and Measuring Instruments*, IEE, London, 1996.

15. Ingle, V.K. and Proakis, J.G., *Digital Signal Processing Laboratory Using the ADSP-2101 Microcomputer,* Prentice Hall, Upper Saddle River, NJ, 1991.
16. New, B., A distributed arithmetic approach to designing scalable DSP chips, *EDN,* August 17, 107, 1995.
17. Fine, B., Considerations for selecting a DSP processor: ADSP-2100A vs TMS320C25, Application Note , Analog Devices, Norwood, MA, n.d.
18. Titus, J., What is DSP all about?, *Test and Measurement World,* May, 49, 1996.
19. Levy, M., EDN's 1997 DSP architecture directory, *EDN,* May 8, 43, 1997.
20. Cushman, R.H., μP-like DSP chips, *EDN,* September 3, 155, 1987.
21. Levy, M., EDN's 1998 DSP architecture directory, *EDN,* April 23, 40, 1998.
22. Hovsmith, S., Adapting hardware to software: productive programming in a multi-core environment, *Information Quarterly,* 6, 14, 2007.
23. Texas Instruments home page; available at http://www.ti.com.
24. Wolf, O. and Bier, J., TigerSHARC sinks teeth into VLIW, *Microprocessor Report,* 12, 1, 1998; available at http://www.bdti.com/articles/tigersharc.pdf.
25. The MathWorks home page; available at http://www.mathworks.com.
26. *DSP Builder Reference Manual,* version 7.2, Altera, San Jose, CA, 2007; available at http://www.altera.com/literature/manual/mnl_dsp_builder.pdf.
27. Mentor Graphics home page; available at http://www.mentor.com.

8

An Introduction to Oscillators, Phase Lock Loops, and Direct Digital Synthesis

Sujeewa Hettiwatte, Coauthor

CONTENTS

8.1 Essentials of Simple Oscillators.. 413
8.2 Principles of Oscillation 414
8.3 Crystal Oscillators.. 415
8.4 Phase-Locked Loop Systems 418
    8.4.1 Building Blocks of PLL Systems... 419
        8.4.1.1 Phase Detector... 419
        8.4.1.2 Low-Pass Filter 419
        8.4.1.3 Voltage-Controlled Oscillator 420
        8.4.1.4 Frequency Divider.. 420
8.5 Clock Systems for Digital Systems and Signal Integrity.................... 420
    8.5.1 Signal Integrity and System Design 421
    8.5.2 Clock Skew and Transmission Line Effects 422
    8.5.3 PCB Interconnection Characterization 423
8.6 DDS Systems and Waveform Generation 423
References 425

8.1 Essentials of Simple Oscillators

Electronic systems designers frequently need to use an independent stimulus to drive other circuits, and in many cases this stimulus is expected to have certain shapes in the time domain and certain frequency characteristics in the frequency domain. For example, in digital circuits, a clock oscillator output gets distributed across many subcircuits, and the distributed clock needs to be very precisely timed with respect to its rising and falling edges.

An electronic oscillator draws DC power from a circuit and creates a repetitive electrical waveform. There are many ways of classifying the oscillators. One way is to classify them according to the components used in

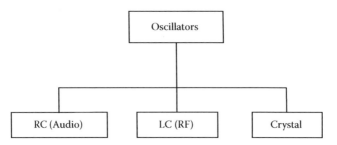

FIGURE 8.1
Classification of oscillators.

the feedback network. Examples are RC, LC, and crystal. Another way is according to the shape of the repetitive waveform they produce, for example, a harmonic oscillator (sinusoidal waveform) or a relaxation oscillator (nonsinusoidal waveforms, such as square, triangular, or sawtooth). Another way of classifying is dependent upon the oscillation frequency, for example, low frequency (1 Hz to 1 MHz) and radio frequency (>1 MHz). In this chapter, which is a summary of some important aspects and approaches to designing oscillators, electronic oscillators are classified according to RC, LC, and crystal, as in Figure 8.1.

The signal source normally found in any analog-type sine wave generator or a function generator is an RC oscillator. Function generators can produce square, triangular, and sawtooth waveforms. A pulse is a square wave generator with a duty cycle of less than 50%; hence, pulse generators can be categorized as function generators. RC oscillators can be further classified as phase shift, Wien bridge, and twin-T.

In high-frequency applications, LC oscillators are used. The signal source in a radio frequency (RF) signal generator is normally an LC oscillator. LC oscillators can be further classified into Hartley, Colpitts, and Clapp types [1].

Crystal oscillators produce very accurate and stable frequencies. However, oscillation frequency tends to change with the age and operating temperature of the crystal. There are a few compensation schemes used in practice to stabilize the frequency, and the crystal oscillators are normally named after the compensation scheme used. Crystal oscillators are discussed further in Section 8.3.

8.2 Principles of Oscillation

The principles of oscillation in an electronic circuit can be described using a positive feedback network, as shown in Figure 8.2. The amplifier and feedback network shown in Figure 8.2 have open-loop gains of A and β. The closed-loop gain is given by

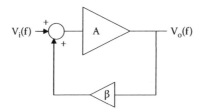

FIGURE 8.2
Positive feedback network.

$$A_f = \frac{A}{1-A\beta}.$$

(8.1)

The condition for oscillations is

$$A\beta = 1\angle 0°.$$

(8.2)

In other words, the loop gain must be equal to unity and the total phase shift around the loop must be 0°. However, in practice, to initiate oscillations, $A\beta > 1\angle 0°$, and once oscillations are initiated, Equation 8.2 has to be satisfied.

8.3 Crystal Oscillators

Crystal oscillators are usually made of quartz. Quartz crystals exhibit piezo-electric properties, which are utilized in the building of the crystal. The frequency of resonance of quartz is dependent upon the direction of cut in the crystal with reference to a Cartesian system of coordinates. These cut directions are named BT, BC, FT, AT, and so forth [2]. Piezoelectricity also works in the reverse mode; that is, if an electrical potential is applied to a quartz resonator, it will deform. Thus, to obtain a stable resonance frequency, part of the electrical signal produced by the resonator is fed back electrically to re-ping the crystal. This allows the resonator to sustain oscillations on its resonant frequency [2].

The typical equivalent circuit of a crystal resonator is shown in Figure 8.3. The complex impedance of the resonator at frequency f can be expressed by Equation 8.3, where f_s is the series resonance frequency given by Equation 8.4 and f_p is the parallel resonance frequency given by Equation 8.5:

$$Z = \frac{f_s^2 - f^2 + jf\frac{R_1}{2\pi L_1}}{-R_1 f^2 + j2\pi f C_o \left(f_p^2 - f^2\right)}$$

(8.3)

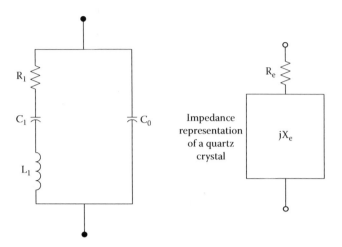

FIGURE 8.3
Equivalent circuit of a crystal resonator.

$$f_s = \frac{1}{2\pi\sqrt{L_1 C_1}}$$ (8.4)

$$f_p = \frac{1}{2\pi}\sqrt{\frac{C_1 + C_o}{L_1 C_1 C_o}}$$ (8.5)

Typical values [3] for a crystal oscillator having an f_s = 2.015 MHz are R_1 = 100 Ω, L_1 = 520 mH, C_1 = 0.012 pF, and C_o = 4 pF. This gives a parallel resonance frequency of f_p = 2.018 MHz. This shows that series and parallel resonance frequencies are close together, and they differ by only a few tens of kilohertz. Figure 8.4 and Figure 8.5 show the reactance versus frequency responses for a typical crystal oscillator. It should be noted here that at f_s, the total reactance of the crystal is not zero. However, this reactance is very close to zero, and for most practical purposes it can be taken as zero. Normally, a

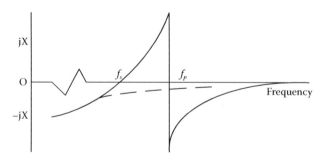

FIGURE 8.4
Reactance versus frequency.

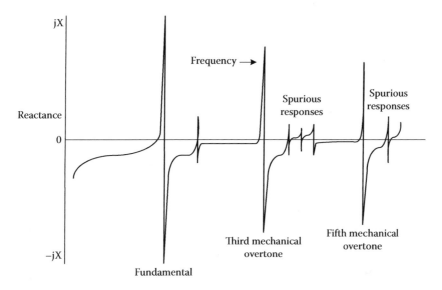

FIGURE 8.5
Reactance versus frequency with overtones.

crystal is operated at a frequency between f_s and f_p, so that reactance is either zero or inductive [3]. To obtain higher resonance frequencies, the overtones (third or fifth in Figure 8.5) are used.

As mentioned earlier, even though crystal oscillators are very accurate and stable, the accuracy and stability of a crystal oscillator might change with operating temperature and age. Therefore, a few compensation schemes are used in practice. Crystal oscillators are normally categorized according to the compensation scheme used. A few popular schemes are

- Temperature-compensated crystal oscillators (TCXO)
- Oven-controlled crystal oscillators (OCXO)
- Digitally controlled crystal oscillators (DCXO)

There are also crystal oscillators that do not use any compensation scheme. These are called uncompensated crystal oscillators.

Temperature-compensated crystal oscillators normally use a voltage-controlled oscillator with a variable capacitance (varactor diode) and a thermistor-resistor network to compensate for changes in frequency. A thermistor-resistor network generates a temperature-varying voltage that alters the varactor's capacitance to cancel out the crystal's thermal drift. Almost all TCXOs use AT-cut crystals, which have a superior thermal stability over a wide temperature range. The temperature-compensated crystal oscillators are hard to manufacture consistently with a temperature stability better than 1 ppm [4]. In oven-controlled crystal oscillators, temperature stabilization is achieved by placing the crystal in a proportionally controlled oven. Such oscillators tend to be bulky, expensive, and power hungry, but they have extraordinarily good stability [4]. OCXOs with stabilities in the range of ±5 ppb (parts per billion) have been reported in the literature [5].

Digitally controlled crystal oscillators try to achieve the stability of OCXOs and the low current drain of TCXOs. There are basically two types of DCXOs based on the compensation scheme used: direct compensation and indirect compensation. In direct compensation, oscillator frequency is electronically tunable via a varactor diode inserted in the feedback network. A compensating voltage is generated that tracks the characteristic frequency-versus-temperature drift of the crystal, pulling the oscillator back to nominal frequency over the design temperature range [4]. In indirect compensation, the oscillator is free to run at its natural frequency, regardless of the temperature. The compensated output is derived by subtracting as many oscillator pulses as necessary to maintain a constant output frequency as the temperature changes [4]. There are advantages and disadvantages associated with direct and indirect compensation schemes used in DCXOs. A more detailed account of DCXOs can be found in Breed and Fry [4].

8.4 Phase-Locked Loop Systems

A common method of generating RF oscillation is by using a phase-locked loop (PLL). PLLs are commonly found in areas such as communications, wireless systems, digital circuits, and disc drive electronics. A PLL can be used for frequency synthesis, jitter reduction, skew suppression, and clock recovery [6]. Jitter is the phase shift of digital pulses over a transmission medium. Skew is the difference in time caused by a signal traveling from a source (transmitter) to the receiver using two parallel paths. The block diagram in Figure 8.6 shows the basic building blocks of a PLL.

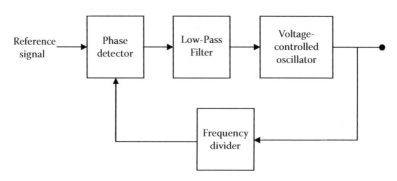

FIGURE 8.6
Basic building blocks of a PLL.

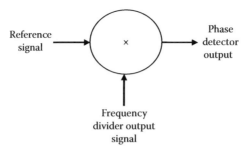

FIGURE 8.7
Multiplier as a phase detector.

8.4.1 Building Blocks of PLL Systems

The PLL contains four building blocks: a phase detector, a low-pass filter, a voltage-controlled oscillator, and a frequency divider.

8.4.1.1 Phase Detector

The phase detector (PD) compares a periodic input signal (the reference signal in Figure 8.6), which is normally a sine wave or a square wave, with the frequency divider output signal. The PD output voltage is proportional to the phase difference between the two signals. In the locked state, the PD output is a DC value proportional to the phase difference between the reference signal and the output from the frequency divider.

In designing a PD, the following properties are of particular interest [6]:

- What is the input phase difference range for which the characteristic is monotonic?
- What is the response to unequal input frequencies?
- How do the input amplitude and duty cycle affect the characteristic?

A PD can be implemented as a multiplier, as shown in Figure 8.7. However, one drawback of this implementation is that the locking range is small [7].

Another implementation of a phase detector is as a phase-frequency detector, shown in Figure 8.8 [7]. This circuit employs two D-type flip-flops and an AND gate. f_r is the reference signal and f_y is the signal from the frequency divider output.

8.4.1.2 Low-Pass Filter

The low-pass filter suppresses high-frequency components in the PD output, allowing the DC value to control the voltage-controlled oscillator frequency.

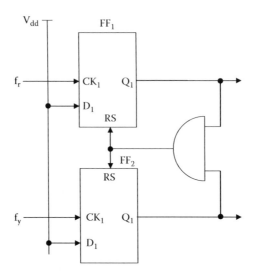

FIGURE 8.8
A phase-frequency detector used as a phase detector [7].

8.4.1.3 Voltage-Controlled Oscillator

The output frequency of the voltage-controlled oscillator (VCO) is a mono-tonically increasing function of the input voltage. In the locked state of the PLL, the output frequency is equal to N times the frequency of the reference signal, where N is the value on the frequency divider.

The most widely used VCO is a varactor, which is a reverse-biased junction diode whose capacitance monotonically decreases with reverse bias. Varac-tors made of silicon are used up to a few gigahertz, and those made of gal-lium arsenide (GaAs) can be used up to 40 GHz or more [7]. Some microwave VCOs use a yttrium–iron–garnet (YIG) resonator instead of a varactor.

8.4.1.4 Frequency Divider

The output of the frequency divider is a signal with a frequency equal to the VCO output frequency divided by N. The simplest frequency divider is a T (toggle) flip-flop. T flip-flops can be cascaded to obtain higher-order fre-quency dividers.

8.5 Clock Systems for Digital Systems and Signal Integrity

Clock signals have the highest frequency in any digital circuit and are heav-ily loaded, because they are distributed to a number of components in a cir-cuit. The timing margin is the amount of excess time (or the slack) available

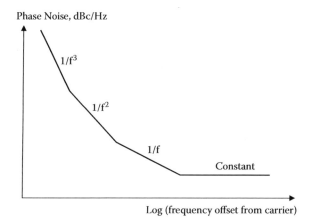

FIGURE 8.9
Tangential approximation of a VCO phase noise shape [7].

in a clock cycle for a logic transition. A system with a higher timing margin can operate at a higher speed than one with a lower timing margin.

Phase noise represents short-term, rapid, and random changes in the phase of a signal. It is normally measured at an offset from a carrier frequency. By definition, phase noise at a given frequency offset (f_m) from a carrier (f_c) is defined as the ratio between the noise in a 1-Hz bandwidth f_m Hz offset from f_c Hz and the carrier, f_c Hz. The phase noise is usually expressed in logarithmic units (dBc/Hz). As an example, in a VCO, phase noise can usually be approximated by the graph in Figure 8.9 [7].

Skew arises in clock systems due to parallel transmission of a signal between a source (transmitter) and a receiver, with different arrival times. The most common cause of skew is due to the difference in the effective lengths of the two transmission paths connecting the source and the receiver. To lower the skew in clock systems, a clock buffer can be placed in a PLL.

The International Telecommunication Union defines jitter as "short-term variations of the significant instants of a digital signal from their ideal positions in time" [8]. Timing jitter is the phase shift of digital pulses over a transmission medium. A PLL-based clock recovery circuit can be used to regenerate data and thereby reduce the timing jitter [6].

8.5.1   Signal Integrity and System Design

In electronics, signal integrity means the signal is unimpaired with regard to functionality. For example, a system with good signal integrity has data that arrive early enough to guarantee setup timing requirements are met and clocks that make only one logic transition per physical transition. Digital systems are tolerant of many signal-integrity effects, including delay mismatches and signal ringing. However, sufficiently large signal-integrity problems can cause systems to fail or, worse yet, work only intermittently.

TABLE 8.1

Definitions of Signal Integrity and Related Terms

Term	Description
Signal integrity	Signal integrity is the ability of a signal to generate correct responses in a circuit. A signal with good signal integrity has digital levels at required voltage levels at required times.
Crosstalk	Crosstalk is the interaction between signals on two different electrical nets. The one creating crosstalk is called an aggressor and the one receiving it is called a victim. Often, a net is both an aggressor and a victim.
Overshoot	Overshoot is the first peak or valley past the settling voltage—the highest voltage for a rising edge and the lowest voltage for a falling edge.
Undershoot	Undershoot is the next valley or peak.Undershoot is the second peak or valley past the settling voltage—the deepest valley for a rising edge and the highest peak for a falling edge.
Skew	Signal skew is the difference in arrival time of one signal to different receivers on the same net. Skew is also used to describe the difference in arrival time between the clock and data at a logic gate.

One common signal-integrity problem is false clocking, where a clock crosses a logic threshold more than once on a transition due to ringing on the line. Some parameters related to product design and signal integrity are described in Table 8.1.

8.5.2 Clock Skew and Transmission Line Effects

The most difficult problem in high-speed design is clock skew. In a digital system, the clock must be distributed to every IC that is operating at the processor speed. Because these ICs can number 15 or more in a typical system, designers are expected to use buffers with the crystal clock reference. Clock skew is the time difference between edge transitions for different buffered outputs.

Buffers generate three types of clock skew: intrinsic, pulse, and extrinsic. Intrinsic skew is the difference between outputs on the same IC. Clock buffers also generate pulse skew. Pulse skew is the difference in propagation delay between a low-to-high transition and a high-to-low transition for a single buffer output. Extrinsic skew is external to the buffers, and it is the time difference that occurs when one sends a fast rise time signal on board traces that have different lengths and capacitive loads.

To compensate for various types of clock skew–related delays, designers have to use transmission line principle–related design approaches. PC board transmission lines come in two flavors: strip lines and microstrip lines. The clock driver must drive a line's characteristic impedance (Z_0), which is a function of trace geometry and the board's dielectric constant (ε_r). For details, see Johnson and Graham [9], Gallant [10], Johnson [11,12], and Sutherland [13].

8.5.3 PCB Interconnection Characterization

As the performance requirements for modern computer and communications systems grow, the demand for high-speed PCBs also increases. Speeds as fast as 1 Gbit/s are expected to be supported by standard PCB technologies, with the rise times of these signals being as fast as 100 ps.

At these speeds, interconnections on PCBs behave as distributed elements, or transmission lines, and reflections due to impedance mismatch are typical signal-integrity problems that board designers encounter in their work. Connections between layers and connectors on a board create discontinuities that distort the signals even further. To accurately predict the propagation of the signals on board, designers need to determine the impedance of the traces of different layers and extract the models for board discontinuities. For details of planning for signal integrity and identifying the bandwidth limitations of PCB traces, see Johnson [11,12].

8.6 DDS Systems and Waveform Generation

Direct digital synthesis (DDS) is a technique by which a signal is generated in the form of a series of digital numbers and converted into analog form by a digital-to-analog (DAC) converter. It is a convenient way of generating a waveform without going into detailed specifications of individual circuit components, such as the capacitors and inductors, usually associated with oscillators generating such waveforms. A DDS-based waveform generator offers the advantage of a larger library of standard waveforms because the waveforms are stored digitally and loaded into waveform memory as needed. A typical DDS-based waveform generator is capable of providing a choice of sine, square, triangular, and sine^2 pulses as well as other continuous waveforms. To test a CD player, the sine^2 pulse most accurately simulates the error voltage that results from a mechanical shock in the control system of the CD player [14].

A fundamental DDS system is shown in Figure 8.10. In this simplified model, a stable clock drives the address counter, which in turn steps through different locations (addresses) in the programmable read-only memory (PROM). The PROM stores one or more integral number of cycles of sine wave or any other arbitrary waveform. As the address counter-steps through each memory location, the corresponding digital amplitude is loaded onto the register. The register content is then loaded to the DAC, which generates an analog waveform. The low-pass filter smoothes this analog waveform to remove any discontinuities.

Because a DDS system is basically a sampled data system, issues involved in sampling, such as quantization noise, aliasing, and filtering, have to be

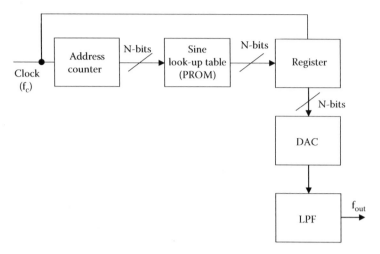

FIGURE 8.10
A fundamental DDS system.

considered in designing such a system [15]. A major problem with the simple
DDS system shown in Figure 8.10 is that the final output frequency can be
changed only by changing the reference clock frequency or by reprogram-
ming the PROM, which is rather inflexible. A much more flexible DDS system
used in practice is shown in Figure 8.11. The operation of this DDS system and
a detailed account of waveform generation can be found in Kularatna [16].

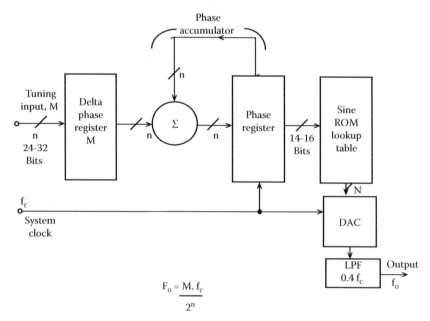

$$F_0 = \frac{M \cdot f_c}{2^n}$$

FIGURE 8.11
A more flexible DDS system.

References

1. da Silva, E., *High Frequency and Microwave Engineering*, Butterworth-Heinemann, Oxford, 357, 2001.
2. Carr, J., Crystals made clear 1, *Electronics World*, September, 780, 1999.
3. Frerking, M.E., *Crystal Oscillator Design and Temperature Compensation*, Van Nostrand Reinhold, New York, 20, 1978.
4. Breed, G.A. and Fry, S., *Oscillator Design Handbook*, 2nd ed., Cardiff Publishing, Englewood, CO, 12, 1991.
5. Travis, B., OCXO touts industry's best stability, *EDN*, March 18, 14, 1999.
6. Razavi, B., *Monolithic Phase-Locked Loops and Clock Recovery Circuits: Theory and Design*, IEEE Press, New York, 1, 1996.
7. Bianchi, G., *Phase-Locked Loop Synthesizer Simulation*, McGraw-Hill, New York, 23, 2005.
8. International Telecommunication Union publications; available at http://www.itu.int/publications.
9. Johnson, H. and Graham, M., *High Speed Digital Design*, Prentice Hall, Upper Saddle River, NJ, 1993.
10. Gallant, J., 40 MHz CMOS circuits send designs back to school, *EDN*, March 2, 67, 1992.
11. Johnson, H., Planning for signal integrity, *Electronic Design*, May 12, 26, 1997.
12. Johnson, H., Transmission-line scaling, *EDN*, February 4, 26, 1999.
13. Sutherland, J., As edge speeds increase, wires become transmission lines, *EDN*, October 14, 75, 1999.
14. Barker, D., Function generators test digital devices, *Electronic Design, Test and Measurement*, December, 89, 1993.
15. *Mixed Signal and DSP Design Techniques*, Analog Devices, Norwood, MA, 2000.
16. Kularatna, N., *Digital and Analogue Instrumentation: Testing and Measurement*, IEE, London, 2003.

9

System-on-a-Chip Design and Verification

Chong-Min Kyung

CONTENTS

9.1 Introduction 428
    9.1.1 What Is System-on-a-Chip? 428
    9.1.2 Why SoC? 429
    9.1.3 Challenges with SoC Design 430
9.2 Front-End Design Flow 431
    9.2.1 Reference Algorithm 433
    9.2.2 Hardware and Software Partitioning 434
        9.2.2.1 Issues..434
        9.2.2.2 Transaction-Level Modeling-Based Methodology 436
        9.2.2.3 System-Level Design Method..438
    9.2.3 Cosimulation 438
        9.2.3.1 Introduction 438
        9.2.3.2 Cosimulation Tools 439
        9.2.3.3 Cosimulation Issues...440
        9.2.3.4 Cosimulation Example 440
    9.2.4 Coemulation 442
        9.2.4.1 Introduction 442
        9.2.4.2 Benefits...442
        9.2.4.3 Coemulation Tools 443
        9.2.4.4 Coemulation Environment 444
    9.2.5 Emulation...445
        9.2.5.1 Emulation Board Survey..446
        9.2.5.2 Debugging Methodology..449
9.3 Back-End Design 452
    9.3.1 RTL Description and Verification..454
    9.3.2 Logic Synthesis...454
    9.3.3 Static Timing Analysis 455
    9.3.4 Design for Test and Auto Test Pattern Generation.....................456
    9.3.5 Gate-Level Simulation and Power Analysis.................................456
        9.3.5.1 Gate-Level Simulation 456
        9.3.5.2 Power Analysis 457
    9.3.6 Place and Route 458

         9.3.6.1   Partitioning 458
         9.3.6.2   Floor Planning 459
         9.3.6.3   Placement 460
         9.3.6.4   Routing 460
         9.3.6.5   Clock Tree Synthesis and Power Routing 461
         9.3.6.6   Extraction and Verification 461
     9.3.7   Postlayout Simulation 463
9.4   Status and Prospect of Radio Frequency Integrated Circuits (RFICs) ... 463
9.5   Mixed-Signal ICs in SoC 465
Acknowledgments 466
References 466

9.1 Introduction

9.1.1 What Is System-on-a-Chip?

The almost-50-year history of microelectronics indicates that integration leads not only to smaller chip and system sizes but also to greater reliability, lower system power requirements, and plummeting cost/performance ratios. After the invention of the transistor, integrated circuit (IC) technology has progressed through the following stages to reach system-on-a-chip (SoC) concepts:

- Small-scale integrated (SSI) circuits
- Medium-scale integrated (MSI) circuits
- Large-scale integrated (LSI) circuits
- Very large-scale integrated (VLSI) circuits
- System-on-a-chip (SoC)

During the early stages of the IC era, one group of designers created monolithic components using semiconductor technology, while systems designers created systems out of ICs. We are now at a juncture where complete systems can be fabricated on a single piece of silicon. Early examples of SoC concepts are monolithic silicon components for calculators, modems, and cellular phones. A systems perspective with some historic experiences is presented in Frantz [1], based on the experiences at a semiconductor company such as Texas Instruments.

During the last two decades, CMOS has developed into an amazing process technology for integrating various IC components. Today, SoC denotes a CMOS system consisting of both hardware and software implemented on a single chip. The hardware includes not only such digital functional blocks as processor cores and memory blocks but also various analog and RF components. Integrating many process-wise homogeneous and heterogeneous

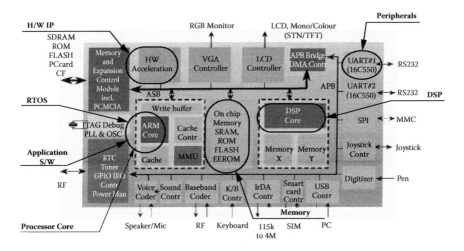

FIGURE 9.1
An example of a SoC consisting of a microprocessor core, DSP core, memory blocks, and other peripheral components.

components on a single chip provides many advantages in speed, footprint, noise, and power consumption. However, these incur more complicated process steps and a larger number of mask layers, which add to the processing expense. Despite the increasing number of mask layers, process complexities, along with shrinking resolution, large-volume products are implemented as a SoC for higher performance, less power consumption, and lower production costs. Figure 9.1 shows an example of a SoC consisting of an advanced reduced instruction set microprocessor (ARM) core, digital signal processor (DSP) core, memory blocks, and other peripheral components.

Among the many components in a SoC, a programmable processor core and on-chip memory are used for controlling the overall function and running the software application. Sometimes a real-time operating system (RTOS) is executed in the processor core. In a multicore SoC, the DSP is also integrated with a general-purpose processor. The DSP is generally used as an application-specific core for various video and audio or similar applications. Other major components in a SoC are accelerating functional units implemented in hardware. Generally, the software application is flexible and easy to maintain but not adequately fast. Some parts of the application need to be implemented in hardware for performance reasons such as timing. Other components in the SoC are peripheral devices that are used for interfacing with external devices. Analog components and opto-/microelectromechanical system (O/MEMS) components can also be incorporated in a SoC.

9.1.2 Why SoC?

The SoC technology has appeared for two reasons: design productivity gap and time-to-market (TTM) need. The progress of silicon manufacturing technology has begun to outpace that of design technology, with the deep submicron regime entering as a main process technology. The resulting design productivity gap implies that designers cannot catch up with available gate sizes by conventional design techniques but have to employ a so-called platform-based design along with a significant reuse of hardware and software blocks which could carry significant intellectual property (IP) elements.

The platform is a set of rules and guidelines for hardware and software architecture, together with a suite of building blocks that fit into the particular architecture. The hardware rules deal with the building of the chip: clocking schemes, power management, interface rules for the blocks, and on-chip interconnect concepts (buses, protocols, etc.). Software rules specify the intermodule communication (data transfer) and general coding rules that allow reuse of software modules across applications. In other words, a platform consists of various predefined elements, such as bus architecture, I/O and intermodule communication protocols, instruction sets, design and verification language, tools, methodology, and operating system, which ease the development of derivative products once the platform is developed.

Philips Semiconductor's Nexperia and Texas Instrument's Open Multimedia Application Platform (OMAP) are examples. They have a dual-processor approach in which an embedded CPU core performs the control functions and simple data processing, together with a second processor—a media core or a DSP that handles streaming data such as audio and video [2].

Another reason for the adoption of SoC is the commercial pressures on TTM. The market dictates that both hardware and software components be implemented quickly in a single device to beat the competition. This situation pushes designers to achieve shorter design cycles. On the other hand, with the cost of fabrication exceeding $1 million for a set of reticles in 130-nm technology, to reduce the cost of the design and development cycle, achieving first-time success is mandatory.

SoC can meet these short TTM needs, because in SoC design, only some components are newly designed, whereas a substantial portion of components are reused as IP as made available from internal and external sources. Because these reused components have been verified in other established products, the verification process as well as the design cycle is reduced. Design approaches using "platforms" allow for the growing complexity related to the design and verification process.

9.1.3 Challenges with SoC Design

Due to shorter TTM and project cost constraints, a SOC project will stumble if the first silicon does not function, even with minor corrections. This means that complete verification is imperative for all SoC design projects. As the gate count increases, the verification portion of the design increases at a steeper rate. In 1996, the verification portion of a design with 300,000 gates

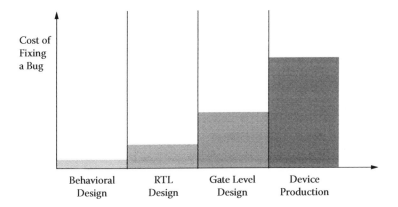

FIGURE 9.2
The cost of fixing a bug increases rapidly as one progresses with the design.

was only 30% to 40%. In 2000, the verification portion of a SoC design with 1 million gates increased to 50% to 60% of the total development effort for a design. The verification time in 2007 and beyond will consume more than 70% of the total design time.

Figure 9.2 shows that the cost of fixing a software bug or a hardware problem increases as the design progresses. In early design stages, a relatively larger number of bugs can be easily found and fixed. As the design progresses, it takes a longer time to locate and fix a bug. Therefore, to reduce the development cost of a SoC, it is necessary to find and fix the bugs at early design stages.

New verification methodologies have to satisfy three conditions: (1) the ability to handle the codesign of two heterogeneous components (i.e., hardware and software components on a single platform), (2) the ability to locate and fix bugs in the early design stages, and (3) minimal time for verification.

9.2 Front-End Design Flow

The SoC design flow consists of many verification steps from a higher abstraction level to a lower abstraction level. At higher levels, the verification can be performed extremely quickly at the algorithmic level. On the other hand, at lower levels of verification, the verification is done exhaustively, considering all implementation details at the penalty of slower speed. Most bugs can be fixed at higher abstraction level verification, but a small number of bugs can survive until the lower abstraction level verification is carried out. Locating and fixing these remaining bugs becomes much harder. Figure 9.3 shows the design flow from the viewpoint of verification. Each design step corresponds to a specific verification methodology.

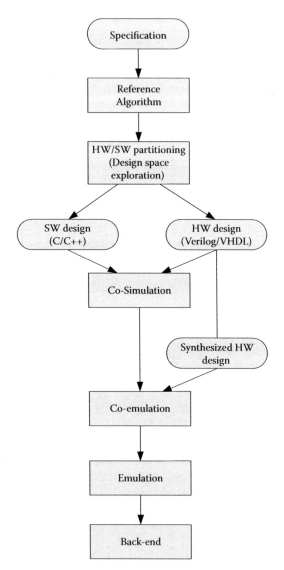

FIGURE 9.3
Typical SoC design flow.

A designer begins with a specification for the application, which becomes the starting point of the SoC design. From the specification, the designer first builds a reference algorithm, which is usually programmed using a high-level programming language such as C/C++. Using the reference algorithm, the designer has to carry out a design space exploration and partition the design into hardware parts and software parts. Hardware parts are generally faster but more costly to implement and less flexible to upgrade. Software parts are generally slower than hardware parts but are easier to implement and modify or upgrade. The designer has to partition the design according to the overall target application.

During this process, high-level system-modeling languages such as SystemC and SpecC are used. Using these languages, some hardware function blocks can be modeled with less effort than the hardware description language (HDL). The hardware characteristics, such as signal, timing, and parallelism, enable more precise simulation. Recent synthesis tools allow transferring a design described in SystemC into register transfer level (RTL) or gate level. Even though handwritten RTL design is better than the synthesized one from the point of view of design optimization, the ever-shrinking TTM requirement mandates the adoption of these tools for complex projects.

After the partitioning step, the designer is aware of the function blocks to be implemented as hardwired logic. Then these hardware function blocks are programmed using HDL. To verify the functional correctness of the RTL designs, software and hardware designs must be simulated together. This is called cosimulation, where the software part runs in the instruction set simulator (ISS), which is either a cycle-level accurate or instruction-level accurate model of the processor core. The hardware part runs on the HDL simulator. An interface channel is required to connect the two distinct simulation platforms, and synchronize, efficient communication between the two is crucial for effective and accurate simulation.

The next step is coemulation. In coemulation, the hardware part is transformed to the lower abstraction level or gate level. The gate-level models are mapped into the FPGA-based emulator to accelerate the simulation speed and to detect any bugs that could not be found in the RTL simulation. Like cosimulation, synchronization and communication efficiency are important requirements in coemulation. In addition, because of the lack of accessibility to signals in the FPGA, special debugging methodologies are required for probing signals. The software part still runs in the ISS, which will be replaced by a real core during the emulation stage. In this emulation environment, the system performs quite similarly to the final product. In order to trace the algorithms running in the core, core-debugging facilities are required. After all these stages, the physical design phase is carried out. In the following subsections, some important requirements of these design steps are described.

9.2.1 Reference Algorithm

Specification of a SoC design can be described with various languages, including plain text, tables, figures, and programming languages such as C/C++. It is just a description of how the target system is operated. It is possible to neither simulate nor synthesize. The reference algorithm is a realization of the specification into a working program code. The reference algorithm allows the designer to evaluate the accuracy of the algorithm and estimate the performance of the final product. The reference algorithm is also utilized as the "model" during the design and verification process.

The reference algorithm is considered the starting point of the SoC design process. From the reference algorithm, design space exploration is begun with hardware and software partitioning. Next is to refine the hardware and software blocks into the lower abstraction level with more details.

9.2.2   Hardware and Software Partitioning

System designers can implement a specific set of functions within a product that satisfies various constraints, such as performance, cost, and power consumption. The choice of implementation architecture determines whether the designers will implement a particular function as a hardware component or as software running on a programmable processor. There are advantages and disadvantages in each choice or a combination. A function that requires heavy computing operations, such as a filter, can be implemented as hardware. On the other hand, a function that is not so complex but requires flexibility can be implemented as software. Decisions in hardware and software partitioning are based on these considerations.

9.2.2.1   Issues

Regardless of the target architecture, there are three main issues that must be resolved in determining the hardware and software partitioning of a system:

- Functional clustering: Cluster the system functionality into a set of tasks.
- Allocation: Allocate the tasks to either hardware or software.
- Scheduling: Schedule the allocated tasks in each of the software or hardware domains to satisfy the specified timing as well as functional properties.

These three problems are interdependent; thus, they must be solved simultaneously to determine an optimal solution. The allocation and scheduling subproblems are known to be NP-hard [3]. (A problem is said to be nonpolynomial time hard [NP-hard] when the problem cannot be solved within time $O(n^a)$, where n is one number of modules being handled and a is a finite number, as there is an exponentially increasing number of possible functional clusters.) Therefore, automated heuristic approaches to solving the hardware/software partitioning problem should be used by the designer to search the solution space. The most common heuristic approaches simply assume a fixed functional clustering and use stochastic search, iterative improvement, and constructive algorithms to solve the allocation subproblems. The tasks need to be scheduled to evaluate the allocations. Stochastic search–based algorithms, such as simulated annealing and genetic algorithms, have yielded near-optimal allocation and scheduling solutions [4].

Most of the optimization problems encountered in electronic design automation (EDA) are NP-hard problems with a large number of variables, with each variable varying in a wide domain. They are usually unsolvable with an analytic approach and therefore have to be solved using a trial-and-error approach, including iterative improvement, the stochastic approach, and the heuristic approach. Iterative improvement accepts good moves while rejecting bad moves; therefore, it can become stuck in a local optimum. (A move is a random trial of value changes of the whole set of variables.) The stochastic approach, on the other hand, sometimes accepts bad moves while always accepting good moves, and therefore has a potential to escape from the local optimum to arrive at the global optimum. This "hill climbing" probability, that is, the probability of accepting bad moves, is proportional to the temperature in a stochastic algorithm such as simulated annealing. The heuristic approach (or algorithm) is one that is based on some domain knowledge of the target problem. It can be used to arrive at a reasonable trade-off, that is, a quite good solution, within a relatively short computing time. The heuristic approach can be used for generating each move in either iterative improvement or the stochastic approach.

To design and verify a hardware/software partitioned system, an adequate design and simulation tool is required. This tool describes the system, which is composed of transaction-level models and signal-level models. In the past, IP designers developed their hardware modules in RTL using Verilog or VHDL, whereas software designers implemented the software part as C or assembly language (based on the hardware IPs already developed). Although the development flow is sequential and requires a number of iterations to fix the bugs between the IP and software, it still worked efficiently because the design complexity was low. However, modern systems contain application-specific hardware and software; therefore, hardware and software components must be codeveloped on a tight schedule to meet the TTM.

To reduce the development time, system-level design languages (SLDLs) have been developed to provide users with a collection of libraries of data types, kernels, and components accessible through either graphical or programmable means to model the systems and simulate system behavior. In the past few years, many SLDLs and frameworks such as SystemC [5], SpecC [6], and SystemVerilog [7] have been developed to manage the issue of complex design.

SystemC provides hardware-oriented constructs within the context of C++ as a class library implemented in standard C++. It enables design and verification from concept to implementation in hardware and software. SystemC provides an interoperable modeling platform that enables the development and exchange of very fast system-level C++ models. It also provides a stable platform for the development of system-level tools. The Open SystemC Initiative (OSCI) is an independent, nonprofit organization composed of a broad range of companies, universities, and individuals dedicated to supporting and advancing SystemC as an open source standard for system-level design.

9.2.2.2 Transaction-Level Modeling-Based Methodology

This section will explain the transaction-level modeling (TLM) methodology, which enables a higher level of abstraction and overall productivity gains.

A transaction is a communication based on a collection of precisely defined signal data. An example is a block memory read or write, which in a hardware simulation is spread over a number of specific input/output (I/O) signals and several clock periods. To interface models to hardware simulations, transactions are decoded by a transactor, which is a data-level translator with buffering capability. For example, a transactor is responsible for converting a set of actual pin signals into a higher-level format, such as converting a repetition of an identical data set into its "count." Another example is converting a binary-coded signal into the corresponding instruction code, thereby reducing communication traffic, and vice versa. A transactor also should be equipped with a timing buffer to meet the timing/bandwidth requirements of two different signal representation levels.

A reference software model may be used for algorithmic refinement or as the basis for deriving hardware and software subsystem specifications. The exact parameters modeled are specific to the system type and application, but the model is typically untimed. To use the untimed model in a higher level of abstraction, TLM-based methodology merges modules described by different languages without precise definitions. These transactions can be used whether the block is SystemC, C, or RTL.

TLM-based methodology enables detailed modeling using a step-by-step refinement process. Without TLM-based methodology, designers need to separately implement each module referred to by the reference software model. Because each module is verified in a different environment, there can be many errors during integration of the entire system. However, TLM-based methodology provides a more effective way to implement the system. In the first step, while the designer has only the specifications of the hardware design and the reference software model, high-level simulation is possible on the software level. In the second step, after designing several modules, the designer models the system by substituting the software model with a verified hardware design. Other parts of the system are still modeled as software models. This step reduces the errors that can occur during integration. In the third step, all the blocks are substituted by verified hardware modules. Figure 9.4, Figure 9.5, and Figure 9.6 describe each step of the TLM-based methodology. After these steps, a verified set of IPs is generated.

Creating the TLM from the software module slightly lengthens the architectural design phase of a project, but it offers several potential benefits. It allows the designers to refine and test software earlier, thereby reducing the overall development time, and achieve more realistic hardware/software trade-offs at an early stage. Furthermore, it enables designers to deliver executable models to customers both for validating the specification and driving changes and for accelerating product acceptance.

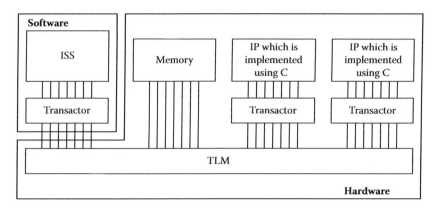

FIGURE 9.4
The first step of TLM-based methodology.

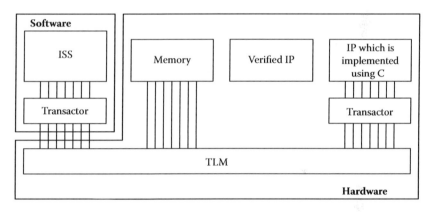

FIGURE 9.5
The second step of TLM-based methodology.

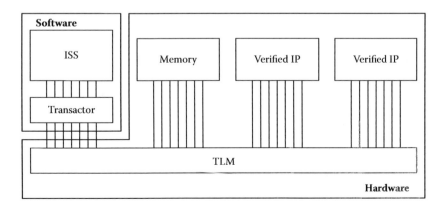

FIGURE 9.6
The third step of TLM-based methodology.

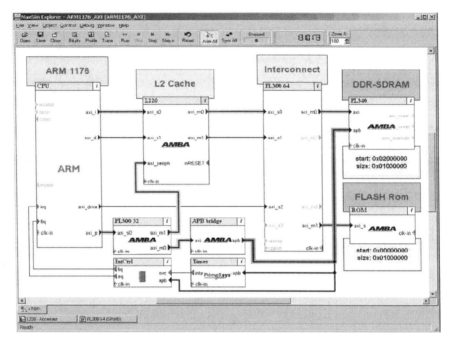

FIGURE 9.7
ARM's MaxSim tool is used to configure an ARM-based SoC. (Courtesy of ARM Limited.)

9.2.2.3   System-Level Design Method

It is not easy to configure a system using SLDLs such as SystemC and SpecC because there are too many components to describe in a system. For easier system implementation using SLDLs, a system-level design approach called electronic system level (ESL) design has been developed. ESL describes a SoC design in a sufficiently abstract and high-level fashion to quickly explore the design space and provide virtual prototypes for system implementation in hardware and software. MaxSim [8] from ARM, shown in Figure 9.7, is one such example. There are other cosimulation tools, such as CoCentric System Studio from Synopsys [9] and ConvergenSC from CoWare [10].

9.2.3   Cosimulation

9.2.3.1   Introduction

Cosimulation refers to verifying that hardware and software parts function together correctly. With hardware–software codesign, it is essential to verify correct functional behavior before system implementation is commenced. The choice of technique in hardware–software cosimulation depends on the availability of models and the performance requirement. Software models can be accurate at every clock cycle or in nanosecond scale. Software models can be the instruction set processor model (or the instruction set simulator

[ISS]), which only guarantees correct modeling of the values in the registers and memory. For asynchronous software–hardware communication, the software can be compiled and simply linked to a hardware simulator. The whole system can be simulated in software by mapping all hardware components into its behavioral software models (e.g., bus functional models). On the other hand, hardware modelers (i.e., hardware emulators) can be used to accelerate the execution of software models.

9.2.3.2 Cosimulation Tools

Many EDA vendors, such as ARM, Synopsys, and CoWare, provide ESL cosimulation tools. These system-level cosimulation tools enable designers to simulate the software part on the microprocessor model along with the hardware part designed in RTL or SystemC. In these cosimulation tools, TLM is widely used for simulation acceleration by a factor of 1,000 to 10,000. Also, their debugging and profiling capabilities provide a very powerful cosimulation environment for system-level architecture exploration and pre-silicon-embedded software development.

ARM's Real View SoC Designer is a toolset for modeling, simulation, and debugging of SoC designs. SoC Designer's virtual prototypes help embedded software developers start the coding of device driver and application software and test them before the RTL description of the hardware part is given. SoC Designer's RTL cosimulation capabilities help the software and hardware integrator find and remove bugs at an early stage. In addition to the extensive library of standard SystemC models, users can choose from processor, peripheral, memory, and bus models from ARM, CEVA, Infineon, LSI Logic, MIPS, PMC-Sierra, and other vendors [11].

CoWare's Platform Architect is a SystemC-based environment for creating product platform and analysis platform architecture. Platform Architect speeds the concurrent design of SoCs with embedded software, enabling users to create and validate SoC designs at the transaction level in SystemC. Features include support for SystemC transaction-level platform creation, architecture analysis, simulation, and debugging, plus optional integration with third-party RTL implementation and verification flows. Together with CoWare's Model Designer and the CoWare Model Library, CoWare's Platform Architect provides a comprehensive system-level design solution available for SystemC [12].

Synopsys's System Studio is a system-level design, simulation, and analysis tool that addresses two critical system-level design areas for innovative SoC applications: algorithms and architectures. Algorithm design is an essential task in signal-processing applications such as wireless telephony, multimedia codecs, DSL, and cable modems. Architecture design involves putting together the right processors, custom logic, buses, memories, and peripherals in order to make effective use of the silicon. SystemC has evolved as the language of choice for this task [13].

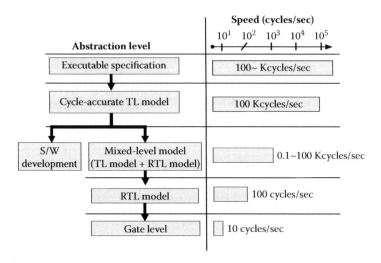

FIGURE 9.8
Typical simulation performance of each design step.

9.2.3.3 Cosimulation Issues

Transaction-level models typically run at least two orders of magnitude more quickly than their RTL equivalents. Simulation speeds of several hundred kilocycles per second for a complete system simulation are readily achievable with TLM compared to several hundred cycles per second in RTL models. This means that by using TLM we can validate a design against more test vectors in the same amount of time than using RTL models, while providing full cycle accuracy. For this reason, TLM is usually used for architectural exploration for system-level design [14].

Figure 9.8 shows the typical simulation performance of each design step. Usually, TLM is gradually refined to the RTL description in a block-by-block approach. As the portion of RTL models in the whole design description increases in the mixed-level simulation, the simulation speed degrades, as the simulation speed of the RTL models is typically more than 100 times slower than that of TLMs. When RTL models are introduced, it becomes practically impossible to run the same group of test vectors used with TLMs (due to slower speeds).

9.2.3.4 Cosimulation Example

Hardware–software cosimulation requires a hardware simulator and a software simulator. In this cosimulation example, MaxSim (from ARM) was used as a software simulator and ModelSim (from Mentor Graphics) was used as a hardware (RTL) simulator. To develop an entire cosimulation environment, it is necessary to design a "transactor" that converts transaction level (TL)

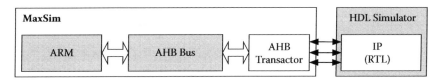

FIGURE 9.9
MaxSim–ModelSim cosimulation environment.

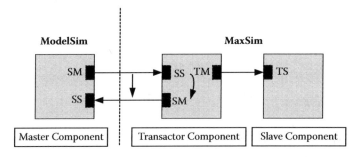

FIGURE 9.10
Signal level (RTL) to TL transactor interface. SM and SS denote signal master and signal slave, respectively, whereas TM and TS denote transaction-level master and slave, respectively.

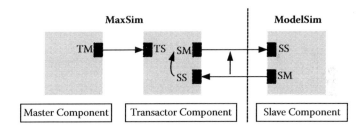

FIGURE 9.11
TL to signal-level (RTL) transactor interface.

to RTL and vice versa. In most cases, the software part is represented in TL to achieve fast execution, whereas hardware is designed in RTL, with its interface represented in signal level. Figure 9.9 shows a block diagram of cosimulation using a transactor that converts the TL advanced high-performance bus (AHB) to signal-level (RTL) AHB.

Figure 9.10 and Figure 9.11 show how a transactor is designed and its operation. The transactor has a signal master (SM) port and a signal slave (SS) port. In Figure 9.10, the transactor receives the RTL signal from the RTL simulator, while the transactor sends TL data to a slave component through the transaction master (TM) port. The opposite direction of data movement, which is from TL to signal level (RTL), is shown in Figure 9.11.

9.2.4   Coemulation

9.2.4.1   Introduction

Coemulation, which is a technique for verifying a given design by mapping it onto a combination of hardware and software, is becoming more and more popular for two reasons: (1) ever-increasing hardware complexity and the resultant excessive verification time and (2) the ever-shrinking TTM requirement. Moreover, embedded system software needs to be verified for its functional correctness long before chips become available. For applications where TTM and project costs are critical, coemulation saves a significant amount of time and reduces the risk of costly hardware design errors. Coemulation addresses one of the most critical steps in the embedded system design process, that is, the integration of hardware and software.

9.2.4.2   Benefits

Coemulation has two primary benefits. First, the coemulation process allows software engineers to interact early with the hardware design. Traditionally, software engineers have been unable to execute software until the hardware became available. Figure 9.12 shows how coemulation can drastically improve the project schedule. Compared with conventional design flow, a virtual chip is expected to generate the ISS of a processor along with the real target environment, so it makes functional models work on real target systems [15]. Figure 9.13 shows the hardware emulation environment using a virtual chip.

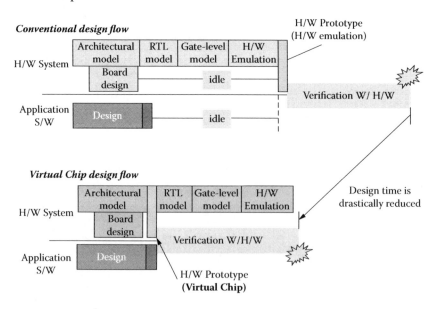

FIGURE 9.12
Project schedule with and without coemulation.

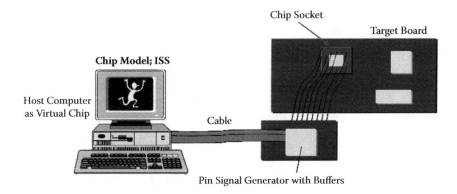

FIGURE 9.13
Hardware emulation using a virtual chip.

A second benefit is that it creates a perfect stimulus that can be used in the final product. Because the coemulation uses the final software design, this stimulus is very useful for the hardware engineers, as it assists them in locating bugs that may occur in a real product environment that are hard to find via simulation alone.

9.2.4.3 Coemulation Tools

Dynalith Systems' Intelligent Prototype Verification Engine (iPROVE) offers easy C-based test stimuli generation on PCI and FPGA execution along with C/HDL coemulation. iPROVE can verify the user's design by mapping the HDL description of each design onto an FPGA card. An extremely versatile built-in logic analyzer (BILA) enables the designer to monitor various signals. Also, a data pumping port (DPP) enables the designer to send and receive data through external connections. Figure 9.14 shows the iPROVE coemulation environment.

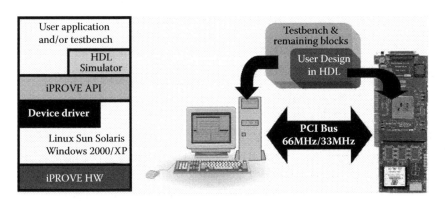

FIGURE 9.14
iPROVE coemulation environment. (Courtesy of Dynalith Systems, Seoul, Korea.)

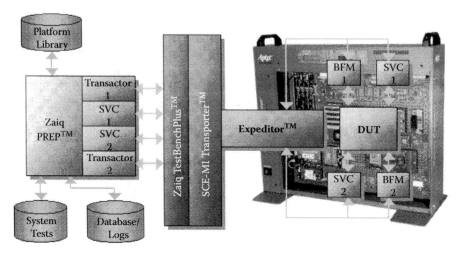

FIGURE 9.15
SoC Validation Lab coemulation environment. (Source: Mentor Graphics.)

The Aptix SoC Validation Lab is a widely used transaction-level coemulation solution (see Figure 9.15). The product can use Zaiq's PREP test development environment and the SystemWare Verification Component (SVC) library based on Aptix's Expeditor coemulation hardware and software. The Zaiq PREP provides system-level tests for system-level models, standard RTL simulation models, or hardware prototypes running on Aptix platforms. The SVC library is a set of predefined transaction-level models for many standard protocols. The Aptix platform lets users develop their own transaction-level coemulation environment.

9.2.4.4 Coemulation Environment

As depicted in Figure 9.16, the coemulation system incorporates two processing engines, a software simulator and a hardware emulator, which require a communication channel for interacting with each other. The communication abstraction level can be cycle level or transaction level. In Figure 9.16a, the interface between the simulator and emulator is cycle level. The wrapper calls the cycle-level application programmer's interfaces (APIs) defined by the emulator to transfer the signal information. In this case, the transactor, which converts transaction-level signal to cycle-level signal, and vice versa, can be described in a high-level language such as C/C++, making implementation of the transactor relatively easy. However, communication between the simulator and emulator should be done on a clock cycle base instead of a transaction base. All signals between the two platforms should be transferred at every clock cycle. This incurs large communication overhead, which can be longer than the actual execution time, which in turn degrades overall performance of the coemulation system.

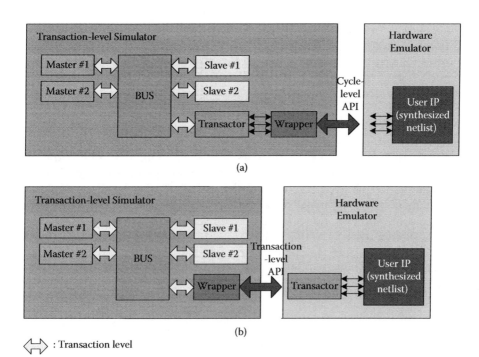

(a)

(b)

$\Longleftrightarrow$: Transaction level

$\longleftrightarrow$: Signal level

FIGURE 9.16
Coemulation environment: (a) cycle-level communication between simulator and emulator; (b) transaction-level communication between simulator and emulator.

Another approach for coemulation is to have the transactor located within the emulator, as in Figure 9.16b. In this case, the wrapper calls the transaction-level APIs to transfer transaction-level information. Data transfer in the transaction level enables efficient communication between the simulator and emulator. However, the transactor design depends on the emulator system protocol, because it should interface with unfamiliar emulator system buses to communicate with the simulator. Consequently, the transactor description is not only time consuming but also error prone. To resolve this problem, some emulators provide transactors for bus protocols such as the Advanced Microcontroller Bus Architecture (AMBA) AHB [16]. On the other hand, because one transaction may occur several cycles in advance, synchronization between the simulator and emulator becomes another important issue.

9.2.5 Emulation

Once cosimulation is finished, target hardware and software need to be further tested to show correct functionality in a real system environment. Although the software's functionality is verified from cosimulation, RTL hardware described by HDL does not guarantee the correct behavior in the final chip and precise and reliable communication with the microprocessor. Although cosimulation uses interprocess communication such as virtual socket for a communication channel between the software and hardware model, it cannot show the situation of the real system. For example, the virtual communication channel cannot prove pin-to-pin communication in a real PCB situation.

Emulation is another methodology to verify hardware functionality. Compared to simulation, emulation is characterized by semiobservability and an expensive development framework. However, emulation will follow the simulation for previously undetected bugs and extremely long executions that cannot be achieved through simulation [17]. Emulation needs duplicate effort, producing additional costs and delaying TTM, but it can shorten the long execution times and show more realistic behavior. To overcome the difficulties of emulation, evaluation boards are used for the target microprocessor, which executes target software, and FPGA for the target hardware. Most off-the-shelf evaluation boards embed an ARM processor as the main processor. In the section below, current off-the-shelf evaluation boards will be described.

Current EDA tools support semicustom chips designed by logic synthesis and automatic placement and routing (P&R). At first, RTL hardware described by HDL should be verified for the correct functionality. By logic synthesis, developers can get a gate-level netlist. The gate-level netlist consists of basic gates, and these are prepared for semicustom designs that most foundry companies support. Hence, if RTL HDL code is synthesizable to gate level, there is no problem with using a semicustom design.

Even though RTL HDL code is synthesizable, we cannot directly proceed to fabricating the target chip. Because the gate-level functionality is not verified, designers cannot verify that the target chip will work correctly after semicustom design using the same code. To fill the gap between RTL HDL code and semicustom design, FPGA-based emulation is needed. After synthesizing RTL HDL code and performing P&R, a gate-level netlist is mapped onto the FPGA and the FPGA emulates the target hardware. The following section describes off-the-shelf emulation board systems and debugging methodologies.

9.2.5.1 *Emulation Board Survey*

Most current off-the-shelf emulation boards have a main processor and an FPGA. A processor-debugging interface allows developers to test the software functionality with the hardware components mapped onto the FPGA. Communication between the processor and the FPGA is mostly performed through a shared bus, such as AMBA AHB [16], CoreConnect [18], or Wishbone [19], with the associated architecture and IPs provided as synthesizable RTL HDL blocks. Emulation boards generally provide the connection between the processor and the FPGA within the PCB. Sometimes, emulation

FIGURE 9.17
ARM's Versatile family Platform Baseboard [21]. (Copyright © 2003 ARM Limited. All rights reserved.)

boards do not have microprocessors but provide baseboards that can be augmented by stacking up various core modules. Core modules include the processor core and have a system connector that is an interface between the baseboards and the core modules. For instance, ARM's RealView Versatile family [20] consists of Platform Baseboards, Core Tiles, Logic Tiles and Interface Tiles.

Figure 9.17 shows an ARM926EJ-S processor-fitted ARM's Platform Board that provides a multitude of input/output (IO) ports, such as serial interface, USB, and Ethernet. Using Platform Baseboard, one can pursue the platform-based design methodology, where SoC developers do not need to re-design the peripheral devices. Versatile's Logic Tile provides FPGAs connected to the AMBA system bus. Therefore, software executed in the ARM processor module can communicate with hardware mapped onto the logic tile's FPGA through AMBA system bus.

Figure 9.18 shows ARM Core Tile which includes ARM processor with on-chip caches, Tightly Coupled Memories (TCM), and debug/trace logic. Core tiles can be stacked on top of Platform Baseboards, and the multiplexed AMBA interface is routed to the Core Tile board header connectors.

Among the features provided by Core Tile, most important are the debuggability and the observability of software and the controllability of hardware. An embedded hardware can be controlled through communication architecture. In other words, hardware behavior is controlled by message passing through the communication bus, while the software debuggability

FIGURE 9.18

ARM core tile for the ARM1136JF-S™ Processor. (Copyright © 2005 ARM Limited. All rights reserved.)

and observability are offered by the Core Tile. For ARM processor's debugging and trace, most Core Tiles have Embedded Trace Macrocell logic and trace connectors (JTAG).

Commercial logic synthesizers provide a gate-level netlist, whereas FPGA vendor tools perform P&R. Because communication between the hardware and software is performed through the system bus, in order to add the device under test to the emulation board system, developers must modify the bus logic function to cover the device under test address map and interconnect other bus masters/slaves.

As shown in the top of Figure 9.19, FPGA in Logic tile provides room for DUT, and developers should prepare the connection to the system bus bridge. HDL codes for bus logic function to connect DUT with the system bus bridge are available from the vendor's technical support.

Modern SoCs require two or more microprocessors to meet performance constraints. Current off-the-shelf emulation systems can support multi-processor or multi-core environments. For example, as shown in Figure 9.18, Core Tile has stackable header connectors. Not only homogeneous processors but also heterogeneous processors can be handled on the emulation boards. For multimedia applications, stream processing operation is more important than the other fundamental operations and hence DSPs are used to accelerate the speed.

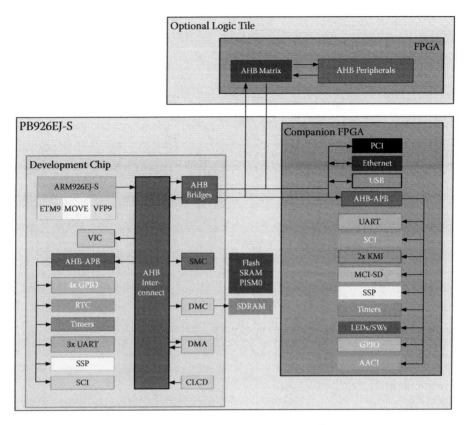

FIGURE 9.19

ARM Platform Baseboard Block Diagram [20]. (Copyright © 2005 ARM Limited. All rights reserved. ARM1176™ is an unregistered trademark of ARM Limited.)

9.2.5.2 Debugging Methodology

To debug and trace software, the designer should know the processor's status and instructions executed within the processor. The processor's general and special registers are accessible through the debugging interface. A typical debugging interface is defined by the JTAG. Boundary scan testing was developed in the mid-1980s as the JTAG interface to solve physical access problems on PCBs caused by increasingly dense assemblies made available by compact packaging technologies [22]. The JTAG provides hardware debuggability and observability to embedded system developers.

As shown in Figure 9.20, boundary scan cells are organized with scan-chained registers, which transfer logic data, and are controlled through the test access port (TAP). TAP control operation is shown in Figure 9.21. I/O signals are controlled via the JTAG probe. Software developers verify the software's functionality by using software debuggers executed in a host PC.

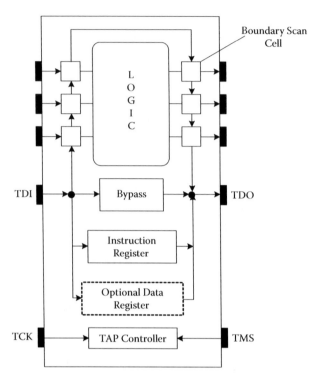

FIGURE 9.20
Boundary scan device.

An in-circuit emulator (ICE) is an invaluable tool for embedded system software developers, where the microprocessor of the target hardware is replaced by the ICE [23]. An ICE can emulate the target processor in real time. The developer loads the application into the emulator and proceeds to run, step, and trace into it. Inside the ICE, a special bond-out version of the same chip resides.

Lauterbach's Trace32-ICD [24] is a debugger for embedded processors and DSPs, such as ARM, MIPS, and ZSP. Using Trace32-ICD, developers can download the application image into an embedded memory in the emulation system and control the operation of processors (i.e., run, step, and trace the application code).

Sometimes, developers need to work within the internal hardware of the processor. For example, in order to debug and optimize the application software, developers may need to know the contents of various registers and memory modules. The software debugger collects information and displays it in the GUI window. Figure 9.22 shows how the debugger displays processor register values and memory contents. Developers can change the values and gather trace information. These features can shorten software development time and help meet TTM requirements.

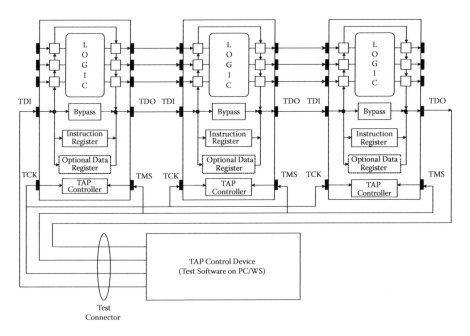

FIGURE 9.21
Test access port control device [20].

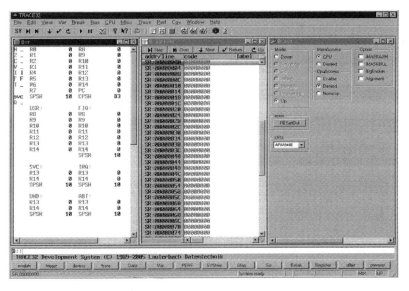

FIGURE 9.22
Trace32 software debugger. (Source: Lauterbach.)

9.3   Back-End Design

In contrast to the front-end (or the prelayout) design process, the actual layout process is usually called back-end design. There are three different approaches to back-end design of a SoC: full-custom, semicustom, and mixed design.

In full-custom design, all design procedures are performed by the designer without using predesigned IPs. Full-custom design was commonly used in early chip design. Today, it is used for designing special applications that require ultrahigh performance, such as DSPs or graphics chips. A good example is the Pentium processor series by Intel. The necessity of fully customizing the circuit and layout design comes from the efficiency limitations due to automation tools, particularly when a high-performance SoC is required. The relatively high cost and development time of the full-custom approach can be justified only if the final chip has a big market and a large profit expectation.

The semicustom design method is the most commonly used method for modern digital circuit designs. This method limits the number of circuit primitives used in the design, and there may be performance degradation compared to a full-custom design. Semicustom design is used if the market is large and there is a quick TTM, especially in digital circuits. There are two types of semicustom design: cell-based design and array-based design. Only cell-based design will be discussed in this chapter. In cell-based design, logical functions like AND and OR are already built as a standard cell that varies from each ASIC foundry company. The set of these cells is called the cell library and also contains information on timing and electrical behavior. Automation tools convert logical functions to cells by using standard cell libraries and information from the HDL. Despite its inferior performance, the semicustom method is widely used in designing core chips of peripheral devices for PC or handheld devices when the tight turnaround time of the market is more critical than performance. Figure 9.23 shows various VLSI designs and layout styles.

Mixed design (full custom and semicustom) is also used for SoC design. In some SoC designs, not only digital circuits but also analog blocks are embedded within the chip for special purposes. Figure 9.24 describes the general steps of semicustom design.

In semicustom design, system-level or high-level simulation is performed first. In this process, high-level languages such as C, C++, and SystemC are used for the design of both hardware and software modules. The motivation for using such high-level languages, despite the difficulty of modeling some hardware-specific features such as parallelism and concurrency precisely, lies in the drastic increase in simulation speed, generally by a factor of 100 to 1000. The second step is the implementation of real hardware through system-level simulation using HDL, considering properties like timing, chip area, and power. After the RTL description, circuits must be verified with RTL simulation using a test bench. If any errors are found, the designer has

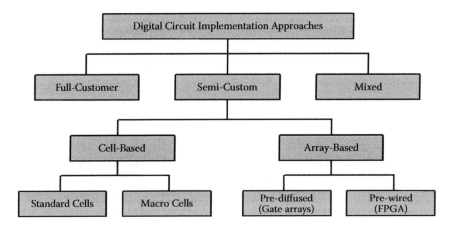

FIGURE 9.23
Various VLSI designs and layout styles.

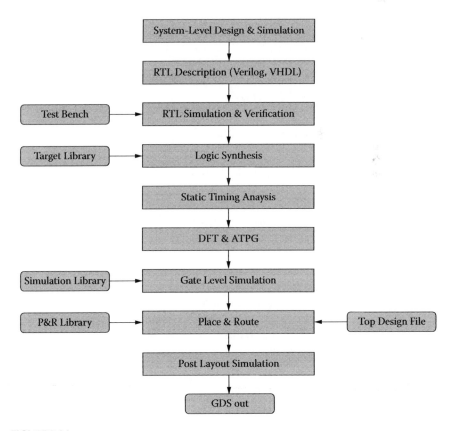

FIGURE 9.24
Semicustom design flow.

to repeat the previous stage to correct the RTL description. Developing an effective test bench is critical at this stage.

Logic functions described in the RTL must be mapped onto the standard cell library as provided by the selected ASIC foundry. This procedure is called logic synthesis, and real hardware blocks such as gates, macros, and memory are implemented at this stage. P&R is done in the next stage. The standard cell libraries of the foundry should be made available in advance. The next tasks include static timing analysis (STA), design for test (DFT), and auto test pattern generation (ATPG), which are covered in subsequent sections.

The gate-level netlist is extracted from logic synthesis and used for gate-level simulation. Because improper propagation delay values are ignored in RTL simulation, gate-level simulation can still indicate some faults, even if RTL simulation is successful. At this stage, if such faults are detected, some modifications are required in either the RTL description or the netlist. Thus, it is important to design a circuit accurately at the RTL description level.

The next stage is placement of the gate-level netlist and routing, which requires the design file containing power, clock, and I/O pad information. Final simulation (or postlayout simulation) has to be done after the layout is completed. Postlayout simulation takes the parasitic delays (due to wires) into account. The graphical data system (GDS) file containing actual layout information will be available after this stage. Fabrication is performed in an ASIC foundry with this GDS file. In the following sections, static timing analysis, gate-level simulation, and back-end design will be discussed, followed by power measure, DFT, and ATPG.

9.3.1   RTL Description and Verification

Compilation of a design into its RTL description is usually done using a commercial compilation tool such as Verilog-XL (Cadence), NC-Verilog (Cadence), VCS (Synopsys), and ModelSim (Mentor Graphics). Timing verification is carried out using value change dump (VCD) format, with commercial programs such as Virsims (Synopsys), SimVision (Cadence), and UnderTow (Veritools). This procedure only verifies the behavior of the software.

9.3.2   Logic Synthesis

The next step in the back-end process is logic synthesis, where the hardware described in the RTL is converted to the gate-level netlist. This procedure converts the HDL description into basic gates or flip-flops. To design a SoC, logic synthesis must be carried out again just as it was done in the FPGA, because logic synthesis in the SoC uses standard cell libraries that contain timing and electrical information for standard cells. Logic synthesis must be done again (in addition to the synthesis done in the FPGA) because SoC design also requires better performance in terms of timing, area, and power compared to an FPGA, even though the emulation based on an FPGA could save effort in this stage. Logic synthesis is composed of three steps:

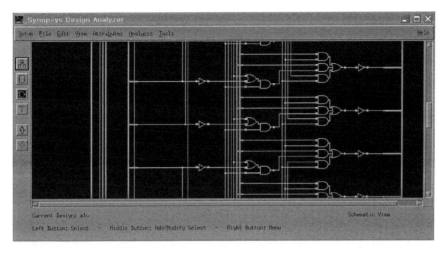

FIGURE 9.25
An example of a 32-bit adder logic synthesis result.

1. Translation—converts the RTL hardware described in the HDL into gate level.
2. Optimization—optimizes the gate-level hardware. Optimization is performed for different metrics, such as area, timing, and power, using different methods such as multilevel optimization algorithms, resource sharing, and scheduling.
3. Mapping—selects the most appropriate gate in the standard cell library according to the specifications (such as timing, area, and power).

Designers have to repeat these steps until the target design satisfies the given specification in timing, area, and power. Figure 9.25 is an example of a graphical result that was synthesized in gate level after describing it in HDL with Design Compiler (Synopsys). If the result does not satisfy the specification after synthesis, modifications in the RTL design or in the gate-level are unavoidable. The margin should be 10% to 20% tighter than the actual specifications because parameters like timing, area, and power can easily deteriorate throughout the back-end process. This step has to be done accurately because it is the beginning of real SoC design.

9.3.3 Static Timing Analysis

There are two basic methods for analyzing timing characteristics during back-end design: static timing analysis (STA) and dynamic timing simulation (DTS). In STA, the system checks the requirement of all possible paths in parallel. The advantages of STA are speed and the ability to simultaneously report multiple critical paths. STA can significantly reduce design iterations due to its high speed and wide coverage. STA is also used to check the states of setup/hold timing violations, multicycle paths, zero-cycle paths, and illegal gated clocks of designed circuits. It can be used to check combinational feedback loops, multiple clock domains, asynchronous signals, boundary I/O delay and slopes, and so on. However, STA can handle only synchronous circuits and is generally less accurate than DTS.

In DTS, the test vector created by the designer is used to obtain the timing behavior. The advantages of DTS are accuracy and variable design styles. DTS has two disadvantages. One is the long simulation time, and the other is the possibility of missing critical paths, depending on the completeness of the test vectors.

9.3.4 Design for Test and Auto Test Pattern Generation

These steps are needed to check for manufacturing errors during chip fabrication. Testing is the final process that verifies the accurate functional (and parametric, when analog blocks are involved or when the SoC is to be binned according to such parameters as maximum clock frequency) behavior of each fabricated chip, using functional test vectors specified by the designer. However, it is not easy to generate all the required functional test vectors or apply all the prepared vectors to each fabricated chip. As the time for testing each chip is reflected in the chip cost, the efficiency of testing, as defined by the fault coverage (i.e., the percentage of manufacturing faults that can be detected divided by the test vector length), is very important in the design of testing scenarios. To make testing easier, DFT is often incorporated in the synthesis step, which inserts a scan path apart from the functional logic path. ATPG generates test vectors according to the given fault coverage, typically based on stuck at 1 or 0 fault models.

9.3.5 Gate-Level Simulation and Power Analysis

9.3.5.1 Gate-Level Simulation

Gate-level simulation is the last verification step before proceeding to the back-end steps (fabrication). It provides information about the delay, area, and power consumption of the final design with a simulation of the gate-level netlist extracted from previous steps, using the simulation models of the standard cell library. Because the RTL simulation does not consider the delay, area, and power associated with actual gates, gate-level simulation provides specifications and characteristics reasonably close to the final hardware.

Gate-level simulation provides more accurate timing analysis to confirm the hardware operation than STA, which is done only at the gate-level netlist. A more accurate timing analysis can be made by identifying which gates or blocks of gates are actually invoked by the given input stimuli and which I/O pairs are used within each multi-input (and multi-output) combinational logic gate, as the propagation delay can be different according to the I/O port combinations of each gate.

Gate-level simulation can be used to reduce power, because it provides any slack for all paths and information about the delays of each net. This may lead to the use of techniques such as dynamic voltage scaling (DVS) [25] and clustered voltage scaling (CVS) [25] to reduce the power consumption of the SoC.

9.3.5.2 Power Analysis

Power analysis is an important process in the design of a SoC. As the complexity of the logic increases, the power consumption increases. The power consumption of a logic block is based on three factors: quiescent (internal) power, switching power, and leakage power. The first two elements account for the largest share of the power consumption. Internal power and leakage power use the values defined by the standard cell libraries, whereas switching power consumption is analyzed using the transition of each net of hardware. The gate-level netlist is mainly used for power analysis because it is precise and efficient, as it is difficult to analyze power at the RTL and system level.

Figure 9.26 illustrates this process. To proceed with the gate-level power analysis, we need (1) the gate-level netlist, (2) test vectors, (3) gate delay information according to the standard delay format (SDF), and (4) the switching activity interchange format (SAIF) information. SAIF is divided into two subsets, the forward SAIF (or library SAIF) and the backward SAIF (back-annotated SAIF). The forward SAIF contains information on the logic switching behavior of each library cell (i.e., when and how each output port of the library

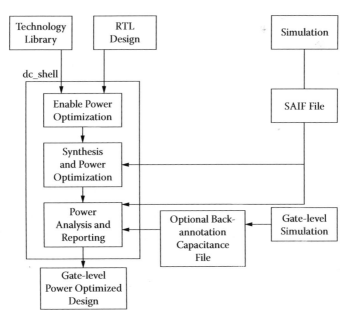

FIGURE 9.26
Power analysis flow. [Used by permission from Synopsys, Inc. All rights reserved.]

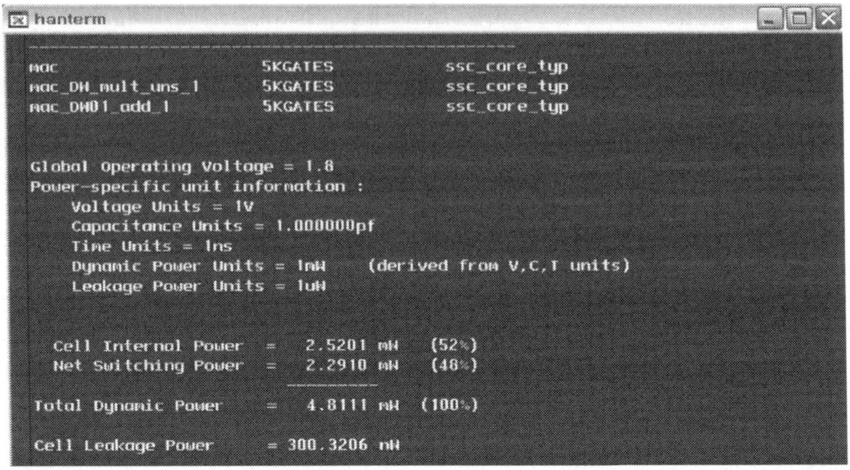

FIGURE 9.27
Power compiler result.

cell is affected (switched) by each different combination of input signals). The library SAIF is generated from the state-dependent, path-dependent (SDPD) analysis of each library cell in the gate-level netlist. Power consumption can vary with various operation modes (state dependent) and various input-to-output paths (path dependent). The backward SAIF, generated through the gate-level simulation, contains actual switching activity information that is used to calculate the power consumption due to dynamic switching.

Figure 9.27 shows the real power consumption estimated by the Synopsys power compiler by the power analysis process in Figure 9.26. In Figure 9.27 we can identify the different components of total power consumption, divided into switching power, internal power, and leakage power.

The analysis of power at the gate level has an accuracy of 80% to 90%. The loss of 10% to 20% accuracy is because the P&R has not yet been processed, which makes the arrangement of components imprecise, and the information on wire lengths is not reflected.

9.3.6   Place and Route

The netlist or the circuit representation is created after the logic synthesis in back-end design. In the P&R process, which is the final design stage, we translate the circuit representation to a geometric representation (layout). Figure 9.28 illustrates the details of the back-end design process, including partitioning, floor planning, placement, routing, extraction, and verification.

9.3.6.1   Partitioning

Partitioning divides the total circuit of the system into subcircuits or subsystems called blocks. Partitioning provides the area of each block and the possible shape of the blocks. In addition, it provides the number of terminals that each block requires, making the floor-planning process complete. The results of the partitioning process are used in other steps, such as partitioning-based placement.

Partitioning improves the speed of the design without breaking the functionality of the original system. It also simplifies the routing task. The system is partitioned to minimize the interconnection among blocks—the min-cut problem. The process is aimed at minimizing delay. There are many algorithms to do partitioning. Group migration algorithms and simulated annealing/evolution algorithms are the common ones.

Group migration algorithms partition randomly at first, adjusting the components among blocks to minimize the interconnections between blocks as well as the delay. This algorithm is very efficient. Simulated annealing/evolution algorithms often provide the lowest-cost solutions but with a larger computation overhead than other algorithms.

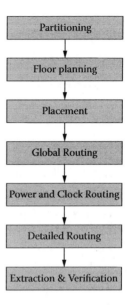

FIGURE 9.28
Back-end design flow.

9.3.6.2 Floor Planning

Floor planning determines the locations and shapes of partitioned blocks. In addition, it determines the location of each I/O pad, the number and locations of the power pads, and the location and type of clock distribution. Floor planning is not a uniform operation and is iteratively based on the designer's intuition, experience, and the engineering process. The final task of floor planning is to minimize the total chip area and delays and to place highly connected blocks physically close to each other.

Floor planning has two stages: generation of the floor plan and evaluation of the floor plan for cost efficiency (see Figure 9.29). It is an iterative process to achieve the most efficient floor plan.

After generation of the floor plan, we estimate the efficiency using a cost function. The primary variables that influence the cost function are the area (bounding box of the floor plan) and the total wire length. Secondary variables are the critical path delays, noise, heat dissipation, etc. The area is determined

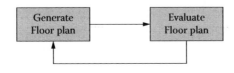

FIGURE 9.29
Iterative approach for floor planning.

by the designer, allowing the approximate wire length to be calculated. Finally, the floor plan is chosen, minimizing cost, wire length, and dead space.

The designer should define the position of the pads before floor planning. Also, if any memories (synchronous RAM, asynchronous RAM, ROM, dual-port RAM, etc.) are contained in the design, the designer must create the memory cells aligned with the ASIC foundry requirements using a memory generator. Using these memory cells, the designer will have to perform the RTL, gate simulation, and P&R.

9.3.6.3 Placement

Placement arranges the logic cells within the allowed region as guided by the floor plan. Good placement avoids overlap between the cells and makes them easier to wire. In addition, placement should minimize variables such as power dissipation, all performance-critical net delays, crosstalk between signal paths, interconnect congestion, and total interconnect length. Also, blocks must be placed as closely as possible for less silicon area and smaller interconnection delays.

There are different types of placement, according to the type of placement objects, such as a cell or a block. Standard cell placement deals with only cells of the same height and different widths, whereas building block placement deals with blocks of arbitrary sizes and shapes. There are also gate array placement and FPGA placement, both of which correspond to mapping each logic block in the logic design onto a gate array cell or a proper number of configurable logic blocks (CLBs) of the FPGA.

Figure 9.30 shows the placement of a small SoC based on the OpenRISC 1200 processor offered by OpenCores (http://www.opencores.org). This is generated using Verilog.

9.3.6.4 Routing

Routing connects a wire between pins. In general, routing can be classified as global routing or detailed routing. For clock and power routing, clock tree synthesis (CTS) is used.

Global routing allocates routing regions for each net instead of actually connecting pins belonging to each net. Detailed routing generates an actual geometric layout for all nets based on the global routing result.

First, global routing partitions the entire routing space into routing regions, as shown in Figure 9.31. (This is basically defining the regions.) It allocates regions to each of the pins without any overlap and then checks for any possible timing violation (region and pin assignment). Detailed routing can be classified as channel routing and switchbox routing (see Figure 9.32).

The purpose of channel routing is to reduce the number of tracks and vias. In switchbox routing, the routing requirement comes from four directions (compared to two, i.e., top and bottom in channel routing), which must be satisfied by routing within the available switchbox region.

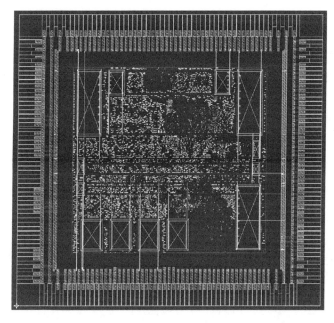

FIGURE 9.30
Placement result of OpenRISC-based SoC. (Courtesy of Dynalith Systems, Seoul, Korea.)

9.3.6.5 Clock Tree Synthesis and Power Routing

In spite of noise or delays due to wires, the clock should ideally operate with zero rise times or delays (zero clock skew). Also, despite an uneven distribution of capacitive load, each clock port in the logic block and flip-flops must be provided with clock signals of equal amplitude and timing. Clock tree synthesis (CTS) is executed to achieve zero clock skew. The designer should be able to compare the clock distribution before executing CTS and the clock tree after executing CTS. In power routing, it is critical to supply the same voltage to all V_{dd} ports in the chip. Also, for noise minimization, we need to consider placing as many power and ground pins as required by the distribution of the load.

9.3.6.6 Extraction and Verification

This is the final step within the P&R stage and checks the layout for correctness. This provides the design rules for every ASIC foundry. For example, there is a minimum length between metal layers and a minimum size of contacts for the circuits to operate accurately. The design rule check (DRC) checks the design rules, and this stage must be carried out very accurately. After passing the DRC stage, we can extract the files, including the wire delay (e.g., the SDF file) and layout information, to proceed with chip fabrication

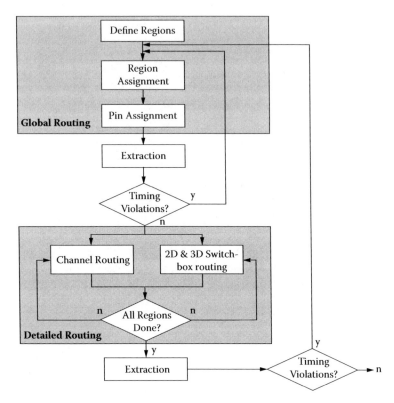

FIGURE 9.31
Routing flow.

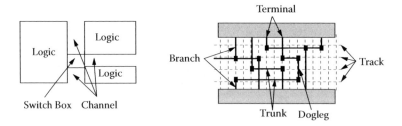

FIGURE 9.32
Switchbox and channel.

(e.g., the GDS file) and simulation (Verilog file). From the P&R stage, the design should proceed to postlayout simulation to check the functional correctness of the chip as a whole.

9.3.7 Postlayout Simulation

Wire delay information is not considered in gate-level simulation because it executes before the P&R stage. However, the actual wire delays can adversely affect the timing of a design. From the postlayout simulation, the designer can extract the file including wire delay information based on the resistance and capacitance of wire. Although the wire delay of the actual chip may be slightly different from the extracted information, the designer can obtain a reasonable estimate to see how the timing is affected by wire delays.

If there is no significant timing error found by the simulation with the extracted information and gate-level netlist, the layout information can be sent to an ASIC foundry for SoC fabrication. If an error is identified, the RTL or gate netlist should be modified and the above stages repeated. For this reason, back-end design takes more time than front-end design.

9.4 Status and Prospect of Radio Frequency Integrated Circuits (RFICs)

There has been an increased demand for RF transmitter and receiver ICs for mobile communications systems and wireless systems such as code division multiple access (CDMA), global system for mobile communication (GSM), personal communications services (PCS), digital cellular system (DCS), wideband CDMA (WCDMA), wireless broadband/worldwide interoperability for microwave access (Wibro/WiMax), global positioning system (GPS), wireless local area network (LAN), dedicated short-range communication (DSRC), Bluetooth, Zigbee, ultra-wideband (UWB), and radio-frequency identification (RFID). Current RFICs can be categorized mostly based on application standards, IC technologies, and transceiver architectures. Within the last decade, many application standards have been introduced for wireless mobile and communications systems. Complexity, cost, power dissipation, and the number of external components have been the primary criteria in selecting transceiver architectures. As IC technologies evolve, the relative importance of each of these criteria changes, allowing new approaches that once seemed impractical. Mixed-signal CMOS and silicon-germanium (SiGe) BiCMOS are major technologies used to integrate transceivers in a single chip. Recent RF CMOS technologies have feature sizes ranging from 0.18 μm and reaching 0.13 μm, 90 nm, and 65 nm.

Direct conversion architectures, which have no image frequency problems, are quite common these days in transmitters and receivers. Because an image rejection filter is not necessary in the architecture, they are easy to implement on a single chip. In handheld systems, recent developmental attempts were made to replace the conventional RF circuits as antenna front-ends by digital circuit blocks. These are digital RF circuits, which must have novel transceiver architectures can readily be integrated with other digital circuits in CMOS.

The low-noise amplifier (LNA) plays a key role in receiver performance and overall system performance. Its main function is to provide enough gain to overcome the noise of subsequent stages and add as little noise as possible to the signal. Many CMOS LNA designs have demonstrated low noise figures and high gain. Some applications, such as ultra-wideband, need very wide band characteristics or high linearity. The first stage with noise input matching must have a very low noise figure.

Mixers are used to provide frequency conversions in transmitters and receivers. These need nonlinearities, and usually a second-order nonlinearity, to up-convert or down-convert frequencies, which are easy to implement using CMOS. Because mixers also produce unnecessary frequencies, filters must be placed at the input and output of the mixers. Many new transceiver architectures, such as polar transmitters and digital receivers, eliminate the noisy and inefficient circuitry in the circuit.

A transmit–receive switch, an essential part of RF circuits in transceiver systems, selects between the transmitting and the receiving path to the antenna. This is usually implemented with Gallium-Arsenide (GaAs) pseudomorphic high electron mobility transistor (PHEMT) devices for mobile communication systems such as GSM and EDGE. Low insertion loss in the on state and high isolation in the off state are the key requirements.

Power amplifier (PAs) is a key element of transmitters. PAs consume a lot of power and their efficiency is important in battery-powered portable systems. State-of-the-art PAs for mobile communication systems are made of GaAs heterojunction bipolar transistors (HBTs), and attempts have been made to implement these with CMOS and integrate with control circuits. Output-matching circuits are conventionally implemented on a package because on-chip passive elements have low Q values. There are two kinds of RF PAs: linear and switching. Linear PAs (class A, B, AB, and C) have very low power efficiency but good harmonic performance. Switching PAs (class E, D, and F) have very high power efficiencies but inferior harmonic performance. Therefore, one must use a linear PA in spite of its very low power efficiencies when one must amplify nonconstant envelope signals. Most of the digital communication systems with optimized spectral efficiencies have nonconstant envelope signals. Polar transmitters utilize switching PAs to amplify nonconstant envelope signals.

Much research has been carried out to develop efficient PAs. Smart PAs control the bias of gate or base power transistors and increase the power efficiencies at low power. A DC-DC converter for the PA is used with special control to adjust the bias of drains or collectors. Architectures for direct RF synthesis have been proposed and implemented.

Frequency synthesizer, which consists of a voltage controlled oscillator and a feed back loop, generates pure high-frequency carriers at wanted frequencies. Since voltage controlled oscillators are susceptible to the phase noise, there have been a lot of work to reduce the phase noise. A careful design of a CMOS oscillator allows the phase noise to be comparable with the bipolar transistor oscillators. Also new architectures were introduced for larger tuning range and smaller chip size. The modulus control in the loop allows selecting a wanted frequency. Integer N architecture has integer modulus of the divider in the loop. In this case, the loop bandwidth is limited because the input reference frequency must be equal to the channel spacing. In fractional N synthesizers, the output frequency can vary by a fraction of the input frequency, allowing the latter to be much greater than the channel spacing. Recently introduced all digital phase locked loops (ADPLL) meet most of these requirements.

9.5 Mixed-Signal ICs in SoC

As the performance and complexity of SoCs increase, there are various functions that must be implemented in the mixed-signal IC domain. These include circuits for data conversion, power management, communication interface, and clock generation. Because of their flexibility and performance, SoCs must have a communication interface for data rates up to a few tens of gigabits per second. This requires serial link technology, whose core consists of high-performance phase-locked loop (PLL). SoCs should also provide flexibility in handling various types of data, whether it is analog or digital, and data conversion blocks need to be integrated. For efficient computation and power management, it is necessary to provide multiple power supplies in addition to versatile clock generators that provide a wide range of clock frequencies, from megahertz to gigahertz. Wide-range clock generators are usually designed using a PLL or a delay-locked loop (DLL).

As the integration level and process technology advances mixed-signal circuit design not only faces demand for higher performance but also must confront multiple design challenges. One of the most compelling issues comes from the reduced supply voltage in advanced CMOS technology. Because supply voltage limits the signal power, it is difficult to increase the SNR of the mixed-signal circuit with an advanced CMOS process that uses a less-than-1-V supply. It will be increasingly difficult to design core analog circuits such as op amps, comparators, and voltage-controlled oscillators in such situations. Another design challenge comes from the noise that arises from the digital circuitry. Switching noise coupled to the supply, ground, and substrate degrade the performance of mixed-signal circuits. Although there have been efforts to reduce the noise by isolating the sensitive circuits from the noisy environment, or even canceling the switching noise, an effective systematic solution has yet to be found.

From the standpoint of systematic design and verification methodology, designing mixed-signal circuits is still primitive compared to designing digital circuits. Mixed-signal circuit design and verification cannot be automated as easily as digital circuits [26]. Because of the different accuracies required for analog and digital subblocks, simulation and verification of the overall mixed-signal system can be very time consuming. Different operating frequencies are also an issue. For example, PLLs have subblocks that operate at different clock frequencies with more than an order of magnitude difference. Hence, fast simulation and verification of the circuit become extremely difficult.

In order to overcome these design challenges, one of the clear trends in mixed-signal circuit design is to exploit digital techniques as much as possible. Because CMOS scaling is clearly more advantageous to digital circuits than analog circuits, digital signal processing techniques that assist analog functions can overcome many of the design challenges. These include techniques such as digital calibration and oversampling. With the help of ultrafast switching times of advanced submicron CMOS processes, all-digital techniques are gradually being introduced into the analog/mixed-signal domain. Recently, all-digital PLLs [27] and all-digital analog-to-digital converters [28] have emerged as strong candidates to replace the traditional mixed-signal designs for future CMOS processes.

Acknowledgments

The author likes to acknowledge the kind contributions made by Prof. Songcheol Hong, Prof. Seonghwan Cho at the Department of EECS at the KAIST, and graduate students at the VLSI Systems Lab in the Department of EECS at the KAIST, Ki-Yong Ahn, Sangheon Lee, Seonpil Kim, Kyungsu Kang, Sangkwon Na, Jaemoon Kim, and Jungsoo Kim.

References

1. Frantz, G.A., System on a chip: a system perspective, *International Symposium on VLSI Technology, Systems, and Applications*, Proceedings of Technical Papers, 1, 2001.
2. Classen, T.A.M., System on a chip: changing IC design today and in the future, *IEEE Micro*, May–June, 20, 2003.
3. Garey, M., and Johnson, D., *Computers and Intractability: A Guide to the Theory of NP-Completeness*, W.H. Freeman, San Francisco, 1979.
4. Wong, D.F., Leong, H.W., and Liu, C.L., Simulated annealing for VLSI design, Kluwer Academic Publishers, Norwell, MA, 1988.
5. Open SystemC Initiative home page; available at http://www.systemc.org.
6. SpecC System; available at http://www.cecs.uci.edu/~specc.

7. SystemVerilog home page; available at http://www.systemverilog.org.
8. RealView SoC Designer; available at http://www.arm.com/products/Dev-Tools/SoCDesigner.html.
9. System Studio; available at http://www.synopsys.com/products/designware/system_studio/system_studio.html.
10. CoWare Platform Architect Overview; available at http://www.coware.com/products/platformarchitect.php.
11. ARM home page; available at http://www.arm.com.
12. CoWare home page; available at http://www.coware.com.
13. Synopsys home page; available at http://www.synopsys.com.
14. Lee, J.-G. and Kyung, C.-M., A predictive packetizing scheme for reducing channel traffic in transaction-level hardware/software co-emulation, *IEEE Transactions on Computer-Aided Design of Integrated Circuits and Systems*, 25, 1935, 2006.
15. Kim, N., Choi, H., Lee, S., Lee. S., Park, I.-C., and Kyung, C.-M., Virtual chip: making functional models work on real target systems, *Proceedings of the 35th Conference on Design Automation*, 170, 1998.
16. AMBA System Architecture; available at http://www.arm.com/products/solutions/AMBAHomePage.html.
17. Hassoun, S., Kudluqi, M., Pryor, D., and Selvidge, C., A transaction-based unified architecture for simulation and emulation, *IEEE Transactions on Very Large Scale Integration (VLSI) Systems*, 13, 278, 2005.
18. CoreConnect Bus Architecture; available at http://www-03.ibm.com/technology/ges/semiconductor/power/licensing/coreconnect.
19. SoC Interconnection: Wishbone; available at http://www.opencores.org/projects.cgi/web/wishbone/wishbone.
20. RealView Versatile Family; available at http://www.arm.com/products/Dev-Tools/VersatileFamily.html.
21. RealView® Integrator™ Family; available at http://www.arm.com/miscPDFs/17086.pdf.
22. Boundary-Scan Technology; available at http://www.jtag.com/main.php?cm=p8_1___.
23. In Circuit Emulators; available at http://www.algonet.se/~staffann/developer/emulator.htm#What%20is%20an%20ICE%20?.
24. TRACE32; available at http://www.lauterbach.com/main.html.
25. Piguet, C., *Low-Power CMOS Circuits: Technology, Logic Design and CAD Tools*, CRC Press, Boca Raton, FL, 2005.
26. Chou, E.Y. and Sheu, B., Nanometer mixed signal system on a chip design, *IEEE Circuits and Devices*, July, 7, 2002.
27. Staszewski, R.B., Muhammad, K., Leipold, D., Hung, C.-M., Ho, Y.-C., Wallberg, J.L., Fernando, C., Maggio, K., Staszewski, R., Jung, T., Koh, J., John, S., Deng, I. Y., Sarda, V., Moreira-Tamayo, O., Mayega, V., Katz, R., Friedman, O., Eliezer, O.E., de-Obaldia, E., and Balsara, P.T., All-digital TX frequency synthesizer and discrete-time receiver for Bluetooth radio in 130-nm CMOS, *IEEE Journal of Solid-State Circuits*, 39, 2278, 2004.
28. Watanabe, T., Mizuno, T., and Makino, Y., An all-digital analog-to-digital converter with 12-μV/LSB using moving-average filtering, *IEEE Journal of Solid-State Circuits*, 38, 120, 2003.

Index

Note: Locators for figures are in italics and for tables in bold.

A

ACPI (advance configuration and power
    interface) specification, 98
active filter networks
    bi-quad, *255*
    multiple feedback, *255*
    unity gain (Butterworth), *255*
ADC, 12-bit
    testing of using a 4096-point FFT, *291*
ADC architectures, 306–322
    flash converters, 308
    integrating ADCs, 309–310
    ranges of resolution and sample
        rates, *279*
    recycling, *315*
    self-calibration techniques, 322
    sigma-delta converters, 315–321
    subranging ADCs (Half-Flash
        ADCs), 311–312
    successive approximation ADCs,
        306–307
    two-step architectures, 313–314
    pipeline-subranging , with a digital
        error correction, *312*
ADC characteristics, amplitude
        quantization on a time domain
        waveform and, *295*
ADC drive amplifier considerations, **340**
ADC errors, 292–298
    differential nonlinearity, 295–296
    gain error and offset error, 297
    integral nonlinearity, 297
    testing of ADCs, 298
ADC gain (bandwidth), ENOB versus
        frequency and, *292*
ADC resolution and dynamic range
        requirement, 287–291
    analog bandwidth and, 289

effective number of bits of an ADC,
        288
    harmonics and, 289
    spurious components and, 289
ADC-amplifier interface, *339*
ADCs (analog-to-digital converters)
    8-bit subranging, *310*
    effective number of bits of, 288
    integrating, 309–310
    interface between serial output
        ADCs, 400
    interface with DSPs, 399–400
    parallel interfaces with, 399
    successive approximation, 306–307,
        307
    testing for its performance
        parameters, *290*
    testing of, 298
ADCs, Σ–Δ(sigma delta)
    block diagram of, 315–321
    key concepts behind, 315–321
address generation units, 385–387
    example: ADSP-21xx Family, 385–386
addressing
    bit-reversed, 381
    circular, *380*
    linear, *380*
ADSP-2100A program sequencer
        architecture, *392*
ADSP-2101's shifter, block diagram of,
        388
aliasing, implications of, 281
    time and frequency effects of, *281*
ALM in a Stratix II device, architecture
        of, *350*
Altera CPLD device, architecture of, *346*
Altera Stratix device, block diagram of,
        351
ALU (arithmetic logic unit), group
        functions, 397

amplification, 211–242
    instrumentation amplifiers, 227–234
    operational amplifiers, 211–226
    video and communications
        amplifiers, 241–243
amplification, nonlinear approaches to,
    235–240
    practical log amps, 239
    root mean square-to-DC converter
        ICs, 230–240
amplifier frequency response, 22–31
    BJT equivalent circuits, 22–25
    BJT small signal operation and
        models, 26
    high-frequency models of the
        transistors and, 26–31
amplifier types
    feedback, and the modification of
        input and output impedances, **14**
amplifiers; *See also* op amps (operational
        amplifiers); SHA (sample-and-
        hold amplifier); VFA (voltage
        feedback amplifier)
    communications, 241–243
    generic types of and simfple
        examples, **212–213**
    noise in, 41–43
    series-series feedback, *18*
    series-shunt feedback, *17*
    series-series topology, *12*
    series-shunt topology, *12*
    shunt-shunt topology 12
    video, 241–243
amplifiers, single pole
    effect of feedback on, *21*
analog bandwidth, 289
    harmonics and, 289
    spurious components and, 289
    spurious-free dynamic range,
        290–291
analog signal, at frequency f_a sampled
        at f_s, *283*
analog-to-digital converters, *see* ADCs
        (analog-to-digital converters)
antialiasing filters, system dynamic
        range and, *284*
aperture, aperture time and, 301–303
aperture jitter, SHA output and, *303*

application specific integrated circuits,
        see ASICs (application specific
        intergrated circuits)
arithmetic logic unit, *see* ALU
        (arithmetic logic unit)
ARM , 200, 437, *447–449*
ASICs (application specific integrated
        circuits)
    advance configuration and power
        interface (ACPI) specification, 98
    powering high-power processors
        and, 94–98
assembly drawing, example of, *71*
assembly instructions, 72

B

back-end design, 452–462
    flow, *459*
    gate-level simulation and, 456–457
    logic synthesis, 454
    place and route, 458
    postlayout stimulation, 463
    power analysis and, 456–457
    RTL description and verification, 454
    static timing analysis, 455
    test and auto test pattern generation,
        456
bandwidth, 211
baseband antialiasing filters, 283–285
bathtub curve, reliability of products
        and, *193*
bill of materials (BOM), 71
bit sizes, for 2.048-V full-scale
        converters, **287**
BJT (bipolar junction transistors)
    base current noise density of, *47*
    equivalent circuits, 22–25
    small signal operation and models, 26
bode plots
    gain and phase margins, 20–21, *21*
    for the loop gain, *21*
BOM (bill of materials) structure,
        typical, *72*
boost converter, *115*
boost converter IC and charge pump
        use of, *96*
boundary scan device, *450*
brick converter, examples of, *95*

bridge amplifiers, *248–249*
bridge circuits, 243–248
    properties and, **245–246**
bridge converters, 129–136
buck converter, *114*
buck topology
    ZCS-QR operation and, **142–143**
    ZVS-QR operation and, **142–143**
buck-boost converter, *116*
bugs, fixing
    rapid increase of as design
        progresses, *431*
burst noise, *42*
Butterworth low-pass response, *254*
Butterworth polynomial coefficients,
    254

C

calculation of the high-frequency 3-dB
    corner frequency ω_H, 34–38
calculation of the low-frequency 3-dB
    corner frequency ω_L, 32–33
California Energy Commission, *see* CEC
    (California Energy Commission)
Canadian Standards Association,
    see CSA (Canadian Standards
    Association)
candidate DSP architecture for IIR/FIR-
    type filtering, *375*
capacitors, 51–52
    in practical circumstances, *52*
CEC (California Energy Commission),
    99
cellular phones, 88, 148
    block diagram of 2005, *89*
centralized power architecture
    distributed power architecture (DPA)
        and, 79–80, *80*
charge division, 329
charge pump converters, 170–172
charge pumps, types of, *172*
circuit, combinatorial, *344*
circuit capacitance, effect of, 41
circuit noise
    bandwidth calculation and, 44–46,
        48–49
circuits
    passive components in, 49–54

circular addressing, 379–380
clock systems for digital systems and
        signal integrity, 420–423
    clock skew and transmission line
        effects, 422
    direct digital synthesis and, 420–423
    oscillators and, 420–423
    PCB interconnection
        characterization, 423
    PLL (phase lock loops), 420–423
    signal integrity and system design,
        421
clock tree synthesis, power routing and,
    461
CMOS analog switches, 265–267
    equivalent circuits for, *269*
    AC performance of, *270*
codes
    binary, *324*, **325**
    one-of-n, *324*, **325**
    thermometer, *324*, **325**
coemulation, 442
    benefits of, 442
    environment, 444, *445*
    tools, 443
complex programmable logic device, *see*
    CPLD (complex programmable
    logic device)
compliance folder management, 68–69
component engineering, 62–64
    component specifications, 64
    preferred component list, 63–64
    reliability of components, 63
    spares requirement, 63
components
    reliability of, 63
    specifications for, 64
    suppliers of, 69
computation instructions: a summary of
        the ADSP-21xx Family, 394–397
    ALU group functions, 397
    MAC functions, 394–396
    shifter group functions, 397
computational input/output registers,
    396
computational loads, approximate
    for decimations-in-time FFT and, **9**
    for direct DFT and, **9**
configurable logic blocks for digital
    systems design, 343–362

design tools for PLDs, 351
FPGAs (field programmable gate
arrays) and, 346–350, 361
overview of a hardware description
language, 352
programmable logic devices, 344–345
pulse-width measurement, 357–360
control loop
of a buck converter, *155*
design, 152–159
converters, switch mode
control of, 140–146
current mode control, 141
hysteretic control, 141–146
voltage mode control, 140
cosimulation, 438–441
example, 440–441
issues, 440
tools, 439
CPLD (complex programmable logic
device), general block diagram of,
345
crowbar protection, overvoltage
conditions and, *187*
crystal resonator, equivalent circuit for,
416
CSA (Canadian Standards Association),
safety standards of, 100
current division, 329
binary, *331*
uniform, *331*
current feedback op amps, 224–226
current limiting methods
constant current limiting, *188*
foldback limiting, *188*
hiccup current limiting, *188*
current mode control, 141, *145*
current-steering architecture, 335
current-steering DAC with an R-2R
ladder, *336*
customer requirements, understanding,
59–60

D

DAC architectures, 333–337
current-steering architecture, 335
intermeshed ladder architecture, 335

ladder architecture with switched
subdividers, 333–334
other architectures, 337
R-2R network-based architecture, 336
resistor ladder DAC architectures,
333
DACs
interface with DSPs, 401–403
interfaces and, 401–403
DACs (digital-to-analog converters),
322–328
data sheet terminology and, 324–328
missing codes in, *326–327*
nonmonotonicity in, *326–327*
parallel input, 401–402
parameters of, *328*
performance parameters, 324–328,
325
using a voltage division technique,
330
DACs, parallel
interface with DSPs, *402*
interfaces and, 401–402
DACs, principles used in, 329
charge division, 329
current division, 329
voltage division, 329
DACs, serial
interface with DSPs, *404*
interfaces and, 403
DAG (data address generator), block
diagram of the ADSP-2101, *386*
data acquisition system interfaces,
337–339
amplifier-ADC interface, 339
signal source and acquisition time,
337–338
data address generator, *see* DAG (data
address generator)
data address units of ADSP-21xx
Family: an example, 385–386
data converters, 277–342
ADC architectures, 306–322
ADC errors, 292–298
ADC resolution and dynamic range
requirement, 287–291
basic SHA operation, 299–304
DAC architectures, 333–337
DACs, 322–328

data acquisition system interfaces, 337–339

DDS systems and waveform generation, 423–425

effects of sample-and-hold circuits, 298

interface with DACs and, 401–403

interface with DSPs, 398–403

principles used in DACs, 329

sampled data systems, 279–286

DC-DC converter techniques

bridge converters, 129–136

flyback converters, 109–128

popular, **84**

selection of, 81–83

SEPIC converters, 136

DC-DC converters

design of, 106–135

forward mode converters, 108

two-transistor forward mode, 129

DDS (direct digital synthesis), 413–426

DDS system

fundamental, *424*

more flexible, *424*

waveform generation and, 423–425

DDS systems and waveform generation, 423–425

debugging methodology, emulation and, 449–451

decimation process, *319*

design management, 73–75

design of DC power supply, 78

approaches and specifications, 79–89

loading considerations, 90–99

off-the-line power supplies, 100–196

power management and, 77–204

specifying DC power supply requirements, 90

design process, 57–76

design management, 73–75

elements of, *59*

product documentation, 70–73

role of external agencies, 67–69

specifications for the design, 59–60

design specifications, 60

sample template for, **61**

dielectric absorption, 304

effect of, *53*

differential amplifier, *214*

BJT implementation, *214*

FET implementation, *214*

transfer characteristics for BJT case, *214*

transfer characteristics for FET pair, *214*

differential nonlinearity, 295–296

differential to single-ended conversion, 216

digital control, 176–184

in a simple buck converter design, *180–181*

digital filters, 370–372

FIR filter, 370

IIR filter, 371–372

digital signal controller, *see* DSC (digital signal controller)

digital signal processing system, *365*

digital signal processors, *see* DSPs (digital signal processors)

digital still camera system components, *85*

digital-to-analog converters, *see* DACs (digital-to-analog converters)

direct digital synthesis, 413

discrete and digital signals, 6–8

discrete Fourier transform, 7

discrete time Fourier transform, 7

fast Fourier transform, 8

discrete time sampling of analog signals, 280

distributed power architecture, *see* DPA (distributed power architecture)

DPA (distributed power architecture), centralized power architecture and, *80*

DSC (digital signal controller)

power management (PM) solution for from Texas Instruments, *88*

SMPS design and, *183*

DSC-based approach for PFC-based full-bridge configuration with dual processors, *184*

DSP applications, examples of, 403–406

DSP architecture, 375–378

basic Harvard architecture, 376

microprocessor architectures and, *368*

modifications to Harvard
architecture, 379
multiple-access memory-based
architectures, 378
SISC architectures, 377
DSP core
SoC (system-on-a-chip), *429*
DSP-based filter implementation, *370*
DSPs (digital signal processors), 363–412
architectural elements of, 381–392
bit-reversed addressing, 381
circular addressing, 379–380
evolution of DSP architecture, 369–378
examples of DSP applications, 403–406
filtering applications and, 369–378
instruction set, 393–398
interface with ADCs, 399–400
interface with DACs and, 401–403
interface with data converters,
398–403
interface with parallel DACs, *402*
interface with serial DACs, *404*
latency and, 367–368
microprocessors and, 366–368
parallel interfaces with ADCs, 399
practical components and developments in the late 1990s, 407
recent technology trends in design,
408–409
recent technology trends in DSP
design, 408–409
special addressing modes, 379–380
sample period and, 367–368
special addressing modes, 379–380
DSPs, architectural elements in, 381–392
address generation units, 385–387
loop mechanisms, 391–392
multiplier/accumulator, 383–384
shifters, 387–390
DSP-to-ADC parallel interface,
generalized, *400*

E

effective aperture delay time,
measuring, *303*
efficiency requirements for external
power supply curves based
on CEC, EPA, and EU
recommendations, *99*

efficiency vs. input voltage for a
compact switching regulator
family, a typical example, *83*
weighted efficiency, 86, **87**
electrolytic capacitors, aluminum
behavior of at different frequencies
and temperatures, *166–167*
electromagnetic compliance, 67–68
Electronic Numerical Integration
and Calculation, *see* ENIAC
(Electronic Numerical Integration
and Calculation)
EMC (electromagnetic compatibility)
cost diagram, *68*
EMI reduction in switch mode
converters, 148–151
emulation, 445–451
debugging methodology, 449–451
emulation board survey, 446–449
engineering services, 66–67
ENIAC (Electronic Numerical
Integration and Calculation), 366
Environmental Protection Agency, *see*
EPA (Environmental Protection
Agency
EPA (Environmental Protection Agency,
99
error amp types usable in switching
power supplies, **156–157**
error minimization, options at system
and device levels, *206*
error sources of an analog switch,
268–270
EU (European Union), 99
European Union, *see* EU (European
Union)
external agencies, role of
component suppliers, 69
regulatory agencies, 67–69
reviews, 69–70
test houses, 69
external power supplies and new
energy standards, 99
extraction and verification, 461–462

F

factory test specifications, 72
fast Fourier transform, 8

feedback
  amplifier types and the modification
    of input and output impedances,
    14
  bandwidth extension, 11–12
  effect on a single-pole amplifier, *21*
  frequency response and, 9–12
  gain desensitization, 10
  input/output impedance
    modification, 11–12
  noise reduction, 10
  reduction in nonlinear distortion,
    11–12
feedback amplifier, *10*
feedback topologies, basic, *12*
FFT algorithm, output of
  bit-reversed addressing and, *382*
field programmable gate arrays, *see*
  FPGAs (field programmable gate
  arrays)
filter implementation
  by DSP techniques, *373*
  in DSPs, 373–374
filter noise bandwidths compared with
  -3 db corner for frequency, *209*
filter responses, ideal cases of, *250*
filter sensitivities to real op amp
  characteristics, **257**
filtering applications and the evolution
  of DSP architecture, 369–378
  digital filters, 370–372
  DSP architecture, 375–378
  filter implementation in DSPs,
    373–374
filters, 249–264
  active and passive, 251–257
  phase response of, *253*
  switched capacitor, 258–264
finite impulse response (FIR) filter, 370,
  371
  implementation by multiplication
    and addition of pointed data, *406*
flash converters, 308
flash or parallel ADC block diagram,
  308
floor planning, 459
  iterative approach, *459*
Fourier transform, 5
  discrete, 7
  discrete time, 7

FPGAs (field programmable gate
  arrays), 346–350
  in practice, 361
  types of, 349–350
frequency divider, 420
frequency response of capacitively
  coupled amplifiers, *28*
frequency response of common emitter/
  common source amplifiers, 32–38
frequency spectrum of a switched
  capacitor filter, *264*
front-end design flow, 431–451
  coemulation, 442–444
  cosimulation, 438–441
  emulation, 445–451
  hardware and software partitioning,
    434–438
  reference algorithm, 433
fundamentals, review of, 1–56
fuses, 102

G

gain desensitization, 10
gain error and offset error, 297
gate driver techniques, comparison of,
  131–132
gate-level simulation, 456–457
  power analysis and, 456–457

H

hard-switching (PWM) techniques,
  resonant techniques and, 137
hardware and software partitioning,
  434–438
  issues, 434
  system-level design method, 438
  transaction-level modeling-based
    methodology, 436–437
hardware description language,
  overview of, 352
hardware emulation, using a virtual
  chip, *443*
Harvard architecture, 366, *377*
  basic, 376
  with dual-ported memory, *378*
  modified, *378*, 379

high-power processors, powering
  ACPI (advance configuration and
    power interface) specification and,
    98
high-speed sampling, 282
hold mode droop, 304, *305*
hold-mode settling time, *301*
hold-to-track transition specification,
  305
hysteretic control, 141–146
hysteretic control and a typical circuit
  Intel VRM (voltage regulator model)
    and, **146**

I

IEC (International Electrotechnical
    Commission), safety standards
    of, 100
IIR filter, 371–372
  simple, *372*
in-amp requirements and
    implementations, *233–234*
inductor-equivalent circuit, *54*
inductors, 53
industry-favorite configurations and
    areas of usage, *106*
input and output impedances,
    modification of
  amplifier types and, **14**
  feedback and, **14**
input current and voltage relationships
    of a rectifier stage, *105*
input imperfections, 215
input line filters for SMPS systems, *151*
input section of an of-the-line power
    supply, *101*
input transients, protection against, 189
input/output capacitors, reliability of,
    189–191
input/output characteristic of an ideal
    3-bit DAC, *323*
inrush current limiting, 103
instruction set, 393–398
  computation instructions: a
    summary of the ADSP-21xx
    Family, 394–397
  other instructions, 398

instruction set groups (using ADSP-
    21xx family as an example), **398**
instrumentation amplifiers, 227–234
integral nonlinearity, 297
integrating converter, typical, *310*
intermediate bus architecture, 81,82
intermeshed ladder architecture, 335
  with one level multiplexing, *335*
internal departments, responsibilities in
    product design, 61–67
  component engineering, 62–64
  design engineering, 61
  engineering services, 66–67
  marketing, 67
  material procurement, 65
  production engineering, 65
  quality assurance, 66
  test engineering, 65
International Electrotechnical
    Commission, *see* IEC
    (International Electrotechnical
    Commission)
International Solid State Circuits
    Conference, *see* ISSCC
    (International Solid State Circuits
    Conference)
introduction to oscillators, phase lock
    loops, and direct digital synthesis
  clock systems for digital systems and
    signal integrity, 420–423
  DDS systems and waveform
    generation, 423–425
  essentials of simple oscillators, 413
inverting CFA configuration, VFA and,
    227
iPROVE coemulation environment, *443*
isolation barriers, required in medical
    environments, 272
isolation techniques, examples of,
    273–274
ISR devices from Power Trends, Inc. in a
    DPA solution, *95*
ISSCC (International Solid State Circuits
    Conference), program committee,
    322

K

Kirchoff's Laws, 3

L

ladder architecture with switched subdividers, 333–334
laptop computers, 148
LDO (low dropout regulators), 160–169
  adjustable circuits, *169*
  block diagrams, *162*
  noise measurements of, *167–168*
  ground currents in, *170*
LDO pass transistor configurations, characteristics of, **163**
LE in Altera Cyclone FPGA devices, architecture of, *348*
load, effect of on stability, *165*
load loci
  for PWM (hard-switching) techniques, *144*
  resonant (soft-switching) techniques, *144*
loading considerations, 90–99
  external power supplies and new energy standards, 99
  powering high-power processors and ASICs, 94–98
log amps, 239
log amps with negative values of *x* input, *240*
logic devices, programmable, 344–345
logic element, basic structure of, *347*
logic synthesis, 454
  32-bit adder, *455*
loop measurements and typical response, *158–159*
loop mechanisms, 391–392
loop gain and the stability problem, 13–21
  bode plots and gain and phase margins, 20–21
  injection transformer setup for loop-gain measurements, *160–161*
  Nyquist plot, 17
  poles and zeros, S-domain, and bode plots, 17–19
low-frequency 3-dB corner frequency ω_L, calculation of, 32–33
low-frequency response, 28–29
low-pass and band-reject filters using a filter plot Excel spreadsheet, *259*

low-pass filter, 419
  key filter parameters for, *250*
LUT-based implementation of $Z = \Sigma m(0,3,4,7,8,11,12,15)$, *347*

M

MAC (multiplier/accumulator), functions of, 394–396
MAC block diagram of the ADSP-2104, *384*
managing project scope changes design management and, 74
marketing, 67
marketing requirements specification, 60
material procurement, 65
MaxSim-ModelSIm , *441*
memory blocks
  SoC (system-on-a-chip), *429*
microprocessor core
  SoC (system-on-a-chip), *429*
microprocessors, merging of with DSPs, 369
Miller's theorem for simplifications of circuits, *37*
mixed-signal ICs in SoC, 465
models, 32–38
modified charge division, *332*
moving average calculation as a simple example of DSP applications, *405*
MP3 players, 88
multiphase converter approach, *174*
multiplier as a phase detector, *419*
multiplier/accumulator, *see* MAC (multiplier/accumulator)

N

noise
  in amplifiers, 41–43
  error considerations and, 208–210
  in op amps, 217
  in passive components, 39–40
  in semiconductors, 41–43
noise bandwidth, circuit noise calculations and, 44–46

noise in circuits, 39–48
  circuit noise calculations and noise
    bandwidth, 44–46
  effect of circuit capacitance, 41
  noise figure and noise temperature,
    47–48
  noise in passive components, 39–40
  noise in semiconductors and
    amplifiers, 41–43
noise reduction, 10
notation used in instruction set of
    ADSP-21xx family, **395**
Nyquist plot, 17
  of an unstable amplifier, *19*

O

off isolation and crosstalk in a typical
    analog switch, *271*
off-the-line power supplies, design of
  achieving high power density, 173
  charge pump converters, 170–172
  control loop design, 152–159
  control of switch mode converters,
    140–146
  digital control, 176–184
  efficiency improvements in switch
    mode systems, 147
  EMI reduction, 148–151
  low dropout regulators (LDO),
    160–169
  PFC in off-the-line power supplies,
    104–105
  postregulation techniques, 174–175
  power supply protection, 184–191
  rectifier section, 100–103
  resonant converters, 137–139
  testing of power supplies, 193–196
Ohm's Law, 3
(op amps) operational amplifiers,
    211–226
  common configurations, **228–231**
  common mode voltage
    considerations in, *220*
  current feedback op amps, 224–226
  differential to single-ended
    conversion, 216
  ideal and practical, *215*
  important parameters, **218–219**

input imperfections, 215
noise in, 217
noise spectral density of, *221*
output stage, 216
single-rail, 217–223
slew rate and settling time, *219*
transition to single-rail , *222*
oscillation, principles of, 414
oscillators, 413–426
  classification of, *414*
  crystal, 415–417
  essentials of, 413
  principles of oscillation, 414
  simple, 413
output stage, 216
oversampled Σ-Δ ADC, *320*
oversampling, *317*
  digital filtering and, *318*
overvoltage and overcurrent protection,
    185–188

P

palm computers, 94
partitioning, 458
passive component tolerances and
    worst-case design, 53–54
passive components
  capacitors, 51–52
  in circuits, 49–54
  inductors, 53
  noise in, 39–40
  passive component tolerances and
    worst-case design, 53–54
  resistors, 50
PCB interconnection characterization,
    423
PDAs, 148
performance parameters and data sheet
    terminology, 324–328
PFC (power factor correction), off-the-
    line power supplies and, 104–105
phase detector, 419
phase-frequency detector used as a
    phase detector [7], *420*
PIC microcontroller-based power
    converter using a half-bridge
    driver IC, *182*

place and route, 458
clock tree synthesis and power
routing, 461
extraction and verification, 461–462
floor planning, 459
partitioning, 458
placement, 460
routing, 460
placement, 460
placement result of OpenRISC-based
SoC, *461*
PLDs (programmable logic devices)
design tools for, 351
general structure of, *345*
PLL (phase lock loop), 413–426
PLL (phase lock loops), 413
PLL systems, 418–420
building blocks of, *418*, 419–420
frequency divider, 420
low-pass filter, 419
phase detector, 419
voltage-controlled oscillator, 420
pole locations, **252**
transient and, *20*
poles and zeros, S-domain, and bode
plots, 17–19
positive feedback network, *415*
possible CMOS latch-up problem and
overcurrents
protection against, *268*
postlayout stimulation, 463
postregulation techniques, 174–175,
175–177
power analysis, 457
power analysis flow, *457*
power compiler result, *458*
power factor correction, *see* PFC (power
factor correction)
practical approach to, *106–107*
power management concepts, 84–89
power supplies, testing of, 193–196
power supply
measurements, *195*
specifications, **91–92**
power supply calibration, *196*
power supply inrush current limiter
techniques
inductor based, *103*
MOSFET-based approach, *103*

resistor relay, *103*
thermistor technique, *103*
power supply protection, 184–191
age-related aspects, 192
overvoltage and overcurrent
protection, 185–188
protection against input transients,
189
reliability of input/output capacitors,
189–191
thermal design, 184
power supply subsystem, overall design
approach to, *93*
powering high-power processors
ASICs (application specific integrated
circuits) and, 94–98
practical implementation of a state
variable filter using a commercial
switched capacitor filter IC,
262–263
preferred component list, 63–64
preprocessing of signals, 205–276
amplification, 211–242
bandwidth, 211
bridge circuits, 243–248
filters, 249–264
general considerations, 207–211
introduction to, 206
noise and error considerations,
208–210
signal isolation, 271–273
signal range, 207
switching and multiplexing of
signals, 265–270
process statement, general form of, *355*
product documentation, 70–73
production documents, 71–72
technical manual, 73
user manual, 73
production documents, 71–72
assembly instructions, 72
bill of materials, 71
factory test specifications, 72
production engineering, 65
programmable logic devices. *see* PLDs
(programmable logic devices)
project schedule with and without
coemulation, 442

projects
  implementing the schedule, 74
  managing closure, 75
  managing costs, 75
  managing personnel, 75
  managing quality, 75
  managing scope of, 73
  managing strategy of, 73
  managing time and, 74
pulse-width measurement system,
    schematic diagram of, *358*
push-pull converter, *120*
PWM converters, quasi-resonant
    converters (QRCs) and, **140**

Q

quality assurance, 66
  product-specific quality plan, 66
  quality plan ensuring performance
    and reliability, 66
  quality plan for product design, 66
quality plan
  performance and reliability, 66
  product design and, 66
quantization noise, for 2.048-V full-scale
    converters, **287**
quantization process, quantization
    error, and timing cycle, *294*
quasi-resonant converters (QRCs), PWM
    converters and, **140**
quasi-resonant principle, 138–139

R

radio frequency integrated circuits
    (RFICs), status and prospect of,
    463–464
reactance
  versus frequency, *416*
  versus frequency with overtones, *417*
rectifier section, 100–103
  fuses, 102
  inrush current limiting, 103
reduced component designs using
    power ICs, *128*
reduction in nonlinear distortion,
    bandwidth extension, and input/
    output impedance modification
    by feedback, 11–12

reference algorithm, 433
regulatory agencies, 67–69
  compliance folder management,
    68–69
  electromagnetic compliance, 67–68
representative flyback converter and
    transformer, *125–126*
resistor ladder DAC architectures, 333
resistor ladder DAC with a switched
    subdivider, *334*
resistors, 50
  at higher frequencies, *51*
resonant converters, 137–139
  comparison between hard-switching
    (PWM) and resonant techniques,
    137
  quasi-resonant principle, 138–139
resonant principle, *138*
review of design release for pilot
    production, 70
review of the test plan, 70
reviews, 69–70
  review of design release for pilot
    production, 70
  review of the test plan, 70
  specifications review, 70
RMS-to-DC conversion, *241*
root mean square-to-DC converter ICs,
    230–240
routing, 460
routing flow, *462*
RTL description and verification, 454

S

Sallen-Key implementations using a
    single op amp, *256*
sample period and latency, *369*
sample-and-hold amplifier. *see* SHA
    (sample-and-hold amplifier)
sample-and-hold circuits
  effects of, 298
  error sources and, *300*
sampled data systems, 279–286
  baseband antialiasing filters, 283–285
  discrete time sampling of analog
    signals, 280
  high-speed sampling, 282
  implications of aliasing, 281

key elements of, *280*
undersampling, 286
sampling ADC quantization noise, *317*
sampling clock jitter, effects on the SNR, *304*
second-order Σ-Δ ADC, *321*
self-calibration techniques, 322
semiconductors, noise in, 41–43
semicustom design flow, *453*
sensor amplifier noise sources, *210*
SEPIC concept, application circuits and, *137*
SEPIC converter, *117*, 136
serial DSP-toADC interface, generalized, *401*
serial input DACs, 403
SHA (sample-and-hold amplifier)
specifications, **301**
waveform, *302*
SHA architectures, 305
SHA operation, basic, 299–304
aperture and aperture time, 301–303
dielectric absorption, 304
hold mode droop, 304
hold-to-track transition specification, 305
SHA architectures, 305
track mode specifications, 299
track-to-hold mode specification, 305
SHA output, effects of aperture jitter on, *303*
shifter group functions, 397
shifters, 387–390
short-circuit and open-circuit time constants for the
sigma-delta converters
block diagram of a Σ–Δ ADC, 315–321
key concepts behind Σ–Δ ADCs, 315–321
signal integrity, **422**
related items and, **422**
system design and, 421
signal isolation, 271–273
signal level (RTL) to TL transactor interface, *441*
signal range, 207
signal source, acquisition time and, 337–338

signals with high dynamic range, analog-processing advantage of, *237*
simple charge division, *331*
simple log converter and transfer function, *238*
simple RC network and its gain and phase plots, *153*
simplified block diagram of a Stratix II ALM[3], *349*
simplified interface between a low-voltage ADC and a signal source, *337*
simplified real op amp showing single-ended output and the differential output, *214*
simulation result for the pulse-width measurement system, *360*
SINAD and ENOB for AD676, *289*
single rail amplifier based on reference source approach, *223*
single-rail op amps, 217–223
SISC architectures, 377
small-signal equivalent of junction diode with noise, *43*
SMPS, digital control concept used in buck and boost mode possibilities and, *178–179*
SNR of a circuit, increasing
with an additional noise-free amplifier at the front end for SNR improvement, *11*
basic amplifier indicating the case of input referred noise, *11*
SoC (system-on-a-chip), 428, 429
back-end design, 452–462
challenges of, 430
defined, 428
design and verification, 427–468
example of, *429*
front-end design flow, 431–451
mixed-signal ICs in SoC, 465
reasons for, 429
status and prospect of radio frequency integrated circuits (RFICs), 463–464
typical design flow of, *432*
SoC design, challenges with, 430
SoC Validation Lab coemulation environment, *444*

software requirements, 60
source resistance, effect of, *338*
spares requirement, 63
specifications for the design, 59–60
    customer requirements, 59–60
    internal departments and, 61–67
    marketing requirements, 60
    software requirements, 60
specifications review, 70
specifying DC power supply
    requirements, 90
spectral density, 44
spurious-free dynamic range (SFDR),
    290–291, *292*
standard second-order filter responses,
    252
state variable filter and a digital
    implementation for variable
    parameters, *258*
static ADC metrics, *296*
static timing analysis, 456
subranging ADCs (Half-Flash ADCs),
    311–312
summary of power supply topologies,
    110–113
switch mode systems, efficiency
    improvements in, 147
switchbox and channel, *462*
switched capacitor converters, *171*
switched capacitor implementation,
    262
switching and multiplexing of signals,
    265–270
    CMOS analog switches, 265–267
    error sources of an analog switch,
        268–270
switching supply, losses associated
    with, *130*
synchronous rectification techniques,
    comparison of, **150**
synchronous rectifiers
    control-driven SR (CDSR), *149*
    self-driven SR (SDSR), *149*
system loads, partitioning of, **87**
system-level design flow in DSP Builder
    [26], *409*
system-level design method, 438
system-on-a-chip. *see* SoC (system-on-
    a-chip)

T

tangential approximation of a VCO
    phase noise shape [7], *421*
technical manual, 73
test access port control device [20], *450*
test engineering, 65
test houses, 69
thermal characteristics of a typical
    power package, **185**
thermal design, 184
thermal noise, in a resistor, *40*
thermal resistance from case to ambient
    with PC-board foil area, *186*
time and frequency domains, 3–5
    Fourier transform, 5
TL to signal-level (RTL) transactor
    interface, *441*
TLM-based methodology
    first step of, *437*
    second step of, *437*
    third step of, *437*
Trace32 software debugger, *451*
track mode specifications, 299
track-to-hold mode specification, 305
transaction-level modeling-based
    methodology, 436–437
transfer characteristics of a 3-bit
    unipolar ADC, *316*
transfer functions, **252**
transformer core reset techniques, *124*
transformers, forward mode
    equivalent circuit for, *124*
transient protection for common and
    differential mode surges, *191*
transistor
    characteristic of, *23*
    common emitter characteristics, *25*
    common source amplifier
        circuit, *38*
        high-frequency equivalent circuit,
            38
        Miller's theorem used to simplify
            the circuit, *38*
        Thevenin's equivalent circuit and,
            38
    large signal model of, *24*
transistor equivalent circuits, 32–38
transistors, high-frequency models of

frequency response of amplifiers and, 26–31

high frequency response, 30

low-frequency response, 28–29

use of short-circuit and open-circuit time constants for the approximate calculations of ω_L and ω_H, 30–31

transmission line effects, clock skew and, 422

two-switch forward converter, *121*

two-transistor forward mode converters, 129

typical n-channel multiplexer and its conditions required for equal on- and off-state errors, *266*

typical power supply with common mode filters, example of, *152*

typical sensing elements and their output variables, **243**

typical simulation performance of each design step, *440*

typical video amplifier characteristics, *242*

U

UL (Underwriter's Laboratories), safety standards of, 100

undersampling and frequency translation, *286*

Underwriter's Laboratories. *see* UL (Underwriter's Laboratories)

unit circle with N = 8, *8*

University of Pennsylvania, 366

user manual, 73

V

VDE (Verband Deutscher Electrotechniker), safety standards of, 100

Verband Deutscher Electrotechniker. *see* VDE (Verband Deutscher Electrotechniker)

VFA (voltage feedback amplifier, 227

VHDL (Very high speed integrated circuits Hardware Description Language)

built-in data types in, **353**

operators for behavior description, **354**

VHDL code

for a modulo-*N* binary counter, *357*

for an N-bit, two-input multiplexer, *359*

for an *N*-bit binary counter, *359*

for a pulse_detect component, *360*

pieces of , to demonstrate the differences between signal and variable assignments, *356*

VHDL description

of a 1-bit full adder using process statement, *355*

general structure of digital systems, *353*

video amplifiers, 241–242

VLSI designs, layout styles and, *453*

voltage division, 329

voltage drop of a MOSFET compared with a Schottky diode, *148*

voltage feedback amplifier. *see* VFA (voltage feedback amplifier)

voltage mode control, 140, *144*

voltage rails, sequencing of

output tracking (simultaneous or coincidental), *98*

ratiometric method, *98*

sequential power-up, *98*

sequential tracking with interlocks, *98*

voltage regulator module under the VID command signals (four or five bits) from the processor, *97*

voltage-controlled oscillator, 420

von Neuman, John, 366

Von Neumann architecture for non-DSP processors, *376*

W

Wheatstone bridge, basic, *244*

Z

ZCS-QR operation, buck topology and, **142–143**

ZVS-QR operation, buck topology and, **142–143**